About Island Press

Island Press is the only nonprofit organization in the United States whose principal purpose is the publication of books on environmental issues and natural resource management. We provide solutions-oriented information to professionals, public officials, business and community leaders, and concerned citizens who are shaping responses to environmental problems.

In 1994, Island Press celebrated its tenth anniversary as the leading provider of timely and practical books that take a multidisciplinary approach to critical environmental concerns. Our growing list of titles reflects our commitment to bringing the best of an expanding body of literature to the environmental community throughout North America and the world.

Support for Island Press is provided by Apple Computer, Inc., The Bullitt Foundation, The Geraldine R. Dodge Foundation, The Energy Foundation, The Ford Foundation, The W. Alton Jones Foundation, The Lyndhurst Foundation, The John D. and Catherine T. MacArthur Foundation, The Andrew W. Mellon Foundation, The Joyce Mertz-Gilmore Foundation, The National Fish and Wildlife Foundation, The Pew Charitable Trusts, The Pew Global Stewardship Initiative, The Rockefeller Philanthropic Collaborative, Inc., and individual donors.

Prairie Conservation

Prairie in wind is wave,
big bluestem and indian grass
on a clean horizon of plain.
Prairie in summer is soft,
undulating greens frosted
with crimson paintbrush,
harbor of upland plover nests
in fresh blades.

Winter pulls up in prairie
cold and harsh,
chunks of ice moon
swallow its warmth
like prairie roses dying
from summer to snow.

Song of prairie is long,
low, and ever leaving.

Sue Samson

Prairie Conservation

Preserving

North America's

Most Endangered

Ecosystem

Edited by

Fred B. Samson

Fritz L. Knopf

ISLAND PRESS

Washington, D.C. • Covelo, California

Library of Congress Cataloging-in-Publication Data

Prairie conservation: preserving North America's most endangered
    ecosystem / edited by Fred B. Samson and Fritz L. Knopf; foreword
    by E. Benjamin Nelson
         p.   cm.
    Includes bibliographical references and index.
    ISBN 1-55963-427-8 (cloth). — ISBN 1-55963-428-6 (paper)
    1. Prairie conservation—North America. 2. Ecosystem management—
North America. I. Samson, Fred B. II. Knopf, Fritz L.
QH77.N56P735   1996
333.74'16'097—dc20   96-11227
                                                               CIP

Printed on recycled, acid-free paper ♻

Manufactured in the United States of America

10 9 8 7 6 5 4 3 2 1

Contents

Foreword

Those of us living in the Great Plains are about to embark on a bold experiment in problem solving. Through a program called the Great Plains Partnership, the people who live and work on the Plains are being given the opportunity to define and create their own generationally sustainable future and, at the same time, to preserve and enhance the biological uniqueness of the prairies, wetlands, and waterways. It is a rare opportunity: solutions will be derived from the people with assistance from the government rather than the government dictating what will be done.

There is much that makes the Great Plains special, from the family farmers whose ancestors homesteaded here to the native wildlife that inhabits the prairie grasslands. The goal of the partnership is to protect and strengthen the economic and the ecological interests through a cooperative effort involving all levels of government, environmental groups, and farming and ranching associations. Solutions will be developed by local communities, but they can seek assistance from the partnership in removing institutional barriers, obtaining resources, and developing appropriate scientific procedures and methodologies. Such public-private partnerships offer a new path for tackling complicated problems, particularly in an era of limited government resources. The money for command and control of single agency government operations is dwindling in recognition of the fact that top-down governing seldom provides the best solutions.

Success will not come easily. In trying to reach a balance between natural resources and economic interests on the Great Plains a "problem" may be viewed differently by different groups or individuals, and that can be part of the problem. How then can we attain a common vision to accomplish prairie conservation with all its resources, natural and human, historic and future?

The solution starts with intent and commitment as offered by the Great Plains Partnership. It continues with a full understanding of the rich bioregional history of the Great Plains. It builds on the best available science—from economic to that describing the unique plants and animals that find their life history requirements nowhere else but on the Great Plains. Finally, it requires that private landowners, who own up to 98 percent of the Great Plains, get involved.

The opportunity is there. And I am confident the people of the Great Plains will pioneer a new path to environmental problem solving across the continent.

E. Benjamin Nelson
Governor of Nebraska
Chairperson, Western Governors' Association's Great Plains Partnership
Steering Committee

Preface

The once great prairies with their fruits and wildlife nourished our nation through its weak infancy. They nourished it again through its reckless and wasteful adolescence. The nation has now reached a maturity which should make it capable of recognizing that the prairie can no longer give that which it does not have, and that as man destroys it, he destroys himself.

Eugene M. Poirot, *Our Margin of Life*

Worldwide grasslands are the most imperiled ecosystem (Worldwide Monitoring Group, Cambridge, England, personal communication). Native grasslands in North America that once extended from Canada to the Mexican border and from the foothills of the Rocky Mountains to western Indiana and Wisconsin have dramatically declined in area. Some states and provinces in North America report less than one-tenth of 1 percent of the historic area of native grassland is intact (Samson and Knopf 1994); others report significant but less severe declines.

In the United States the most endangered ecosystems are typically at low elevations, with easy terrains and amiable climates, fertile soils and abundant natural resources, factors that encourage human settlement and resource use (Noss et al. 1995). These same factors may lead to declines in species and communities and, in some cases, extinction. There is reason to believe endangered ecosystems in Canada share similar ecological and economic traits, and declines in species and communities, particularly those associated with grasslands, are known.

We emphasize this information not to raise alarm nor find fault but to suggest that prairies are a priority, perhaps the highest priority, in the conservation of North American agricultural and natural resources. The intent of this book is to collectively increase awareness, information, and understanding of prairie—the tallgrass, mixed-grass, and shortgrass—of the North American Great Plains. Part 1 emphasizes its unique bioregionalism—a sense of place—and economic value. Part 2 brings a new synthesis to vegetation ecology of prairie and prairie wetlands and provides recommendations in management. Part 3 describes prairie legacies—the animals that evolved on the prairies and that tend to be restricted to the Great Plains. Part 4 is a multi-level approach to prairie conservation. It begins with a North American conservation program, continues with a description of efforts in Canada, the identification of areas important to biological diversity, a state-province plan to protect an area important to biological diversity, important joint ventures in Canada and the United States to protect prairie wetlands, a partnership approach to protect what may be North America's most singular ecosystem—the Nebraska Sandhills—and concludes with an international perspective. The summary provides an overview of current information and future goals.

Field in *Lost Horizon* wrote, "The prairie, in all its expressions, is a massive, subtle place, with a long history of contradiction and misunderstanding. But it is worth the effort at comprehension. It is, after all, at the center of our national identity" (Heat-Moon 1991). We hope this book will provide a major foundation to further understand prairie and to establish it as a focus in conservation of North America's natural resources.

PART 1

Value in Prairie

A Long Love Affair with an Uncommon Country: Environmental History and the Great Plains

Dan L. Flores

The sea, the woods, the mountains, all suffer in comparison with the prairie. . . . The prairie has a stronger hold upon the senses. Its sublimity arises from its unbounded extent, its barren monotony and desolation, its still, unmoved, calm, stern, almost self-confident grandeur, its strange power of deception, its want of echo, and, in fine, its power of throwing a man back upon himself.

Albert Pike, *Journeys in the Prairie*

In the larger context of conserving biological diversity in . . . natural ecosystems in North America, prairies are a priority, perhaps the highest priority.

Fred B. Samson and Fritz L. Knopf, *BioScience*

Before sunrise, summer solstice, 1995.

Atop a sandstone ridge that juts like a ship's prow from Texas's Llano Estacado plateau, my girlfriend and I sit a few feet from a snapping juniper fire in the twilight of a Great Plains dawn, sipping hot coffee and waiting. Off across the blue planes into the opalescent distance to the northeast, the direction of the solstice sunrise, homestead lights that looked last night like tiny fireflies on a gigantic lake have now winked out. Since the coffee was made, car lights flashed once momentarily in the middle of those horizontal planes, then were swallowed up.

Since that moment and for minutes since, the panorama of country laid out before us has given every appearance of a tabula rasa, an empty and runeless slate across which human history is yet to be written. It's an illusion, of course. But as we sit here waiting out the sunrise, looking off at a landscape so vast and flat it seems that we can see to the curve of the earth, we indulge the illusion, pretend that the country below us is not yet Texas and that all the immensity looming beyond is not yet Kansas, South Dakota, Saskatchewan. Nature emptied of the modern human presence and its technological clutter seems for us late twentieth-century folks an instinctive longing that probably has something to do with our hunter-gatherer ancestry. If so, we'd better figure out how to let it go.

But this morning, for a very few moments before the jets begin to drone and the faint grinding of truck gears becomes audible, Catherine and I permit the fantasy. Indeed, we've deliberately enhanced its effect—first by choosing to celebrate solstice at all, second, by doing it in remote Caprock Canyons State Park, one of the new nature preserves beginning to appear on the Great Plains. Too, there's the small gathering of rocks Catherine circled together soon after crawling out of her sleeping bag this morning. With a center cairn, it's a miniature medicine wheel, a kind of rock calendar marking the seasons that past cultures of the Great Plains built on hundreds of overlooks up in the northern prairies. As we see it, the wheel and the flint scattered in the dirt all around us represent tangible ties between us and them in this most anciently occupied of all North American landscapes. It's not that we want to ignore the last five hundred years of history here, but we recognize that the human drama has been happening on the plains for far longer. The truth is that this is roughly the 11,200th summer solstice sunrise people have watched across this ground.

We anticipate a fireball sunrise—the moon rose late in the night as a flaming crescent—and we're not disappointed. From first glint until it has cleared the earthline by a distance of two diameters, this solstice sun drifts skyward as a flattened disk, Martian red through the slate blue, and a skyscape effect I recognize because Georgia O'Keeffe captured it in her art here three-quarters of a century ago. The fireball makes this an easy sunrise to mark, and using the center cairn of our wheel as a sight, we embrace the timeless human ritual of setting down a line of rocks to mark the sun's point of emergence and ascent.

Later, after we've packed back down to the canyon floor, we drive out past a hundred-centuries-old Folsom site that archaeologists have interpreted as a shrine to the long-horned bison. The country that we pretended was an unmarked slate at daybreak is in fact historically rich and diverse. Tule Canyon, where Col. Ronald Mackenzie had fourteen hundred Indian ponies shot to death after his attack on the last big southern Plains village in Palo Duro Canyon in 1874, is the next major canyon to the north. Beyond that is JA Ranch country, where Charles Goodnight and John Adair began the Anglo-American ranching empire on the southern Plains. Atop the Llano Estacado that looms like a flat-topped mountain range to the west lies the quintessential twentieth-century plains adaptation: an agribusiness empire based on mined aquifer water. And when we drive through the tiny burg of Quitaque, Texas, whose economic fortunes are now closely tied to the nearby park and canyon trails, I remember that Frank and Deborah Popper walked these streets a few years ago, ruminating on the future of depressed Plains counties like Briscoe, and on the possible twenty-first-century outlines of Great Plains society.

In the history of American places, the history of human adaptations to landscapes, and the history of ideas about the past as it has played out across ecological regions, the American Great Plains stands in the first rank. Other American regions—New England, the South, the intermountain West—possess their own recognizable historical identities. But the grassy heart of the country, those great rolling sweeps of prairie that commence around the Mississippi River, where the rainshadow of the Rockies is first manifest, and stretch away in a steadily rising sea of undulations to the foot of the mountains 800 km west, has been home to a human history that is both older and apparently more fragile than that of other American regions. What struck American explorers and homesteaders out of the eastern woods as a place that was alien and marginal for Anglo-Americans was in truth—as the poet Walt Whitman would recognize—the most anciently American of all living places. By virtue of its minimalism, the Great Plains is a part of North America that captures more transparently than elsewhere the precarious fingerholds by which human cultures cling to the earth. And the last two centuries of Plains history illustrate very well two additional themes of environmental history: the wholesale assault on nature that some modernist models of living imply and the consequences of that assault both to nature and to human society.

Modernism's assault on nature has fashioned a legacy that is playing out around us as we enter the new century, and it appears to be serving the Great Plains up as an example. If the intermountain West, from New Mexico to Montana, is today running scared of growth and Californication, the Great Plains is a large slice of the West that does not share that particular fear. Despite the occasional story of the random Dust Bowl refugee family returning to calmer soils three-quarters of a century after parts of the Plains lost nearly 20 percent of their population in the 1930s, the Great Plains is facing different worries. Rather than being overrun by the Californian recoil into the hinterland, the Great Plains is hemorrhaging people. It's a long-term process that dates to the 1920s, the high tide of population for much of the rural Plains, but that accelerated sharply in the 1980s, when large areas of West Texas, Kansas, Montana (and, if you want to get specific about it, 38 of 41 of arid North Dakota's counties, 50 of 52 of Nebraska's, and 22 of 23 of Oklahoma's western counties) lost as much as 10 percent of their populations in a single decade (Popper and Popper 1987; Matthews 1992). The losses amount to a hemorrhage because the people who are exiting are the young and the creative, leaving the Great Plains with the oldest population of any part of the country except Florida. Why it's happening, and what the future holds, are questions that point backward as well as forward in time.

For many, the modern Great Plains is a conundrum. A century ago the Plains was the heart of the West, the terrain of Sitting Bull and Crazy Horse and Bill Cody, the setting for Old Shatterhand's adventures in the Karl May novels that have turned a large segment of Europe into aficionados of the West. For a more reliable account of the nineteenth-century Plains, read Lewis and Clark. When they traversed the grasslands almost two centuries ago, the Great Plains was the Serengeti of North America. Now the region is widely and popularly regarded as ugly, monotonous, uninteresting, unromantic. If the Montana Rockies strike Americans as the "last best place," to use writer William Kittredge's phrase, then the Great Plains—home to one-fifth of the American land mass and 6.5 million people—has become the West's last and least appreciated place (Kittredge and Smith 1988; Riebsame 1990; Matthews 1992).

Great Plains space serves as preface to a discussion of how humans and the natural world of this remarkable country have interacted and what the possibilities are of some continuation. In late December 1993, in a single day, I replicated the movement of nineteenth-century Anglo-Americans across the continent from East to the West. I started where I started originally, where my ancestors' bones lie buried in the damp woods along Bayou Pierre in some of Louisiana's most remote eighteenth-century cemeteries, where the only names on the crumbling tombstones are Floreses and Lafittes, and then retraced in twelve hours the journey it took American pioneers eight weeks to accomplish.

Driving up out of the shrouded bayous and misty pineywoods, where the rainfall approaches 114 cm annually, within 96 km of the Louisiana border I had imperceptibly gained higher oak country with the first scatterings of little bluestem meadows, which seemed to become larger and more savanna-like with each mile. By the time I was two-thirds of the way to Dallas I was out of the woods and in the Blackland Prairie, the country broadening into a true grassland aspect, the gently undulating prairie crests from 1 to 5 km distance, one from the other, with riparian tree growth in every declivity. Human sprawl around Dallas–Fort Worth had all but obliterated the north-south finger of oak woods once known as the Eastern Cross Timbers, but through the signs and city-edge suburbs, all of a sudden there was the Grand Prairie, a windswept, midheight grassland built on Cretaceous ocean terraces that a century ago was home to the first big herds of bison and wild horses as one moved west. Beyond that the Western Cross Timbers, a wider finger of oak and junipers that straddles the Red River for 320 km in both directions, still strike the modern traveler as a phenomenon of the continental transformation, although this woodland belt, too, is rapidly being cleared away.

And so on into nightfall, the country climbing higher and spreading out and drying out—annual precipitation down to 50 cm and dropping past Dallas—the trees steadily dwarfing, until with my headlamps lighting Highway 114, I realized that the trees were all but gone when the lights of small towns began to appear miles away in the spreading pancake of flatness. At this point I knew I'd come out, that I was no longer in country but on it. Approaching the shortgrass High Plains, where rainfall drops to under 50 cm a year and emigrants watched wagon wheels shrink loose on their axles, I'd floated out of choppy swells onto a smoother sea, had come to the surface for far-seeing. A memory came at me from twenty-five years before, from some of my first trips west onto the Plains and the excitement of spacious—western!—topography, the wonder of seeing those distant town lights 30 and 40 km away and having no idea how the country was going to look in the light of day. It's a feeling I've lost with more than a decade of living on the Plains, but that feeling of euphoria at spaciousness is another epiphany for humans, with our evolutionary roots on the savannas of Africa.

This country of grass, space, and sky has been regarded as an environmental anomaly, at least in the United States view of the continent, since the time of Lewis and Clark, Zebulon Pike, and the Long Expedition of 1820. Of course environmental anomalies, or normalities, result from experiential and cultural biases. Spaniards out of the Meseta or Hispanos up from Mexico returned from their early *entradas* onto the Llanos del Cibolos (Plains of the Buffalo) with the impression of a country that was densely inhabited by nomadic Indians and that possessed great potential as pasture for *rancherias* (Cabeza de Vaca 1984; Castaneda 1984). When they arrived, Comanches on the southern Plains and Salish peoples on the northern sweeps—both groups from farther west—found the Plains superior to the country they had known, a hunter's paradise. But Anglo-Americans out of the woods of northern Europe and the Atlantic Seaboard have the reputation for having long found the Plains alien, deficient, and peculiarly hard to love in its natural state.

One wouldn't have imagined so from reading some of the first U.S. accounts of the region. In 1806, for example, William Dunbar, the Natchez, Mississippi, scientist who was involved in President Jefferson's early exploration of the Southwest, wrote of the Great Plains:

By the expression plains, or prairies . . . it is not to be understood a dead flat without any eminences. . . . The western prairies are very different; the expression signifies only a country without timber. These prairies are neither flat nor hilly, but undulating in gently swelling lawns, and expanding into spacious valleys, in the center of which is always found a little timber, growing on the banks of brooks and rivulets of the clearest water. . . . Those who have viewed only a skirt of these prairies speak of them with a degree of enthusiasm, as if it were only there that nature was to be found truly perfect; they declare that the fertility and beauty of the vegetation, the extreme richness of the valleys, the coolness and excel-

lent quality of the water found everywhere, the salubrity of the atmosphere, and above all, the grandeur of the enchanting landscape which
this country presents inspires the soul with sensations not to be felt in
any other region of the globe. (Dunbar 1806)

Dunbar was not alone in this kind of glowing reaction to the Plains. Mountain
man, poet, and later Arkansas judge Albert Pike, who traversed the Llano Estacado in 1832, was moved to similar rapture by the prairies, as were later poets,
writers, and artists like Walt Whitman, Mari Sandoz, Willa Cather, and Georgia
O'Keeffe (Pike 1969; Whitman 1983; Turner 1989; Flores 1991b; Bloemink
1995). But almost from the beginning of Anglo-American encounters with the
Plains, explorers and observers who felt obliged to respond to the grasslands' potential for transformation into the *agricultural* landscapes that buttressed the expanding American frontier tended to regard the Plains as a place of deficiencies.
Zebulon Pike, who traversed the Plains in 1806–1807 and spoke of the country
as "sandy desarts" (Pike 1966), and Stephen Long, whose party crossed the
southern Plains during August of a drought year in 1820, gave the Plains an identity that it has never entirely relinquished as the "Great American Desert." Long
put it this way: "The plains are almost wholly unfit for cultivation, and, of course,
uninhabitable by a people depending upon agriculture for their subsistence"
(James 1906). Even the national necessity of replacing Great American Desert
imagery with imagery of the same region as a Garden of the World in the mid-
nineteenth century never quite erased the stigma of these popular initial impressions (Smith 1950; Blouet and Lawson 1975).

The historians Walter Prescott Webb and James Malin summarized those so-
called deficiencies in major works of Great Plains history half a century ago. Because of the rainshadow effect of the Rockies, the Great Plains has been regarded
as deficient in moisture; because this semiarid quality made it a grassland, it was
seen as deficient in trees. Its rivers were shallow, seasonal, unsuitable for navigation (Webb 1931; Malin 1984). To Webb, these environmental deficiencies
spurred pioneers to technological adaptations: Colt revolvers, barbed wire, wind-
mills, sod houses, and so on. And for Americans whose nature aesthetic had been
shaped by painters of mountain landscapes and whose evolving nineteenth-century sense of the nation's cultural worth often depended on the presence of monumental and vertical scenery, the sere, sunlit savannas—especially after the great
wildlife herds were extirpated and the Indians shunted off to reservations—came
to be regarded as aesthetically deficient, too (Runte 1987; Nash 1983; Novak
1979).

Environmental and conservation history on the Great Plains has been much
shaped by these historical impressions of deficiency. By the early twentieth century, the Plains that Dunbar and Pike and many others—Col. Richard Dodge, for

instance—had rhapsodized over had been stripped of much of its wildlife; much of its expanse had been homesteaded, fenced, and in a word, privatized. Increasingly even the vast grasslands were being lost, plowed to plant wheat, cotton, sorghum, and with that transformation the Plains lost its romance and came to be seen in our own time as a precariously dry and boringly flat version of the midwestern farm belt (Riebsame 1990; Evernden 1983).

No matter that the Plains actually continued to harbor dozens of grassland complexes, and that the widespread imagery of flatness is belied by the plethora of mountain islands found in the region's grassy sea (the Wichitas, the Black Hills, the Big Horns, the Sweetgrass Hills, the Bearspaw Mountains, the Crazies, the Big and Little Belts, the Highwood Mountains, the Turtle Mountains, for example). Or that there were still wildlife-rich rivers, potholes, and playa lakes. Or that the Badlands of the northern Plains and the sandstone canyons along the Llano Estacado of the southern Plains have turned out to be catalysts for artistic creativity and inspiration (Flores 1989, 1991b; Broach 1992). As geographer Bret Wallach pointed out in the journal *Landscape* a few years ago, twentieth-century Americans have not so much had our aesthetic sense filled by the Great Plains as we have come to deny that the Great Plains is aesthetic or much worthy of preservation (Wallach 1985).

If history is the measure by which regional inhabitants understand their adaptation to place, then it is clear why most citizens of the Great Plains appear not much concerned about their future. The history they've been served up—including Walter Prescott Webb's 1931 classic *The Great Plains: A Study in Institutions and Environment*, supposedly still the most frequently read book about the Plains—has been essentially pioneer celebration at the conquest of nature and the subjugation of the native inhabitants. But the new western history, a significantly more critical approach, and especially the new field of environmental history, make for rather more useful literature if the purpose of history is not just celebration but to ground us in place and give us some perspective and context, and some powers of discernment.

History used critically can be a very useful tool, but its limitations are obvious. It shares with the sciences some ability to analyze cause and effect, to explain why things have come to be the way they are—but unlike the hard sciences, history is not predictive. Its insights are a one-way street, and the signs point backward. The path of history, as Stephen Jay Gould demonstrated so nicely with his study of the Burgess Shale in *Wonderful Life* (Gould 1991), is a layer cake of contingencies, each choice resting on hundreds and thousands of earlier choices. While it is possible to read direction from the tapestry of those choices, the ability to infer the future diminishes the farther ahead we try to look. Who can say now what new matrix of circumstances, knowledge, values, or technology will affect the decisions extending our future? About the best we can say is that we have a reasonable idea where we've come from, and we're pointed somewhere, but we don't have much of an idea what is out there or where we're going.

The historical tapestry indicates that there are elements of the Great Plains environment that have been critical to the human experience here, and may remain so. These are best expressed by looking at the long story of human adaptations to the Great Plains—a story that stretches back at least 11,200 years and undoubtedly carries lessons for present and future inhabitants.

The story of sequential human cultures interacting with the Great Plains environment demonstrates some interesting patterns. One pattern that lends weight to the arguments of those who believe that late twentieth-century Plains culture may have overshot environmentally is simply this: Because the Plains environment is made fragile by cyclical drought, the hold human societies have had here has often been disruptive and tenuous. No part of the continent invites such easy human environmental alteration, yet can collapse so quickly and completely under that wooing as the Great Plains (Madole 1994; Riebsame 1993; Finley 1990).

Held against the eastern woodlands or the Rockies, the Great Plains is ecologically a simple system, with fewer of the safeguards built into more diverse systems. This is what makes it deceptively fragile. Thus with few exceptions—a several-thousand-year bison-hunting continuum on the northern Plains appears to be the major one—the big picture of human interaction with the Plains centers around a series of ecological crashes and simplifications, several of them with profound consequences for both the natural world and the human societies that depended on them. In fact, human interaction with the Plains has been so transformative that far from a howling wilderness, for at least eleven thousand years, the Great Plains has been the environment that those early Spanish accounts so honestly portrayed—an occupied landscape much shaped by human activity, especially human-set fires (Denevan 1992; Pyne 1982).

The most far-reaching Great Plains ecological collapse since men and women arrived on the continent seems to have been the first one. Except for the Antarctic and a few isolated islands in the Pacific, the Americas were one of the last human-free places in the world twenty-five thousand years ago. And all over the world, the principal result of human arrival in such places was ecological extinction and diaspora (Ponting 1992). In the Americas the Pleistocene extinctions saw thirty-two genera (and many dozens of species) of mammals disappear, evidently largely as a result of the intrusion of highly skilled human hunters from Siberia into an ecological setting that had never before experienced human hunting pressures. This extinction crash that peaked around ten thousand years ago not only eliminated almost all of the American-evolved megafauna, leaving only half a dozen grazer-browsers of mostly Eurasian derivation on the Plains, it also eliminated the so-called Clovis and Folsom cultures involved in the crash. A set of human cultures (the Paleolithic peoples) who occupied the Plains for three thousand years—two hundred times longer than we have—failed when the giant animals they had hunted were gone (Martin and Wright 1967; Martin and Klein 1984). The crux is that the bison, elk, and other specific animals we have learned to associate with the historic Plains were in fact the products of extinctions in which early humans were involved.

This ancient phase of Plains history was finished off a millennium later by another major disruption, this one primarily climatic: the great two-thousand-year drought of the Altithermal, when the climate warmed and dried subtly, but enough, in a country so delicately balanced between precipitation and evaporation, to cut plant diversity by as much as 50 percent on the southern Plains. Human adaptation evidently could not keep pace with this climatic change. While the northern Plains saw a succession of bison-hunting specialists for thousands of years after the Pleistocene crash, the southern Plains were all but abandoned during the Altithermal (Johnson 1987; Bryan 1991). A sensible strategy of both animals and people on the Great Plains, abandonment was also the response of the Plains Villager and Plains Woodland cultures of a thousand years ago, which had pushed their riverside farm villages far out onto the Plains, then fell back to the tallgrass prairie perimeter when another significant drought struck around A.D. 1300 (Wedel 1953).

A more recent lesson of fragility and marginality is the ranching collapse of the 1880s. What this one seems to demonstrate is that even in a country that evolved with large herds of native grazers like bison, a policy of land privatization coupled with commercial grazing based on exotic animals was capable of bringing on short-term cultural ruin and long-term ecological alteration. Although traditional ranching historians argue that a variety of human adaptive responses (fencing, irrigated hay pastures, winter feeding, better stock) emerged from the Big Dieup, modern environmentalists have questioned the grazing adaptation in light of a much-transformed grassland ecology on the Plains over the last century (Worster 1992). Plains ecology has been more grazing resilient than that elsewhere in the West simply because for millennia the Plains has been essentially an immense pasture, home to large wandering herds. But the spread of juniper, sagebrush, and mesquite (mostly due to rancher intolerance for the natural fire ecology), the invasion of exotics (Russian thistle, or tumbleweed, is just one of several dozen serious ones), riparian destruction, and significant alteration of grass species composition all spring from the rancher adaptation (Box 1967; Jordan 1994).

Another major pattern in the long-term environmental history of the Great Plains has been the tendency for Plains cultures to respond to the minimal but apparently limitless resources of the great grasslands with narrow economic specializations. The Paleolithic big game hunters inhabited the richest, most diverse Great Plains environment men and women have been privileged to see, a North

American version of today's Masai-Mara of East Africa. But so far the archaeological record indicates that their response to that wetter, lusher, more diverse environment of a hundred centuries ago was a narrow hunting specialization on the giant, lumbering animals of the late Ice Age. Similarly, when the Pleistocene extinction crash and a stressed climate regime produced that huge biomass of a single species, the modern bison, Plains cultures increasingly specialized in hunting bison. While this adaptation seemed to work rather well on the northern Plains for at least a couple of thousand years (and a two-thousand-year adaptation would have to be called a successful one), the alteration in technology and economic outlook that accompanied the arrival of Europeans on the continent undermined it. Indeed, after the year 1700, many Indian cultures along the periphery of the Plains abandoned horticulture or gathering altogether and became bison specialists, hunting increasingly for the market economy. The result of this pressure and other, external ones, was yet another great environmental crash, and the demise of still another way of life (Flores 1991a).

There are some contrasting examples provided by history on the Plains, but their lessons may be hard to accept for modern inhabitants. The buffalo lifestyle of the series of Archaic cultures on the northern Plains, cultures like the Oxbow, McKean, Pelican Lake, Besant, Avonlea, and Old Women's, offers one positive historical example of a Plains sense of place predicated around nature (Bryan 1991). Another is the legacy of the Archaic peoples of the southern Plains who moved in after the Altithermal ended and resided in the region for almost forty-five hundred years (Hester 1976; Schlesier 1994). Although their story remains one of the least understood in the archaeological record, the evidence so far is that these Archaics occupied the grasslands longer and more successfully than anyone else before or since. Their secret is instructive, but there was nothing magical about it. Consciously, as an act of policy, they kept their numbers small. They were mostly generalist hunter-gatherers whose economies were diverse, so that their effect was spread across a wide range of resources. Although they certainly did alter their environment, using broadcast fire that suppressed shrubs and brush and enlarged the areal extent of the grasslands, their real genius seems to have been not so much altering local bioregions as adapting to them. If cultural longevity and integration with intact natural ecosystems make up the essential environmental criteria for human success, then the Archaic lifeway is probably our species' most successful model on the American plains. We ought to find out as much as we can about it, and study it well.

That all these patterns appear to continue into our own time folds us like layers into the fabric of history. Narrow economic specializations, such as the monocrop dry farming of wheat or the dripping of irrigation water onto skinned prairies revegetated with exotic crops such as sorghum and cotton, possess none of the safeguards of diversity inherent in the Archaic lifestyle. And according to most gaugers of Great Plains health, mounting internal pressures on the key resources of soil and water, coupled with external pressures from the market and from global climate change, keep these modern economic specializations on the familiar, slippery slope of Plains human history, with periodic ruin grinning up at us through bright teeth.

That the Plains remains a fragile and uneasy place to live even with modern technology has been seared into the American consciousness by the Dust Bowl of the 1930s. Triggered as so many times in the past by drought, the Dust Bowl appears in history as the logical consequence of five decades of the most massive human ecological transformation of the Plains since the Pleistocene. During the preceding half century the Plains had been debuffaloed and dewolfed. Now much of it was degrassed. With a series of dry years, first the northern Plains, then the southern Plains, basically collapsed in a nightmare of erosion and dust storms—seventy major ones on the southern Plains in 1935. The ecological collapse and epic human abandonment of the Dust Bowl have become the great historical experience, equivalent to the Civil War, of the modern Great Plains (Worster 1979; Riebsame 1986). It is also a historical experience that many traditionalists have used to argue against doomsday projections for the Plains, as an example of how hard times acted as a natural selection process to fix more appropriate institutions in place (Riebsame 1993).

There are observers of the Great Plains today—not many, most of them on the periphery, it is true, and sad to report very few of them serving the taxpayers well from university positions on the Plains—who believe that the Great Plains is facing yet another watershed. There are worrisome doubts about the success of modern adaptations to the Plains, fears that in fact we haven't done much adapting at all but have simply imposed patterns from other places, other visions. From Texas to Kansas and Colorado, half a century of irrigated agribusiness has increased the human carrying capacity of the Plains twenty times beyond what it was under Comanche hegemony. But it has drained the resource on which this society rests—the fossil Ogallala Aquifer—so much that in Texas alone the wells have stopped pumping on 20 percent of the acreage irrigated just ten years ago. On the southern Plains the aquifer's expected remaining life span is estimated at no more than twenty to fifty years (Green 1973; High Plains Associates 1982; Kromm and White 1985, 1994; Worster 1994). After that, no more water. Compared to the forty-five-hundred-year life of the Archaics, that kind of life span may not be enough for future archaeologists even to bother with.

The litany of problems is not short. Ecologically the Plains is a mere shadow of the vast grassland that throbbed with a diverse and enormously numerous wildlife right down to 125 years ago. Oklahoma Indian friends of mine say that sometimes in their Saturday night peyote ceremonies, the reverberations of all those migrations, all those animals, all that life drama, can still be heard echoing across the Plains. The tallgrass prairie, whose areal extent has declined as a result of agriculture 82 to 99 percent since 1830, has suffered the greatest disaster of any ecosystem on the continent. But the mixed-grass and shortgrass systems have also been devastated by agriculture, particularly on the northern Plains and in the Canadian provinces, where the amount of native grasslands remaining ranges from 28 percent (North Dakota) down to 19 percent (Saskatchewan). In Texas a mere 20 percent of the shortgrass plains country still has native vegetation (Samson and Knopf 1994; Lesica 1995). The effects of natural systems dismantling are graphic in the part of Texas I know best; only 3 percent of the native grassland remains in Lubbock County (Flores 1990). The widely touted Conservation Reserve Program (CRP) of 1985, which hoped through federal rents to restore half the croplands on the Great Plains to grass, unfortunately has a mixed record. It did return grasses, but especially on the northern Plains much of the restoration was done with the exotic, crested wheatgrass. And CRP regulations even encouraged a certain degree of breakout of remaining native prairie (Mitchell 1988).

The results of this kind of stunning wholesale alteration are predictable. It's not just the highly visible and charismatic species—the bison that have been reduced to a shadow of themselves, and plains grizzlies and lobo wolves and Eskimo curlews and Merriam's elk and Audubon bighorn sheep that have been entirely erased from the scene—but today 55 grassland species are threatened or endangered in the United States, with a whopping 728 Plains candidates up for those listings! Great Plains bird species, in fact, suffered a sharper population decline (25 to 65 percent in the 1980s) than any other single group of continental species (Samson and Knopf 1994).

Those perceived aesthetic deficiencies that have so colored Plains history—and, by the twentieth century, the fact of privatization—have also figured prominently in the disgraceful lack of Great Plains parks and preserves in the national park system. In 1834, when the artist George Catlin issued his call for "a great nation's park" in the West, it was the Great Plains he had in mind (Catlin 1973). But in fact, the Great Plains remains today the most underrepresented region in the entire National Park Service (NPS) system (Miller 1984). Through the end of homesteading in the 1930s, the National Park Service found proposed Plains parks like Badlands and Theodore Roosevelt (in the Badlands of North Dakota's Little Missouri River) and Palo Duro Canyon in Texas not sufficiently "monumental" compared to the parks of the Far West. The three existing Great Plains parks at the time—Sullys Hill in Nebraska, Platt in Oklahoma, and Wind Cave in South Dakota—totaled fewer than 12,140 ha, and the NPS did all it could to lose them. One National Park Service investigator reported Sullys Hill to lack even "comic" value (Shepard 1995).

It was a struggle to get any significant Great Plains parks at all. In the 1920s the ecologist Victor Shelford and the Committee of Ecology of the Grasslands pressed for large Great Plains preserves based on ecology rather than monumentalism, studied eleven sites, found four more than acceptable, and one (spanning

290,000 km² in Nebraska and South Dakota) was even submitted to Congress (Shelford 1933). But the National Park Service fumbled the ball, and it kept turning down other proposals. A 386,000-km² park enclosing the Palo Duro system at the head of the Red River in Texas was turned down in 1931 (Flores 1993). Badlands in South Dakota was first proposed as a park in 1909, and did finally become a 60,700-ha national monument in 1939. Initially the National Park Service thought North Dakota's Little Missouri Badlands "too barren" for a park, and local ranchers opposed the idea vociferously. But rancher opposition swirled away with the Dust Bowl out-migration, and the National Park Service finally got Theodore Roosevelt Memorial Park in 1947. Both of these Badlands preserves became full-fledged national parks with the Omnibus Parks Bill of 1978, and Badlands got enlarged to nearly 101,000 ha (Flores 1990). In the 1980s Great Plains grassland preserves were augmented with Saskatchewan's creation of its Prairie National Park (Miller 1984; Friesen 1985).

If the Great Plains as a whole remains pathetically underprotected ecologically, the central and southern Plains are almost entirely so. Citizens of places like Texas and Kansas are today among the most divorced of all Americans from any kind of connection with regional nature. With midheight grassland ecology represented by existing parks on the northern Plains, however, the pressing need in the future is for large preserves in the shortgrass High Plains, and in the tallgrass prairies. Efforts to create a tallgrass prairie national park have stumbled along for two decades now, but the Nature Conservancy's recent 12,140-ha acquisition in northeastern Oklahoma may at last serve as a core for tallgrass protection (Hall 1962).

As serious reappraisal of the burden of modern Plains history has begun, voices like those of Donald Worster, Bret Wallach, and Frank and Deborah Popper have called the modern agricultural experiment on the Great Plains a tragic mistake, in the much-quoted phrase of the Poppers, "the largest, longest-running agricultural and environmental miscalculation in American history" (Popper and Popper 1987, 1988). Of course, nobody wants to hear that what their grandfathers did was a mistake. Scarcely anyone finds cause for rejoicing in the prediction that maybe the Indian Ghost Dancers were right about the whites disappearing and the buffalo returning. The Poppers needed six deputies as bodyguards when they talked about their Buffalo Commons in McCook, Nebraska, and when Frank Popper was invited by the citizens of liberal Missoula to discuss the Big Open idea in Montana, his appearance had to be called off because of alleged threats on his life by Montanans from the Plains part of the state (Matthews 1992).

In 1991, in an article in *Great Plains Research*, Colorado geographer William Riebsame explored the use of evolutionary adaptation models to explain how American society might respond to environmental change on the Great Plains of the future. Riebsame contrasted adaptation (the emergence of new characteristics better fitted to changed circumstances) to what systems theorists call resiliency, or a system's tendency to rebound to its previous characteristics following a disturbance. Along with Donald Worster, Riebsame thinks that many of the institutional changes that have resulted from modern society's response to various Plains crises have actually been resilient rather than adaptive. Contour farming, listing, center-pivot irrigation, all these have been merely technological refinements of the status quo, Riebsame argues, which have enabled Plains society to avoid making adaptations to nature's parameters on the Plains (Riebsame 1991; Worster 1979).

The environmental future on the Great Plains is complicated by predicted global climate change, expected by some geographers to proceed at a rate ten times faster than humans have ever experienced (about ½ to 1°C per decade). The southern and central Plains have been singled out as among the continent's regions that will be hardest hit by global warming (Worster 1994; Glanz and Ausubel 1984). Viewing the future of the Great Plains in the context of aquifer drawdown, rapid climate change, and an emerging awareness by Plains people that troublesome changes really are taking place, it seems obvious to some of us that the Great Plains is on the threshold of what could be a major paradigm shift. Whether this shift is brought on by an eventual disillusionment with the continuing pattern of cyclical ruin or by more planned, truly adaptive strategies built around the natural systems and cycles of a restored grasslands, future society here is almost certainly going to look very different.

What form might a new adaptation to the Plains take? There are some models. Phil Burgess, of Denver's Center for the New West, has recently published a study titled *A New Vision of the Heartland: The Great Plains in Transition* that posits what Burgess calls an "urban-archipelago" society, a high-tech oasis country with population islands supported by a service economy. As for the rural lands in between, Burgess has so far been mute (Shepard et al. 1994). Perhaps Wes Jackson's Kansas Land Institute has an answer for that in Jackson's search for a native grass (maybe a genetically manipulated grama grass) that could replace wheat as a commercial crop (Jackson 1980). The Poppers' Buffalo Commons and Bob Scott's Big Open, both of which envision reintroducing the native Plains fauna (including predators) in enormous wildlife refuges, may be more romantic visions, but they hold out a promise of sustainability based around the Plains' ancient ecological base (Shepard 1995). Depending on one's politics, these ideas either sound like either Eden or disaster.

My own suspicion is that the Great Plains, which after all was the country's western experiment with privatization, might learn something useful from the model presented by the Mountain West. As in the Rockies, there could be a consolidation of the human population in a few choice locations near interstates or in favored service-hub areas. While there might be some rural water importation and agricultural cropping in select locations, and private ranching (perhaps increasingly with buffalo, the grazer evolved to the country) would continue on a reduced scale, the great experiment that privatized virtually all of the Great Plains would have to be reversed. The process initiated by the Dust Bowl with the federal reacquisition of 4.5 million ha of homestead land ought to be reimplemented, and I think it will. In the future the Plains almost surely will feature far more publicly owned and managed lands, some used principally for grazing, but most for restoring the natural ecological diversity that has proved capable of weathering natural change on the Plains so well. The basic structure of the Plains economy ought to rest on the human interaction with restored nature. There's a name for this kind of model, of course: eco-tourism.

The key to a nature-based sense of place on the American Great Plains thus lies in a democratizing and ecologically restorative land system. With the return of bison to tribal lands on many of the Great Plains' Indian reservations, the Indians might—and they should—lead the way. For the rest of Plains inhabitants it's going to be far harder, and it is possible that it will require another great collapse on the order of the Dust Bowl to ratchet Plains society to its next stage (Flores 1994).

With fewer national parks and preserved ecosystems than any other region in North America, the modern Great Plains is today struggling to create a sense of place based not so much on herefords, furrows, and big farm machinery, as on Plains nature. A growing number of Plains writers, artists, and photographers have been trying heroically in recent years to rescue a place-centered society predicated on the old Plains natural world. Given the transformations of the last century, the process has far to go (Evans 1988; Flores 1989; Flores and Winton 1989; Broach 1992; Howarth 1993; Shepard 1995).

On the other hand—as Catherine and I intuited watching summer solstice from the rim of the Llano Estacado—the peculiar trajectory of Great Plains human history may make that connection increasingly more possible in the future. There may well be fewer of us living here a century from now—maybe 3 million instead of 6.5 million. But we are likely to occupy a more complete Great Plains than today, one sufficiently restored that nature and humans can once again intertwine in a genuine Plains sense of place—a sense of place that modernism so trampled in its rush to remake the Plains into something it wasn't.

CHAPTER 2

The Economic Value of the Prairie

Jeffery R. Williams
Penelope L. Diebel

What Is Value?

The value of the prairie stems from its relationship to people. Values arise from human wants and desires. As long as society views the prairie as beautiful or a source of knowledge, it will have value (O.J. Reichman 1987). However, benefits of the prairie are not limited to beauty and knowledge. Value is derived from the land in its natural state, but for many, in a developed state. Development of the prairie means losing some or all of those services and benefits derived from its natural state and gaining those benefits extracted from the prairie in a modified state, such as agriculture. Value can thus be derived from consumption of goods extracted from the environment or from services that flow from the prairie. Value may vary when one differentiates between private and social benefits.

An example of this difference in values is the recent disagreement over the valuation of Pacific northwest forest resources for timber and spotted owl habitat. The private value of timber land can easily be derived in a market transaction that results in the sale or leasing of land to harvest timber. This private value is a function of the earnings derived from selling the harvested timber in the marketplace. However, a significant number of people in the larger society believe the value is higher for spotted owl habitat. Although this is an oversimplification of the issue, values derived outside of a market transaction were found to be greater or more important than those easily derived in the timber land market. The end result was the protection of selected forest resources for spotted owl habitat. The importance of this result is in the recognition that market price does

Prairie Conservation
Island Press (Washington, DC • Covelo, CA)

not always equal value. Value instead is a measure of all benefits, some quantified by market price and others not as easily quantifiable. These conflicts are eventually judged by public institutions and, therefore, a measure of social benefits influences the valuation and decisions regarding natural resources such as the prairie.

The Social Concept of Benefit

The value of benefits are the gains in individual and societal welfare because of the use or existence of a resource or services associated with the resource. To receive the greatest welfare from the prairie, society must compare the values received from any change or use of the resource with the values given up by modifying the resource (Freeman 1993). Values are determined by people not by physical laws of nature. Benefits are a function of human desires and satisfaction derived from the resource. Referred to as utility in economic theory, satisfaction is difficult to measure. Individuals can rank satisfaction obtained from different resources and services although it may be difficult for an individual to evaluate the tradeoffs involved in the ranking process. Comparing satisfaction obtained from a resource across individuals is a more significant problem. For example, two individuals could rank a hike along a scenic trail as their most satisfying outdoor recreation activity, but it is not easy to compare their individual levels of satisfaction. Expressed as economic value this comparison becomes easier if society agrees it can be monetized.

Economic Value

Although human valuation of benefits received is a complex issue. No other common measure of value besides economic or monetary measures has widely accepted approval to evaluate trade-offs. Individual social and economic activity is constrained by our natural resource base and our economic resources. Society is then forced to make trade-offs using economic value. Two types of economic value are discussed: market and nonmarket values, as well as noneconomic values.

Market Values

Market prices are used as a proxy for value. Demand curves, estimated from market data, trace the relationship between the price and quantity of a resource desired at that price. This relationship is shown in Figure 2.1. The intersection of the demand curve (D) with the supply curve (S) reveals the market price (P). In the case of the prairie the supply curve is essentially vertical. Additional prairie cannot be produced even if increased demand for the prairie (D to D_p) raises the market price to P_p. A reduction in the supply of prairie (S to S_p) has the same result if demand (D) is constant.

The market price of prairie is a private valuation measure that reflects how much it would cost the purchaser to obtain ownership rights from another. How-

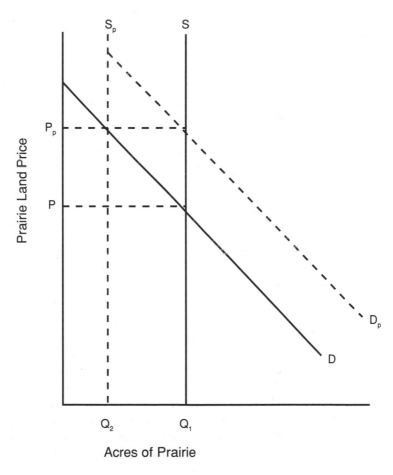

Fig. 2.1 Supply and demand of prairie land.

ever, this price is not a reflection of all the prairie values that may be lost if the prairie is converted to another use. If, for example, a conservation organization were to purchase some prairie land at the market price they would also consider how valuable the land is in meeting their objectives to preserve unique parcels of prairie. These values are not necessarily the same.

Historically, market sales of prairie land have been for conversion to other uses such as intensive grazing, crop production, and nonagricultural development. Therefore, the supply of prairie land and the services derived from its natural state have been decreasing. To estimate the value of an additional loss of prairie acreage to development, it would be necessary to estimate the demand (D_{ps}) and supply (S_{ps1}) curves of prairie services such as recreation, as is displayed in Figure 2.2. The upward sloping supply curve S_{ps1} represents the supply of recreation services available from prairie sites at different costs. This can be used to estimate the impact of a shift in the supply curve due to a loss of prairie land and recreational services (S_{ps1} to S_{ps2}) on the market prices paid for these recreational services, and consumer and producer surpluses.

Consumer surplus is the amount one would pay for the service, over the amount actually paid rather than do without the service. It is represented by the area below the demand curve but above the price P_1 (acP_1). Producer surplus is the area above the supply curve but below the price and is the surplus of payments above the cost of providing the service (cdP_1). As the price rises to P_2 due to a declining supply (S_{ps1} to S_{ps2}), consumer surplus is reduced (abP_2) and producer surplus falls (beP_2). Changes in these surpluses are a measure of change in welfare. These welfare measures could be estimated for any marketed service the prairie supplies such as grazing rights, plant materials including hay, wildflowers and their seeds, as well as recreational opportunities such as camping and hunting. However, the net result of this welfare analysis of the prairie to society is ambiguous as long as the land is usable in some form. Even though prairie land and the supply of prairie services decreases, the land is not consumed. Although certain characteristics of the land, such as its ecological and physical characteristics, and values associated with those characteristics may be irreversibly destroyed, the land is only converted from one use to another. Prairie can be used for other purposes that create positive changes in consumer and producer surpluses in the nonprairie land services markets. Even if these surplus measures of prairie services could be estimated accurately, they are not completely adequate because they represent only the value derived from the private market valuation of benefits; the flow of services from the prairie that can be priced in a market. Market prices coordinate demand and supply to provide optimal levels of resource services. However, the public good characteristics of the prairie cause the market values to be too low.

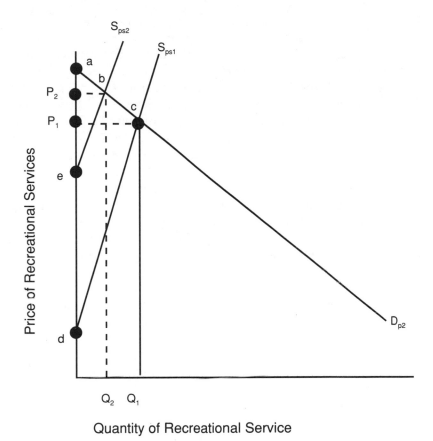

Fig. 2.2 Supply and demand of recreational services from prairie land.

Nonmarket Values

A prairie provides services not directly exchanged in markets, including unpriced recreational, ecological, and aesthetic contributions to society. These services are external to the private market decisions, where prairie land is bought and sold and their value is not captured in the market price. As previously stated this is due to the public good characteristics of the prairie. The first characteristic of a public good is that a person's use of a resource does not directly influence use by someone else. This characteristic is sometimes referred to as consumption indivisibility or nonrivalry. The second is that one person cannot exclude another person from using the resource. This type of resource is often referred to as a high exclusion cost resource. This is true of some of the services a prairie provides such as a scenic view. One of the principle reasons it is not valued in a market is that it has high exclusion costs. To prevent someone from enjoying a prairie sunset vista unless they pay for it would be prohibitively expensive. Often, such services are facilitated by the federal, state or local government, where only a nonexclusive fee is charged for visiting scenic areas. Because these benefits are not exclusive, it is difficult to assign a price or value to them. The price of the service may be low or even free therefore resulting in a high consumer surplus. Dixon and Sherman (1990) note that "Since they cannot be excluded from enjoying the benefits (at least not without cost), there is no incentive for these users to reveal how much these benefits are worth to them." (28)

Because some of the services a prairie generates do have public good characteristics, particularly high exclusion costs, there is potential for free riders to exist in any voluntary prairie preservation or conservation program. Suppose an organization sets out to collect donations to establish a prairie preserve. Because of the public nature of the resource, any or all of the following may happen. Some individuals may contribute because they either expect to derive direct benefits from using the preserve or they obtain indirect, altruistic benefits from contributing. Others may view their potential contribution as having a small impact and assuming others contribute, they expect to receive direct and indirect benefits whether they contribute or not. Others may view their contribution as a small price to pay for the potential benefits, but they also recognize benefits will go to others who do not contribute. Therefore, they decide not to contribute. Some individuals may be unsure of the benefits they will receive, but realize if they do not contribute it may not reduce their chances of receiving benefits, so they do not contribute. The individuals who do not contribute, but still receive benefits from the participants actions, are called free riders. When free riders exist, the optimal quantity of prairie conservation may not take place. As a result, its value is underestimated in the market transaction.

Noneconomic Value

Some may argue noneconomic values exist and to place a dollar value on the existence of an endangered or rare species of wildlife or plant is unethical and impossible. However, economists generally contend that all of these values are economic because they arise out of concern, even if only indirectly, for human welfare. Still others would argue that the market and economic analysis discounts the future and believe it is morally inappropriate to do so. These individuals believe values concerning the future cannot be monetized into an economic analysis. Although economists generally assume that value is a humanistic and not a naturalistic term, and the previous values can be monetized, two areas of questionable monetization are acknowledged.

The first is uncertainty. Society values resources to a certain extent because of inadequate information about the interaction of the natural environment with the human condition. Preservation or conservation of resources may occur because society is uncertain of how valuable they will be in the future. This uncertainty makes the resources difficult to value accurately, because they are conserved or withheld from the market. There is an expectation, that the discounted present value of their future price is higher than the current market price, thus, creating an incentive for conservation. Some individuals may be interested in contributing to the preservation or protection of a resource even if they never use it or see it. The resources are valuable in and of themselves simply because they exist at the present and are protected for the future. An example would be the satisfaction a city dweller receives from knowing a land preserve exists, possibly a prairie preserve, even though they never plan to visit it.

The second area of questionable monetization is irreversibility. If resources are not conserved and the modification of them is irreversible, then they may not exist when a valuable service is discovered. This is often one of the arguments for conservation of the rain forest, where many medicinal species exist and many unknown species may have already disappeared. Species of future importance may also exist in the prairie. The characteristics of nonrenewability and irreversibility can influence resource values. These values may appear to support noneconomic validation, but they are based on human characteristics and are essential economic values that need to be considered in any analysis.

Economic values, as suggested, are human centered. However, some authors have suggested noneconomic valuations could be based on biological or ecological relationships. Kahn (1995), provides an overview of these ideas. A much more important classification of values to the economist is the classification that distinguishes between Use and Non-use.

Use Value of the Prairie

Use values are those derived from direct human interaction with the resource (Taff 1992) and can be classified into consumptive and nonconsumptive. Consumptive uses are those in which some output of the prairie is essentially consumed or converted to another use. These include, but may not be limited to, the following.

Grazing livestock

Harvesting native or cultivated plants

Hunting wildlife

Nonconsumptive uses are services people obtain from the resource in situ or "natural state," and are not related to direct use of some product of the prairie. These include, but are not limited to, the following.

Recreational activities such as hiking, bird watching, and photography

Educational activities

Erosion control and water quality enhancement due to the benefits prairie grasses and native plants provide in their "natural state"

Research activities

Nonuse Value of the Prairie

Nonuse values are those associated with more intangible uses of the environment, such as the aesthetic benefits or the satisfaction derived from the existence of environmental resources (Khan 1995). Nonuse values may include the following.

Existence and option

Aesthetics

Cultural-historical and sociological significance

Ecological or biological mechanisms

Biological diversity

Existence values are independent of resource use. Nonusers receive value in knowing the resources exist even though they never plan to view or use them. Freeman (1993), provides an overview of the literature that discusses why existence values may occur. Option values are a more narrowly defined type of existence value. They occur because individuals may want to preserve a right (reserve an option) for themselves or their heirs to use the prairie resource in the future. When the value can be attributed to an option for future generations, it is also referred to as bequest value. Existence and option values may also be purely altruistic in nature. Individuals value the prairie because others in the current or future generations value its use. These values appear to be difficult to measure or observe. Freeman (1993) suggests that if total value of the resource exceeds use value the difference can be attributed to option and existence values.

Aesthetic values can be attributed to the enjoyment one receives from viewing a prairie. Cultural-historical values are more difficult to define. For example, prairie sites throughout the United States are associated with Native American religious, ceremonial, and historical values. Since colonization, American historical sites have been preserved throughout the prairie because of their social value. Portions of the Sante Fe trail are preserved because of existing wagon ruts made in the historical crossing of the Great Plains by settlers. Opposition to development of natural resources is often associated with saving a species of animal or plant, but development projects have also been opposed because of a site's historical and cultural significance.

Ecological or biological mechanism values stem from the fact that land resources, including the prairie, may provide significant wildlife habitat and environmental services important to society. Prairie resources may contribute to reduced water and wind erosion of soil. Native grasses, recently reestablished on thousands of acres of Conservation Reserve Program ground in the Great Plains have significantly reduced erosion and provided substantial wildlife habitat. Ribaudo et al. (1990) estimate that approximately $8.8 million of off-site damages accrue in the United States from soil erosion. This does not include the value of lost soil fertility. The erosion control services of the prairie produce off-site benefits miles away from the prairie site. For example, streams running through prairie areas are more likely to have less sediment in them and can handle intensive rainfall more efficiently, eliminating the need for artificial flood control impoundments downstream. These impoundments would have substantial impacts on the species of fish populations inhabiting the streams. Some may value natural habitats such as a natural river canal, just because they are undisturbed. Others place value upon the species of fish that flourish in the natural watercourse for human use in consumptive sport fishing or nonconsumptive use such as viewing or nonuse including aesthetic, existence, and option values.

Biodiversity, the stock of genetic material found in natural ecosystems, may be valued for its existence or for its potential use in pharmaceutical and agricultural breeding. An ethical view also exists under which the preservation of prairie biodiversity is morally correct and unrelated to economic issues. Both the areas of ecological mechanisms and biodiversity values are often associated with an "ecocentric" ethical framework. Ecocentric theory states that nonhuman nature is capable of being inherently valuable, which is independent of human valuation. In fact, the Leopold "Land Ethic" may be interpreted to say that human life is only worthwhile insofar that it preserves the global ecosystem. These ethics theories go beyond the supposition that value is a human construct (Pearce and Turner 1990).

Some of the value of biodiversity is associated with the option for future use. This diverse pool of genetic material may be very valuable regardless of whether economic or moral definitions are used. Crosson (1992), believes that expanding agricultural production will require intensive efforts by plant and animal breeders who must have access to a broad range of genetic material to develop more resistant and productive varieties. The genetic information provides direct and indirect inputs for plant breeding programs, development of natural products including pharmaceuticals and drugs, and increasing applications in biotechnology (Sedjo 1992). The genetic information of plants and animals in the prairie environment may prove to be as important as those in the rain forest. As Sedjo states "Species that have no current commercial application, contain no useful natural chemicals, or are as yet undiscovered, nevertheless may have substantial value as repositories of genetic information that may someday be discovered and exploited" (26).

An example of these values is contained in the purchase of prairie lands for public use. The National Park Trust, a nonprofit organization dedicated to acquiring and protecting nationally significant natural and historic properties, purchased the 10,894-acre Spring Hill / Z-Bar ranch in 1994 for $440 per acre. Located in the Flint Hills region of Kansas, the ranch is to become a tallgrass prairie preserve and working ranch. The net present value of benefits of the prairie services over a foreseeable period of time would be at least equal to the purchase price of $440 per acre. However, in this case it is likely the discounted net benefits are greater than the purchase price because then private use benefits alone would have been significant. Once the prairie preserve is open to the public not only will benefits be derived from cattle ranching but other consumptive use benefits will accrue. In addition, nonconsumptive use benefits such as those derived from educational and recreational uses will occur, and nonuse values will be preserved including existence, aesthetic, cultural-historic, and biological diversity values.

Techniques for Valuing the Prairie

Although measuring natural resource values requires economic techniques, it also involves knowledge of the scientific principles of the physical and biological interactions in the environment and their impact on human welfare. In addition, the human impact of resource management decisions on the environment must be well understood. Any estimates of natural resource values are, therefore, influenced not only by the economic models and methods used to estimate value, but also by our basic understanding of environmental relationships and how humans influence them. A lack of scientific knowledge can lead to faulty assignment of values to resources.

The classification of value estimation techniques depends on whether the method yields monetary values directly or inferred through some technique based on individual behavior. Direct estimations are based on observations of competitive or simulated market prices or by direct questioning of people to determine their value of a resource, thus creating a hypothetical market. Indirect methods of estimation can also be based on observed market behavior, but the behavior does not directly involve the purchase or sale of the resource and must be derived from an individual's response to prices of related goods and services or other economic signals.

Use values can be measured through both direct and indirect techniques. The measure of nonuse value is more difficult and many believe can only be measured by a few of the available techniques. Since direct methods measure both use values and nonuse values they will be discussed first. A discussion of indirect methods for estimating use values, specifically nonconsumptive uses will follow.

Direct Methods of Valuation

Direct methods for valuing resources involve the observation and use of price information determined in actual or simulated markets and hypothetical techniques such as contingent valuation and conjoint analysis which rely on survey information.

Consumptive Use Values

The most direct technique is to observe market determined prices for services or goods extracted from the resource. The price of prairie land discovered through transactions in the real estate market reflects the opportunity cost or the owner's willingness to accept a certain value to sell the land on the part of the owner and a willingness to pay a particular value by the purchaser. The purchaser must assume that the discounted value of services at present and in the future will be at least equal to their maximum willingness to pay value. This willingness to pay value can also be derived through analytical methods for consumptive uses. Suppose a prairie owner is using the prairie to graze buffalo to produce buffalo meat. An analysis could be constructed to determine the value of having additional prairie range available to the producer. This discounted value would be the value of an additional acre of prairie for use in raising buffalo, the maximum value the producer would be willing to pay for an acre of prairie to raise buffalo.

Consumptive use values can be estimated directly by studying market demand characteristics for livestock grazing, harvesting of prairie hay, and other prairie products such as seeds and plants, and wildlife hunting. Consumer surplus can be estimated under different prices using market demand curve estimates for these consumptive use products. For example, if additional prairie is converted to housing or commercial developments then it is expected the rental rate or fees to harvest products or access other prairie sites will increase. This will generate a loss in consumer surplus. In addition, the supply curves for these products will adjust and affect producer surplus estimates. If the prairie itself is not lost but is damaged in a way that reduces its productivity and supply of a service, a measure of the change in both producer and consumer surplus for that service of the prairie should be evaluated.

The most widely used technique is contingent valuation. Instead of relying on market information, survey methods are used to assess the willingness of people to pay for the right to use or consume services or products derived from the prairie. The assumption is that responses to a hypothetical market (e.g., in a survey) are comparable to responses in an actual market.

Use of surveys to obtain consumer responses makes contingent valuation vulnerable to various types of error such as strategic bidding, design bias, hypothetical bias, and operational bias. Strategic bias occurs where participants do not reveal the true value they place on a resource. This can be caused by the free rider problem. Individuals realize their access to the resource will not be diminished even without their payment. Their revealed value may be lower than the actual value they would pay if not paying would exclude them from the resource. The design of the survey, to include how participants believe they will "pay" for the resource (e.g., taxes, entrance fees, etc.) is important. Hypothetical bias arises because participants realize an actual market for the prairie does not exist and answer falsely. Finally, operational bias may exist because the contingent valuation method is dependent not only on the structure of the questions, but also on the level of knowledge the surveyed individuals have about the services generated by the prairie, both now and in the future.

An important question is whether this or any other method can properly reflect the future, because future generations are precluded from providing their value estimates. For example, prairie land of marginal value for dryland crops has been broken out by farm managers for irrigated cropping in parts of the Great Plains. As irrigation water availability declines and irrigation costs increase, this land will be of little use for agricultural production. According to Sexton (1980), it will be difficult to stabilize these soils and severe soil erosion may result. Would these farm managers have acted the same way if they had known of the problems that would occur thirty to fifty years later? Uncertainty about future consequences and future needs make conservation decisions difficult. Similarly, how does society use its dollars to signify important prairie services to protect if they do not know of the future consequences of not protecting them?

Conjoint analysis, another direct technique, is designed to establish preference and value of resources. Conjoint· analysis determines a preference function for different levels of characteristics associated with a good. Kahn (1995) indicates that as long as one of the preferences is price, it is possible to use the preference function to derive the willingness to pay for changes in the levels of the other characteristics measured.

Nonconsumptive Use Values

Nonconsumptive use values, such as recreational and educational activities, can be measured using direct techniques similar to those for use values. Market prices can be used to infer the prairie's value to the extent related private markets exist for activities such as hunting and fishing on the prairie. Contingent valuation or conjoint survey techniques can also value nonconsumptive use of the prairie not revealed in the market. For example, to derive a value for a quality prairie experience, levels of quality could be defined by grouping characteristics associated with quality such as the number and variety of plants, birds, and other wildlife. Different combinations of these characteristics are associated with an access fee or admission price. Respondents are asked to rank the bundles they most prefer. The preference ordering indicates the preference to pay for different bundles of quality.

Nonuse Values

The nonuse values attributed to existence, option, aesthetics, cultural, historical, ecological, and biodiversity values cannot be measured using market-oriented techniques because these services are not traded in markets. Although environmental organizations undertake activities to protect areas such as the prairie, neither their level of contributions nor their activities in the market can adequately represent nonuse value estimates. Contingent valuation is important in determining nonuse values where methods that rely indirectly on market values cannot be used. Survey questions could be formulated to determine how much individuals would be willing to pay to protect the prairie in general. The response to such general questions of prairie value would likely include values associated with consumptive use, nonconsumptive use and nonuse.

Indirect Methods of Valuation

Indirect methods of valuation rely on what people do as opposed to what they say they will do. Actual or hypothetical market observations can be indirectly used to measure values based on people's actions. The hedonic price approach, travel cost approach, and replacement value approach are discussed.

Consumptive Use Value

If the consumptive use of a resource service or good is not available in the market, one way to value it would be to determine the cost of replacing these services with a substitute service. An example is the value of prairie hay no longer available for harvest and sale in a local market. The replacement cost associated with a substitute forage would be one way of establishing the use value of the native grass. If a grazed prairie was converted to row crops, the cost of feeding the displaced cattle, either through feed or renting pasture, is another example. This method is only satisfactory if a reasonable technical substitute is available for the lost resource.

Nonconsumptive Use Value

Hedonic price studies use market price information to indirectly measure the value of on-site characteristics that are not consumed, such as proximity to a lake or a view. Hedonic price studies examine differences in market prices from market to market that can be attributed to different characteristics of a similar resource or service. Property values can be statistically estimated with the use of real estate sales data as a function of location and environmental amenity characteristics that may include characteristics associated with living next to a prairie. The value associated with living next to a prairie that is a nonconsumptive use could be estimated in this way. It is an incomplete measure because it only values living next to the prairie, not the value of the prairie itself. If the prairie becomes damaged or is developed for additional housing units, the change in market values attributed to those environmental changes can be measured with this technique. The design of an estimated hedonic price model is very important and difficult. It must be designed so the effects of all characteristics are accounted for separately; sufficient market data must be available; and the public must be able to differentiate the actual physical differences among the characteristics (Mitchell and Carson 1989).

The valuation of a prairie preserve as a recreation site can be established with the travel cost approach. This method develops a total demand curve for a site as a function of distance traveled. These travel costs serve as a proxy for variation in price (entrance or user fees). Price and consumption (visitor days to the site) are used to form the demand curve. Travel cost methods basically derive a willingness to pay for the recreation experience that is a nonconsumptive use. Travel cost models, including the discreet choice, demand system, and hedonic travel cost approaches can be designed not only to value access to resources but to value the quality of the resource experience as well (Mendelsohn 1993). However, travel cost methods may not account for substitute recreational experiences or sites and does not reflect existence values. Nor do they indicate the willingness to accept value for loss of a site that could include specific nonuse values. As Zerbe and Dively (1994:413) state "The willingness to pay of the Sierra Club, for example, to preserve old growth forest in the Northwest may be large but it is likely to be very much smaller than the willingness to sell (willingness to accept payment) of the Sierra Club if they owned the rights to the old growth forest." This would most likely be true for the NPT and the Z-Bar ranch as well.

Replacement costs can also be used to value nonconsumptive uses of a prairie. An example would be the cost of replacing a recreation experience. If a prairie park is no longer available to recreationists, the value of the site could be established by determining what the annualized per capita cost of establishing and using an alternative recreation site that would provide a similar recreation experience. This technique can only provide approximate values because the alternative services from alternative sites may not be perfect substitutes. For example, suppose a section of prairie has been traditionally used as a buffer strip between cropland and a stream. Assume for this example that this strip has been well maintained, has not been damaged by the use of pesticides, provides valuable soil erosion control, and contributes to water quality benefits received by downstream users of the water course. If the buffer strip were to be damaged by pesticides, converted to agricultural production, or used in such a way that it no longer provided these benefits, what would its replacement costs be? The cost of constructing soil erosion control facilities to protect the water resource or of developing treatment mechanisms to repair damage to the water resource could be considered the value of the prairie in providing the erosion control service. In other words, the value of this specific service from the prairie is equal to the cost of providing this service with substitute methods.

Conclusions

The importance of the economic value of the prairie stems from the basic question of whether remaining prairie lands should be preserved and prevented from being converted to nonprairie use. As one approaches this question the issue of the cost and benefits of protecting the prairie and, therefore, its value cannot be avoided. Determining economic value almost always means estimating a monetary value. We live in a world of scarce resources. One of these scarce resources is financial resources. Because many of the services that are derived from the prairie are public in nature, use of limited government budgets often enters the picture in the decision to protect or influence the use of prairie resources whether they are owned by private individuals or not. If the prairie's values are not determined, then the benefits and costs of policies which affect prairies will not be considered properly in government budget allocations. Although economists have developed many techniques for determining values of natural resources, each has shortcomings. The failure of these techniques to accurately estimate value are rooted in the inability to isolate the value for a particular resource from all other economic, political, and social issues. Issues that may never be satisfactorily represented. Although many people are averse to monetizing the value of natural assets, such as prairie, the issue of whether it should be done or not is being overshadowed by the issue of how a better job of accounting for environmental assets in our society's economic decision-making process can be accomplished. There is also concern that the quantity and quality of environmental characteristics influence the health of the macroeconomy and human well-being. In other words, use, abuse, conservation, and preservation of resources are inescapably linked to the performance of the economy and human welfare. Authors such as Daly (1991), and Boulding (1966), have presented the idea that accounting for the value of the environment and determining the optimal size of the economy that can be sustained by the environment are important. To do this requires more formal valuation of environmental resource services.

In light of this situation there are several conclusions which can be drawn.

1. Value is based on human definitions and not "natural" value.

2. Monetization of all value is difficult and inexact, but inevitable.

3. Basic measures include use values (for which we have excellent and poor market values) and nonuse values (for which there is no market value).

4. Human welfare is affected by the valuation of resource services such as those obtained from the prairie but not valued in markets.

5. Measures of nonmarket values are contestable estimates.

6. The task of valuing the prairie is difficult and no two groups will arrive at the same value due to the influence of individual cultures and beliefs.

7. Improved techniques of resource valuation for decision making and accounting of resources in economic indexes is critical to conservation of resources including the prairie.

PART 2

Prairie Ecology

Prairie Ecology–The Tallgrass Prairie

Ernest M. Steinauer
Scott L. Collins

Tallgrass prairie was the dominant presettlement vegetation type in the eastern third of the Great Plains. It occupied approximately 60 million ha (Samson and Knopf 1994) and extended from southern Texas to southern Manitoba (see fig. 3.1) (Küchler 1985). Tallgrass prairie formed a relatively narrow east-west band except in its central portion where the prairie peninsula extended eastward to the Illinois-Indiana border. There is a gradual transition from tallgrass prairie into mixed-grass prairie to the west, deciduous forest to the east, and boreal forest or aspen parklands to the north. These boundaries have shifted historically in response to changing climates and fire frequencies. In this chapter we briefly outline the tallgrass prairie environment, vegetation, historical development, and the role of disturbance, including human impacts, and end with a recommendation for a landscape level management scenario.

Tallgrass prairie is the most mesic of the Great Plains grasslands. Annual precipitation ranges from 60 cm in the northwest to 100 cm in the east and southeast, most of which occurs during the growing season (Vankat 1979), though late summer droughts are common. Climates in tallgrass prairie are continental, with seasonal temperature extremes ranging from -35°C to 45°C.

Vegetation and Ecology

Tallgrass prairie vegetation is dominated by C_4 grasses including big bluestem, switchgrass, indian grass, and rough dropseed. (C_4 plants initially fix carbon

Prairie Conservation
Island Press (Washington, DC • Covelo, CA)

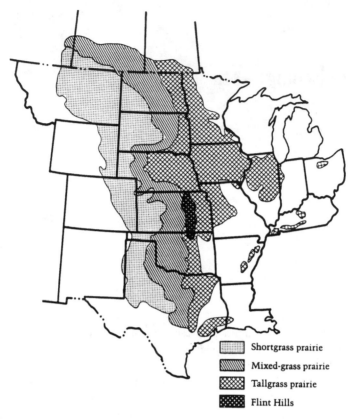

Shortgrass prairie
Mixed-grass prairie
Tallgrass prairie
Flint Hills

Fig. 3.1 Prairie distribution at the time of European settlement (adapted from Reichman 1987).

dioxide into a four-carbon compound rather than three-carbon compound as in C_3 plants.) Mid- and short-statured C_4 grasses exist as subdominants under the canopy of the tall grasses or as dominants on shallow or dry soils or in successional seres; they include little bluestem; sideoats, hairy, and blue grama; western wheatgrass; and buffalo grass. The C_3 graminoid component in tallgrass prairie increases in importance from south to north. Common C_3 graminoids include Scribners panicum, porcupine grass, junegrass, Kentucky bluegrass, and several sedge species. Though graminoids make up the bulk of production in tallgrass prairie, species richness and diversity result from an abundance of forb species (Howe 1994a). The forb component is also more responsive to disturbance and changing environmental conditions (Glenn and Collins 1990; Steuter et al. 1995). Woody species are locally abundant especially in riparian and other areas protected from fire.

Tallgrass prairie is a recently derived vegetation type (Axelrod 1985). Axelrod suggests that a woodland-grasslands mosaic occurred on the Great Plains prior to the Pleistocene. During the Pleistocene the eastern Great Plains was covered by glacial ice in the north and forest in the south. It was not until the Holocene that conditions conducive to continuous grasslands existed on the Great Plains. Extensive use of fire by aboriginal peoples may be at least partly responsible for the development of tallgrass prairie, especially in the prairie peninsula region (Steuter 1991). Evidence for the recent development of the tallgrass prairie flora includes the virtual lack of endemic species and the strong affinities of the tallgrass prairie flora with those of eastern and southwestern North America. Despite its recent origins, tallgrass prairie vegetation is well adapted to current environmental conditions and the historic disturbance regime (Mack and Thompson 1982; Milchunas et al. 1988).

Only about 4 percent of presettlement tallgrass prairie remains (see table 3.1) (Samson and Knopf 1994); most of the remainder was plowed for agricultural purposes soon after European settlement. Large tracts of tallgrass prairie remain in the Flint Hills of eastern Kansas and northeast Oklahoma and on glacial moraines in northeast South Dakota. In general, these areas have steep terrain, rocky soils, or both, making them unsuitable for crop production.

Table 3.1

Estimated original and current area and percent of original area of tallgrass prairie on a state and provincial basis

State/Province	Historic area (ha)	Current area (ha)	Decline (%)
Manitoba	600,000	300	99.9
Illinois	8,900,000	930	99.9
Indiana	2,800,000	404	99.9
Iowa	12,500,000	12,140	99.9
Kansas	6,900,000	1,200,000	82.6
Minnesota	7,300,000	30,350	99.6
Missouri	5,700,000	30,350	99.5
Nebraska	6,100,000	123,000	98.0
North Dakota	1,200,000	1,200	99.9
Oklahoma	5,200,000	N/A	N/A
South Dakota	3,000,000	449,000	85.0
Texas	7,200,000	720,000	90.0
Wisconsin	971,000	4,000	99.9

Source: Adapted from Samson and Knopf 1994.

Throughout the remainder of the former range of tallgrass prairie, only small, scattered fragments remain embedded in nongrassland landscapes. We can gain some indication of the size distribution of tallgrass prairie remnants using the Nature Conservancy's (TNC) tallgrass prairie preserves system as a sample (see table 3.2). Of the 101 TNC tallgrass prairie preserves, 71 are less than 100 ha in size, while only 4 are greater than 1,000 ha (Wayne Ostlie, the Nature Conservancy, personal communication). The small size of most tallgrass prairie remnants results in increased edge effects and likelihood of invasion by undesirable or exotic species, low genetic diversity in local populations, and increased extinction rates. In addition, these remnants are typically isolated, reducing the possibility of gene flow among remnants or recolonization of locally extinct species from neighboring remnants.

Prairies have been extensively studied by the ecological community and have had significant impacts on theory development in community ecology (Clements 1916; Gleason 1926). However, much of tallgrass prairie was extirpated prior to extensive ecological study. Questions as fundamental as presettlement vegetative composition (Clements 1936), extent of grazing by large herbivores (Roe 1970; Bamforth 1987), and fire frequency and season remain (Howe 1994b). Therefore, appropriate management and conservation objectives for tallgrass prairie are neither clear nor straightforward.

Species composition and diversity patterns in tallgrass prairie are structured by forces operating across a wide range of spatial and temporal scales. Evolution of component species, postglacial species migrations patterns, climate, and large-scale topographic patterns constrain species distributions at the largest spatial and longest temporal scales. At intermediate scales, species distributions are constrained by local topographic patterns, drought, fire, and large mammal herbivory (Collins and Steinauer, in preparation). At still smaller scales, local pat-

Table 3.2

The size distribution and total area in each size class of tallgrass prairie preserves owned by the Nature Conservancy as of March 1995

Size class (ha)	Number in size class	Total area (ha)	Percent of total area
<10	14	59.0	0.2
$10-10^2$	57	2291.9	7.7
10^2-10^3	26	7962.2	26.6
10^3-10^4	3	5715.7	19.1
$>10^4$	1	13920.2	46.5

Source: Data provided by Wayne Ostlie, the Nature Conservancy.

terns of immigration and extinction (Glenn and Collins 1992), interspecific competition (Grime 1973; Tilman 1985), and a suite of small-scale disturbances (e.g., small mammal mounds [Platt 1975], harvester ant hills [Coffin and Lauenroth 1990; Umbanhower 1992]) affect the structure of grassland communities. All these factors interact in a complex fashion across scales to increase overall heterogeneity (Allen and Hoekstra 1992).

Intermediate-scale phenomena, such as fire and grazing, occur over areas and at return intervals easily or economically controlled by managers. Large-scale, slowly changing forces (e.g., climate, topography) are beyond managerial control but should be carefully considered when establishing a reserve or reserve system. In addition, many small-scale factors (e.g., insect herbivory) may be difficult to control directly because of the large number of individual events constituting such phenomena or because of our inability to target specific entities (e.g., with chemicals). However, managers can often exert at least some control over small-scale phenomena by manipulating intermediate-scale phenomena (Allen and Starr 1982). For example, the relative abundance of cool- versus warm-season plant species in tallgrass prairie is influenced by fire season (Howe 1994a). Steuter et al. (1995) found that the abundance and distribution of pocket gophers, a small, territorial herbivore, in a sandhills prairie was constrained by land-use patterns of bison, a larger, free-ranging herbivore. It follows then that the outcome of management practices will be constrained by local climate, topography, and the available species pool and will likely differ somewhat among areas.

Presettlement tallgrass prairie was a disturbance-prone system. It burned frequently; a variety of both large and small herbivores existed; and droughts occurred periodically. It is becoming increasingly clear that disturbance is a critical component in maintaining the plant and animal communities and ecosystem processes of tallgrass prairie.

Natural disturbances operate with characteristic size, frequency, and intensity and interact with evolved vegetation characteristics to determine community structure within the environmental context of a given area (Mack and Thompson 1982). Interactions among disturbances, as well as among disturbance and other structuring variables (e.g., topography), resulted in a continuous range of patch types, each with a complex and unique disturbance history. In addition, the disturbance regime in tallgrass prairie developed within the context of a large, continuous grassland landscape. Many tallgrass prairie remnants, on the other hand, are relatively small, isolated, and managed in fixed units to which a limited subset of the natural disturbance regime is applied, usually at regular intervals. The result is a much simpler landscape mosaic.

Because of the importance of disturbances in the development of tallgrass prairie, we briefly review the considerable body of literature dealing with the role of fire, grazing, drought, and small mammal soil disturbance (herbivory) in community structure and dynamics of tallgrass prairie. We also discuss mowing, even though mowing has no presettlement analog.

Fire

Fire played a key role in the formation and maintenance of the tallgrass prairie (Wells 1970; Axelrod 1985). Estimates of historical fire frequency in tallgrass prairie range from two to five years, but fire return intervals probably varied widely (Bragg 1982). Fire frequency initially increased following European settlement but then decreased as fire control efforts were instituted and the prairie landscapes became increasingly dissected by roads, fields, and other artifacts of civilization.

Lightning is the chief natural ignition source of prairie fires. Lightning-ignited fires occurred from March through December but were most common in mid- to late summer (Bragg 1982). The spatial extent of summer fires may have been limited by the large amounts of green matter and high humidity at that time. However, large summer fires probably occurred during drought periods. Up to half of all fires in tallgrass prairie may have been set by Native Americans, including dormant-season fires set to attract herbivores and to protect encampments from wildfires (Moore 1972; Higgins 1986). These fires likely burned large areas due to dry conditions and the lack of green matter.

Fire has a number of important impacts on tallgrass prairie vegetation. Fires, especially those occurring during the growing season, restrict woody and other fire-intolerant species to riparian zones and other protected areas, though precipitation throughout the tallgrass prairie region is sufficient to support woody vegetation in the absence of fires (Bragg and Hulbert 1976; Anderson 1990). Burning releases nutrients bound in litter and deposits them as ash on the soil surface; however, some may be lost via runoff. Though nitrogen (N) is volatilized by prairie fires, fire creates conditions conducive to free-living N fixers, and soil N is rapidly replenished. In addition, the perennial C_4 grasses that dominate tallgrass prairie remove much of the N from aboveground tissues prior to senescence (Adams and Wallace 1985) and thus lose little N in dormant-season fires. Fire, by exposing the soil surface, temporally increases the potential for soil erosion.

Fire-mediated litter dynamics influence the relative contribution of N and light in limiting production in tallgrass prairie (Knapp and Seastedt 1986; Seastedt et al. 1991). In unburned prairie, litter reduces light levels and, consequently, plant growth (Weaver and Roland 1952), though soil N is abundant. High light and temperature levels following fire increase plant production (Knapp 1984) until limited by depleted soil N. Therefore, production can be increased by fertilizing burned but not unburned prairie.

Fire affects plant species diversity in tallgrass prairie in a complex manner.

Frequent spring burning increases the dominance of C_4 grasses and reduces the abundance of cool-season species (Hulbert 1988) thus reducing overall species richness (Collins 1987, 1992). Many prairies are managed in this manner since high C_4 grass abundance is often equated with high quality, late successional status, and since C_4 grasses are preferred for livestock forage. Occasional spring burning may increase species richness (Collins and Gibson 1990) perhaps by opening space for seedling establishment (Abrams 1988). Summer fires, on the other hand, reduce the abundance of C_4 grasses and increase the abundance of cool-season species (Biondini et al. 1989; Howe 1994a). Because the cool-season component contributes relatively little biomass but includes a large number of species (Howe 1994a), summer burns may increase diversity but decrease overall production. Diversity tends to peak several years postfire as increasing litter depths reduce the competitive abilities of C_4 grasses but still allow for seedling establishment (Gibson and Hulbert 1987). Eventually, sufficient litter accumulates to limit seedling establishment and species richness declines (Gibson 1988).

Prairie fires have been shown to affect a variety of animal communities including ungulates (Shaw and Carter 1990), birds (Herkert 1994b), small mammals (Kaufman et al. 1990), insects and other arthropods (Evans 1988), and earthworms (James 1988). The soil microbial community is also affected by fire (Garcia and Rice 1994). Direct mortality from prairie fires has been documented for small mammals (Erwin and Stasiak 1979; Harty et al. 1991) and arthropods (Warren et al. 1987). However, indirect mortality resulting from changes in the plant community, loss of cover and food items, increased exposure to predation, and alteration of the soil microclimate is usually more important (Kaufman et al. 1990). Because many species are restricted to particular habitats, prairie remnants should be managed with a variety of fire frequencies, including infrequently burned areas.

Herbivory

Herbivores inhabiting the tallgrass prairie prior to European settlement included bison, elk, white-tailed deer, mule deer, and several smaller vertebrate and invertebrate species. The extent to which large herbivores, especially bison, grazed tallgrass prairie is unclear (Roe 1970). Historical reports of extremely large bison herds were mainly from the mixed prairie region (McDonald 1981). The large open areas required to maintain social contact among members of large bison herds are uncommon in tallgrass prairie landscapes, which tend to be dissected by wooded stream courses. In addition, bison preferentially graze short, actively growing forage (Shaw and Carter 1990). While tallgrass prairie forage may be suitable for bison early in the growing season, it becomes less so as it increases in height and coarseness during the growing season, though local grazing lawns may be maintained. A possible scenario for bison use of tallgrass prairie is many small herds in the dormant and early growing season with the majority migrating to mixed-grass and shortgrass regions as the growing season progresses. A similar pattern exists among ungulate herds in the short- and tallgrass regions of the Serengeti grasslands (McNaughton 1984). Bamforth (1987) has suggested that European settlement of the eastern portions of the historic bison range altered bison distribution patterns, concentrating bison farther west. Thus the large herds in the mixed prairie at the time of settlement were possibly a recent phenomena and not indicative of historic grazing patterns. Though the extent of large herbivore grazing in tallgrass prairie is far from resolved, many privately held prairie remnants are grazed by cattle, and grazing by native or domestic herbivores is increasingly a part of the management of tallgrass prairie reserves.

Large herbivores impact grasslands at several ecological levels. Large herbivores alter plant species abundance patterns by selective removal of preferred forage species. Bison diets, for example, consist of up to 90 percent graminoids, while cattle diets consist of approximately 70 percent graminoids (Plumb and Dodd 1993). Selective grazing of graminoids releases forbs from competitive pressure and increases plant species diversity (Collins 1987). Tallgrass prairie species that decrease with cattle grazing include many of the dominant C_4 grasses, such as big and little bluestem, switchgrass, and indian grass, while species that increase with grazing include several short-statured graminoids such as Kentucky bluegrass, Scribners panicum, sideoats grama, buffalo grass, and sedges, and forbs (Weaver 1954; Anderson et al. 1970). Heavy grazing of tallgrass prairie results in degraded range with low native diversity and increased abundance of exotic species (Weaver 1954).

Large herbivores select grazing sites at a variety of spatial scales in order to maximize forage intake (Sneft et al. 1987). Sites selected vary seasonally based on dietary requirements, forage quality, and variations in herd size (Vinton et al. 1993; Steuter et al. 1995). Recently burned areas are often preferentially grazed (Shaw and Carter 1990) and the combination of burning and grazing impacts vegetative composition and diversity to a greater extent than either acting alone (Collins 1987). In addition, grazing, by reducing standing crop and litter accumulation, reduces the likelihood of future fires. By grazing only a portion of potential sites, large herbivores increase grassland heterogeneity.

Other impacts of large herbivores include trampling vegetation (Wallace 1987) and wallowing. Bison wallows are depressions ranging from 3 to 5 m in diameter (Collins and Barber 1985) that are virtually free of vegetation when actively used. Some abandoned wallows are invaded by ruderal species; others retain water and support local populations of aquatic plants (Collins and Uno 1983). Wallows may persist for decades after abandonment. Large herbivores also affect plant distributions by dispersing seeds trapped in pelage (Tonielli et al., in preparation) and via seed ingestion followed by deposition in fecal pats (Janzen 1984; Brown and Archer 1987).

Large herbivores indirectly affect the plant community by altering ecosystem level processes. Herbivory reduces litter accumulation, increases rates of nutrient cycling (Woodmansee 1978), and maintains actively growing vegetation. The litter of C_4 grasses is low in N and decomposes slowly, thus slowing nutrient cycling rates. Herbivore excretory products, especially urine, are high in readily available nutrients (Steinauer and Collins 1995). Herbivores remove nutrients from large areas and deposit them in concentrated form as excrement, creating a patchy nutrient environment (Steinauer 1994). Bison urine patches, for example, increase local forage production, alter species composition, and are more likely to be grazed than surrounding off-patch areas (Norman and Green 1958; Jaramillo and Detling 1992b). In addition, large herbivores transport nutrients across landscapes by differential rates of forage intake and excrement among various habitats (McNaughton 1983, 1985).

The effects of smaller vertebrate and invertebrate herbivores, including belowground herbivores, on tallgrass prairie vegetation has been less studied. Several studies suggest that effects of invertebrate herbivores on grassland vegetation and successional dynamics may be substantial (McBrien et al. 1983; Stanton 1988; Brown 1985). For example, production on a South Dakota mixed prairie increased between 26 and 51 percent when nematodes were controlled with carbofuran (Ingham and Detling 1990). However, Gibson et al. (1990) found little change to the plant community following pesticide application to tallgrass prairie soil. The interactive effects of various groups of mammalian and invertebrate herbivores on each other and on grassland vegetation are also not well studied but appear complex.

One group of small herbivores that has received considerable attention is pocket gophers, principally because of their mound building activity (Reichman and Smith 1985; Inouye et al. 1987). The mounds of gophers and other small mammals provide establishment sites for ruderal plant species uncommon in undisturbed prairie (Platt 1975), thus increasing diversity. This is problematic, however, when the potential colonists include undesirable exotic species. High energy costs associated with tunneling constrain gophers' herbivory patterns (Reichman et al. 1982) and choice of forage items. Steuter et al. (1995) found that gophers and bison in a Nebraska Sandhills prairie effectively partitioned the forage resource: bison consumed mostly widely distributed C_4 graminoids, while gophers consumed tap roots of patchily distributed C_3 forbs.

Mowing

Though mowing was not a part of the presettlement disturbance regime, it is now a common management practice on tallgrass prairie and hence warrants discussion. Mowing controls woody plant encroachment and, if followed by hay removal, removes standing vegetation and litter, exposing the soil surface to increased solar radiation (Hoover and Bragg 1981). This, in some ways, mimics the effects of fire. However, mowing leaves stubble and does not blacken or deposit ash on the soil surface. Both of the latter have been associated with increases in production following burning (Hulbert 1988). As with burning, spring mowing increases the relative abundance of C_4 grasses, while summer mowing increases C_3 species abundance (Hoover and Bragg 1981). Mowing does not appear to adversely affect forb richness. In contrast to burning, mowing hay fields and removing forage may be done many times per year. Hence, long-term summer mowing decreases productivity (Ehrenreich and Aikman 1963) though occasional spring mowing increases production on tallgrass prairie (Hulbert 1969). While burning and grazing occur in a patchy manner, entire pastures are typically mown, thus reducing heterogeneity.

Drought

Moderate to severe droughts occur periodically in tallgrass prairie (Weaver 1954; Tilman and El Haddi 1992). The effects of drought include decreased plant production and diversity. Though predrought production levels may return rapidly with more normal precipitation, effects on diversity may be more persistent. The effects of prolonged drought, such as occurred in the 1930s, include shifts in species ranges, invasion of exotics, and considerable habitat degradation. However, human activities likely exacerbated these effects. For example, under presettlement conditions large herbivores likely migrated from drought-impacted areas, whereas fenced herbivores may severely damage pastures before being removed.

The impacts of other disturbances may be more severe during drought. Production is reduced on burned compared to unburned areas during dry years, the opposite response to normal or above normal precipitation years (Weaver 1954). Growing-season fires during drought are likely larger and more intense, which when combined with moisture stress may have more profound impacts on vegetative composition but may be more effective in controlling woody plant encroachment (Anderson 1990). Grazing impacts also appear to be more severe during drought.

Human Activities

Humans have occupied North America throughout the period of grassland development in the Holocene (Axelrod 1985). Though small populations and limited technology probably reduced aboriginal impacts in general, their extensive use of fire likely had a significant impact on species composition and the spatial extent of tallgrass prairie (Steuter 1991). However, North American grasslands appear to have been well adapted to the impacts of aboriginal peoples at the time of European settlement.

Europeans fundamentally altered the disturbance regime of North American grasslands including the introduction of several novel disturbances. Chief among these is plowing, which maintains grasslands in an early successional condition. There is little evidence that oldfield succession leads to a vegetation type that more than superficially resembles presettlement tallgrass prairie. Most remaining private prairie tracts are managed for economic gain via livestock grazing or hay production and may be treated with herbicides to decrease forb production or seeded with exotic grasses. Even prairies protected for conservation purposes are generally managed with a limited component of the presettlement disturbance regime (Steuter et al. 1990b). Reasons for this may be economic, logistical, or a lack of appreciation of the role of complex disturbance interactions in maintaining prairie diversity.

Management

The small size of many prairie remnants may make the application of more than a subset of the natural disturbance regime difficult if not impossible, and reconstructing a more or less fully functioning prairie landscape may be possible on only the largest preserves. No tallgrass prairie reserves currently exist that are large enough to include the full component of presettlement species. Large predators, such as wolves or grizzly bears, which were extirpated soon after European settlement, are unlikely to be reintroduced. In addition, some prairie remnants contain rare or endangered species, further constraining management options.

Given the large scale and complexity of the disturbance regime in prairie, we argue that attempts to "re-create" or simulate presettlement prairie conditions within a single reserve are unlikely to be successful. Rather, we propose a flexible regional scheme that includes consideration of realistic goals relative to factors such as the size of a given reserve, surrounding land-use, distance to the nearest reserve, management objectives, and available natural and human resources. Specifically, we recommend management of tallgrass prairie remnants with the largest subset of the natural disturbance regime or "analog" possible given prevailing conditions. With this in mind, we offer a hierarchically structured regional disturbance model based on reserve size and distribution within a landscape setting (see fig. 3.2). The framework calls for management approaches based not only on the local reserve but also on the management of other regional reserves and surrounding land-use. For example, grazing may not be an important component of a reserve surrounded by grazed prairie, as in the Flint Hills. In a reserve system of many small fragments (e.g., western Minnesota), different

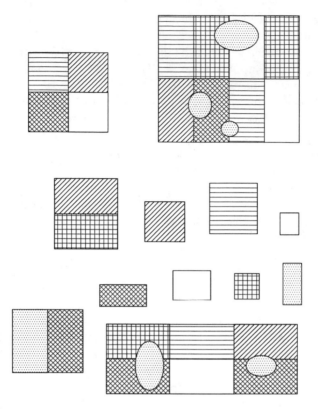

Fig. 3.2 Hypothetical distribution of disturbance types among regional prairie remnants. Boxes represent prairie remnants. Fill patterns represent disturbance types, such as fire frequency, season of burn, mowing, grazing regimes, etc. Larger prairie remnants can potentially contain more than one disturbance type. Distance between remnants not drawn to scale.

fragments could be subjected to different fire frequencies. Thus the management practices of a given reserve are a function of size, goals, surrounding land-use, and management of other reserves in the region.

The largest preserves can potentially include the full component of the natural disturbance regime. TNC's 14,000-ha Tallgrass Prairie Preserve located in the Flint Hills in northeast Oklahoma provides an example. The management plan for this preserve includes reintroduction of free-ranging bison and a fire regime based on biomass accumulation rather than on a set burning schedule (see Steuter et al. 1990b). The goal is to produce a dynamic prairie landscape in which the large-scale forces of fire and bison grazing constrain the more difficult-to-manage small-scale forces. Much of the area surrounding the Tallgrass Prairie Preserve, and the Flint Hills in general, remains in native tallgrass prairie, which helps curb invasion by agricultural weeds and provides a regional native species pool to counteract local species extinction. Even this excellent management plan, however, does not completely simulate presettlement conditions because the bison herd's size and movement are artificially constrained. Nevertheless, this method represents a conscious attempt to incorporate the interactive effects of large- and small-scale disturbance within a large regional context (see fig. 3.2).

On intermediate-size preserves, a subset of the historic disturbance regime could be applied subject to local constraints. Fire regimes could be varied among portions of the preserve, and bison or cattle could be periodically released into temporarily fenced areas. On the smallest preserves only a limited disturbance regime is possible. Many of these preserves are set aside to protect rare species or are near populated areas. Management will therefore have to reflect the constraints imposed by rare species, and mowing may have to be substituted for fire in populated areas. We suggest that to the degree possible all components of the historic disturbance regime and, when necessary, substitutions such as mowing, be applied among the preserves on a regional level (see fig. 3.2). In this manner species dependent upon particular components of the disturbance regime will be more likely to have habitat available.

Of special concern when reintroducing disturbance to tallgrass prairie remnants is loss of native species and invasion by exotic species. Plant populations in natural settings continually fluctuate in size, and local extinctions followed by reestablishment from nearby populations are common (Glenn and Collins 1992). In the current highly fragmented prairie landscape, however, local sources of native propagules are largely absent, and human intervention will likely be necessary to reestablish lost species. Prairie remnants are exposed to a constant influx of exotic species from surrounding agricultural, urban, or forested landscapes. Disturbance may compound this problem by reducing sensitive native populations and by providing establishment sites for exotics (Orians 1986; Parker et al. 1993). Once established, exotics may be difficult if not impossible to remove. The impact of disturbance on local extinction of native species and invasion of exotics needs further study.

Summary

The future of tallgrass prairie does not appear particularly bright. The vast majority of tallgrass prairie has been irrevocably lost to the plow or other human interventions, leaving a few remnants scattered on the landscape. Maintaining these remnants in anything approaching presettlement condition will require considerable expenditures of money and time. Indeed, it may be impossible to stem the gradual loss of species and genetic diversity on the smaller prairie remnants especially in the face of potential global climate changes. The greatest opportunities to preserve tallgrass prairie are likely in the Kansas Flint Hills and the few other areas with extensive tracts remaining. Though our rational minds tell us there is perhaps little hope for long-term preservation of the many small prairie fragments scattered across America's heartland, we find it difficult to release our emotional attachment to these small but beautiful pockets of waving grass. Though the prairie lacks the mass appeal of more spectacular ecosystems such as temperate or tropical rain forests, there are fortunately many prairie enthusiasts. Perhaps with dedicated efforts we can preserve the tallgrass prairie so that those who come after us can marvel at its many wonders.

Prairie Ecology—The Mixed Prairie

Thomas B. Bragg
Allen A. Steuter

The mixed prairie occupies the central third of the North American Great Plains. It is bounded by tallgrass prairie to the east, shortgrass prairie to the west, aspen parkland to the north, and juniper-oak savanna to the south (Küchler 1985). This semiarid land is characterized by seasonal moisture and temperature extremes typical of a continental climate. In the northern mixed prairie, annual precipitation increases from 30 cm in the west to 60 cm in the east; the southern precipitation gradient increases from 40 to 80 cm (Bryson and Hare 1974). Two-thirds of the annual precipitation occurs during the growing season, although regional droughts are common. The west-to-east elevation in the north ranges from 1,130 to 400 m.

During most of the Holocene, mixed prairie uplands and lowlands have been dominated by herbaceous vegetation, with woodlands restricted to isolated buttes, scarps, and riparian habitats protected from fire (Axelrod 1985; Wells 1965). Perennial grasses dominate above- and belowground resources and primary production, but forbs are largely responsible for community diversity. Typically, grasses are represented by tens of species, while forb species number in the hundreds. Distribution and abundance of forbs are also more dynamic and diagnostic of changes in moisture, grazing, and fire regimes than is the perennial grass matrix (Biondini et al. 1989; Steuter et al. 1995). Interestingly, with the exception of blowout penstemon, no widely recognized plant species are endemic to the mixed prairie (Stubbendieck et al. 1993). Blowout penstemon is uniquely adapted and confined to the most actively wind-eroded sites in the Nebraska Sandhills prairie.

Prairie Conservation
Island Press (Washington, DC • Covelo, CA)

Humans have played a major role in the evolutionary history of mixed prairie, largely through their use of fire (e.g., Moore 1972; Higgins 1986). Other important species included the wolf, North American bison, prairie dog, and plains pocket gopher. Historically, mixed prairie formed the central portion of the primary bison range (McDonald 1981), attracting vast summer breeding herds because of the region's openness, high-quality forage, and relatively abundant water (Hansen 1984). The faunal component also included wetland species. The mixed prairie landscape is characterized by broad river valleys with gently rolling interfluvial plains. The regular sequence of rivers flowing through the mixed prairie, the prairie potholes of the Dakotas and Canadian provinces, and the sandhill lake regions of Nebraska formed a dispersed and redundant system of critical water and wetland habitats for a diverse array of migratory and nonmigratory species. The region is a central feature of the Great Plains Flyway, a migratory waterbird spectacle that still rivals that of the great bison herds.

The mixed prairie consists of three types, based on plant community structure and function: northern mixed prairie, sandhills prairie, and southern mixed prairie.

Ecology and Distribution

The evolution of the mixed prairie resulted in biota well adapted to grazing (Mack and Thompson 1982; Milchunas et al. 1988) and fire (Wright and Bailey 1982). The mixed prairie is largely a product of these two forces interacting with regional soils, weather, and climate—particularly periodic drought (Weaver and Albertson 1956). The effects of grazing, whether by bison or cattle, are similar in that standing crop is reduced (see table 4.1). Fire also reduces standing crop (Hopkins et al. 1948) and litter (Willms et al. 1993), thus altering species diversity patterns (Biondini et al. 1989), modifying grazing patterns (Coppock and Detling 1986), and variously affecting animals (Bragg 1995). The interaction of fire and grazing often has different effects than either process alone (Pfeiffer and Steuter 1994). Fire and grazing magnify drought stress on mixed prairie vegetation (Mihlbacher et al. 1989). The adverse effects of drought are most severe on little bluestem, and less so on sideoats grama, blue grama, buffalo grass, and western wheatgrass (Weaver 1968). Dynamic shifts in the plant community also occur with fertilization (Rauzi and Fairbourn 1983), woody plant removal (Vallentine 1980), mechanical disturbance of soil (Haferkamp et al. 1993), and mowing (Launchbaugh 1973). As with fire, mowing tends to reduce production during all but high-rainfall years. Fertilization, mostly used in the northern mixed prairie, generally is not economically feasible.

Table 4.1

Representative standing crop from mixed-grass prairies of the North American Great Plains

Location and Treatment	Standing Crop (kg/ha)		Reference
	Treated	Untreated	
Northern mixed			
Grazed	347	395	Hoffman and Ries 1989
	205–400	228–382	Brand and Goetz 1986
	170–360	268–416	Sims et al. 1978
Burned	265	291	Gartner et al. 1986
	357–403	409	Gartner et al. 1978
Sandhills			
Grazed	182(b)	347(b)	Bragg 1978
Burned	131–393	182–440(g)	Bragg 1978
Southern mixed			
Grazed	182	190	Sims et al. 1978
	242	421	Tomanek and Albertson 1953
Burned	820	861	Nagel 1983
	197(g)	390(g)	Hopkins et al. 1948
	444	216	Adams and Anderson 1978

Note: Treatment includes grazing (g) or burning (b).

Northern Mixed Prairie

The original northern mixed prairie covered approximately 38 million ha in Nebraska, North and South Dakota, and Canada. Plant communities included the wheatgrass-bluestem-needlegrass and the wheatgrass-needlegrass associations of Küchler (1985). Cool-season grasses become increasingly more dominant from Nebraska to Canada. Mesic associations of taller species generally occur on lower slopes, transitioning to midheight and then to shorter species associations on the dry hilltops (Barnes et al. 1983). These grasslands occur primarily on loamy glacial tills and clay to clay-loam soils.

Western wheatgrass is the common denominator of the northern mixed prairie type, even though it is not always a dominant (Gartner 1986). Other common grasses include blue grama, needle-and-thread, green needlegrass, and porcupine grass. Without burning, Kentucky bluegrass and smooth brome, cool-season exotics, increase in the northern mixed prairies (Kirsch and Kruse 1973). Forb productivity ranges from 0 to 40 percent of total net primary production

(Lura et al. 1988) but may vary considerably with heavy grazing (Whitman 1974). Several woody species—western snowberry, fringed sagebrush, and eastern redcedar—have replaced herbaceous species in the region (Wright and Bailey 1982; Kaul and Rolfsmeier 1993).

In general, grazing favors short-statured, or rhizomatous, species, such as western wheatgrass and blue grama, over taller, or bunchgrass, species, such as little bluestem (Mack and Thompson 1982). Shifts in species composition are more a function of grazing intensity and plant species morphology and reproductive mechanisms than whether they are cool- or warm-season species (Ode et al. 1980; Singh et al. 1983; Brand and Goetz 1986; Schacht and Stubbendieck 1985). Grazing decreases litter, but litter accumulation does not appear to limit productivity (Dix 1960). Although heavy grazing or the exclusion of grazing does not increase decomposition (Shariff et al. 1994), moderate grazing increases decomposition and affects soil chemical properties (Dormaar and Willms 1990). Thus, grazing is important in maintaining the ecosystem processes that occurred when large number of bison dominated the Great Plains grasslands. Grasses that decrease with grazing include big bluestem and indian grass. Sedges and leadplant are among other species declining with grazing. Blue grama, ironweed, western ragweed, and curlycup gumweed increase with increased grazing intensity (Branson and Weaver 1953; Brand and Goetz 1986).

Fire was a frequent event in the northern mixed prairie (Higgins 1986). In general, fire reduces standing crop of both cool- and warm-season species during dry years and maintains or increases standing crop in wet years (Engle and Bultsma 1984; Whisenant and Uresk 1990). Fire improves herbage quality and decreases litter (Willms et al. 1980, 1986); it also increases bare ground, allowing more light to penetrate the canopy during the growing season (Dix 1960). Recovery of mulch structure may take at least three years. Reductions in net production and individual species' primary production caused by fire are due to lower plant and soil water potentials on burned sites (DeJong and MacDonald 1975), climatic conditions and the attraction of grazers to recently burned areas (Gartner et al. 1986), site productivity (Dix 1960; DeJong and MacDonald 1975; Whisenant and Uresk 1990), the presence of significant amounts of native warm-season species (Schacht and Stubbendieck 1985), and topographic location. The response to fire also varies depending on the season in which the fire occurs (Dix 1960; Coupland 1973; Engle and Bultsma 1984; Schacht and Stubbendieck 1985; Gartner et al. 1986; Steuter 1987; Whisenant and Uresk 1990; Redmann et al. 1993). Fall burning has the most adverse effect on herbage production, favoring cool-season species over warm-season species. Spring burns decrease some cool-season species (e.g., Kentucky bluegrass and green needlegrass) and increase others (e.g., western wheatgrass, blue grama, and buffalo grass). Although fire may reduce standing crop, community composition and diversity patterns following fire are indicative of a grassland well adapted to its effects (Biondini et al. 1989). The range of variability in plant composition between spring, summer, and fall burns is similar to that caused by annual fluctuations in weather (Biondini et al. 1989). Complete fire suppression results in accumulation of mulch, conditions that favor cool-season exotic species (Ode et al. 1980; Whisenant 1990) and most likely accounts for the active invasion of woody plants in the southern portion of the northern mixed prairie (Kaul and Rolfsmeier 1993).

Sandhills Prairie

Sandhills prairie (Küchler 1985) originally encompassed approximately 7 million ha. The Nebraska Sandhills prairie accounts for approximately 5 million ha of this total, while most of the remainder occurs in central Kansas. The Nebraska Sandhills prairie developed on the largest stabilized sand dune complex in the Western Hemisphere (Bleed 1990). The substrate has often not undergone sufficient change to be classified as a soil, but those that have developed are primarily fine sands or fine sandy loams.

Warm-season grasses dominate primary production, and a distinct community zonation exists based on slope position (Barnes and Harrison 1982). Dominant grasses include prairie sandreed, sand bluestem, big bluestem, little bluestem, blue grama, hairy grama, needle-and-thread, and sand dropseed (Weaver 1965). Sedges are ubiquitous even though they make up only a small component of standing crop. Forbs, such as western ragweed, skeletonweed, and plains sunflower, may represent 10 to 25 percent of the regional species standing crop. In addition to regional variations in dominant species, substantial differences in species composition occur between uplands, slopes, and lowlands (Barnes et al. 1984). Presumably because of recent fire suppression, woody plants are actively invading, especially along the prairie margins (Steinauer and Bragg 1987; Steuter et al. 1990a).

The sandhills prairie has been subjected to large herbivore grazing, at least during more stabilized periods. Bison occurred in the Nebraska Sandhills for at least the last eleven thousand years (Loope 1986). Currently, cattle grazing is the principal use of sandhills prairie. As with other mixed prairie communities, grazing reduces standing crop (Bragg 1978). In general, bunchgrasses are less tolerant of grazing than rhizomatous species. For example, fragmentation of little bluestem plants into scattered clumps with high tiller density occurs with heavy grazing pressure (Butler and Briske 1988). Moderate and heavy grazing also effectively prevent litter accumulation. While this accumulation has no significant effect on overall standing crop, it may affect individual species (Potvin and Harrison 1984). Among the sandhills plant species most heavily grazed are sand bluestem, indian grass, prairie sandreed, blue grama, western wheatgrass, needle-and-thread, and switchgrass. Sandhill muhly and sand dropseed do not appear to be affected by grazing (Bragg 1978). The current bunchgrass composition of sandhills prairie appears dependent on fire exclusion (Pfeiffer and Steuter 1994) since large herbivores intensively graze burned bunchgrasses such as little bluestem.

Historic fires in the Nebraska Sandhills prairie occurred as frequently as every four to five years (Bragg 1986). As in other mixed prairie types, fire causes an initial decline in plant standing crop (see table 4.1) although the decline may not persist longer than one to two years (Bragg 1978), depending largely on weather conditions. Both standing crop and species composition are variously affected by different combinations of burning, grazing, and topographic location. The decline in standing crop is greater with combined fire and grazing than with fire alone (Bragg 1978). Burning also may significantly affect surface stability in the sandhills because fall burns leave the soil surface exposed for several months. A large reduction in bunchgrass composition due to the interaction between fire and grazing may increase the risk of wind erosion (Pfeiffer and Steuter 1994).

Sand lovegrass, sandhill muhly, small soapweed, and sand bluestem are among the species that decline with burning (Bragg 1978). Other species, including Missouri spurge and plains sunflower, increase with burning, as do interstitial forbs (Pfeiffer and Steuter 1994). Sand dropseed cover is increased with summer burns, while the standing crop of larger bunchgrasses is reduced. Rhizomatous grasses maintain or increase their standing crop following fires in years with normal or above-normal precipitation.

Southern Mixed Prairie

The original extent of the southern mixed prairie encompassed approximately 24 million ha. It includes the bluestem-grama and mesquite-buffalo grass associations of Küchler (1985). Soils typically range from loams to clays. A wide variety of warm-season grasses of mid- to short stature increasingly dominate as one proceeds from Kansas to Texas. Shrubs become a significant component on the Rolling Plains of Texas.

The Kansas-Oklahoma component is dominated by blue grama, sideoats grama, western wheatgrass, little bluestem, junegrass, green needlegrass, porcupine grass, Kentucky bluegrass, tall dropseed, Canada wildrye, and sedges (Weaver and Albertson 1956; Wright and Bailey 1982). Forbs make up approximately 25 percent of total standing crop and include locoweed, heath aster, aromatic aster, penstemon, scarlet gaura, annual sunflower, and dotted gayfeather. Dominant invaders include yellow sweetclover, gumweed, and foxtail barley. Diversity is relatively high: 236 vascular plants were recorded in a 259-ha site in southern Nebraska (Nagel 1979).

The Rolling Plains and western Edwards Plateau regions of the Texas component are characterized by a scattered overstory dominated by honey mesquite, with lotebush an important subdominant. The herbaceous component is variously dominated by buffalo grass, sideoats grama, tobosagrass, little bluestem, and Texas wintergrass. Many annual forbs and some annual grasses—bitterweed, Carolina canary grass, and little barley—are abundant during wet winters. Perennial forbs include blanket flower, primrose, lazy daisy, lamb's quarter, butterfly weed, sunflower, Patagonian plantain, nightshade, and scarlet globe mallow (Wright and Bailey 1982). Breaks throughout the Rolling Plains region contain large amounts of redberry juniper (Wright and Bailey 1982). The western Edwards Plateau is similar in composition although common curlymesquite, a stoloniferous shortgrass like buffalo grass, is also prevalent.

Drought and topography affect species composition. Mesic conditions favor taller grasses (e.g., little bluestem and big bluestem), and drier conditions favor shorter grasses (e.g., sideoats grama, blue grama, and buffalo grass) (Albertson and Tomanek 1965; Mihlbacher et al. 1989). Disturbances, such as grazing and wallowing of bison and prairie dog diggings, increase grassland diversity (Collins and Barber 1985).

As in other mixed prairies, most studies indicate that grazing reduces standing crop (Milchunas and Lauenroth 1993), although there are exceptions in which long-term changes appear unrelated to grazing (Mihlbacher et al. 1989). In most instances, mid- and tall grasses decrease with grazing while short grasses, especially buffalo grass, increase as much as 90 percent. While heavy grazing reduces standing crop, moderate grazing may only slightly reduce or even increase production over ungrazed areas (Tomanek and Albertson 1957). With no grazing, however, litter accumulation may cause grass-stand degeneration and reduced production. In the absence of fire, ashe juniper, redberry juniper, and honey mesquite invade grasslands and suppress the herbaceous component, thus lowering forage availability (Wink and Wright 1973; Steuter and Wright 1983).

One effect of the interaction of fire and grazing is that reported by Ring et al. 1985. Their study showed that an area repeatedly grazed throughout the growing season resulted in overgrazed patches within a matrix of lightly to ungrazed pasture. Subsequent fires in these patchy fuels would burn unevenly and result in a patchy burn that has been hypothesized to increase prairie diversity (Biondini et al. 1989).

For the Kansas-Oklahoma component, most of the dominant grass species are tolerant of fire and may require two to three growing seasons to recover (Launchbaugh 1973; Nagel 1983). Summer fires are most detrimental, followed by spring and then by fall burning. Buffalo grass, blue grama, sideoats grama, and Kentucky bluegrass are most severely reduced with spring burning in Kansas and

Oklahoma (Launchbaugh 1964). At least three growing seasons are required for recovery of these species to preburn amounts. Decreases were greatest where litter was heaviest (Launchbaugh 1964). Several broad-leaved plants increase with spring burning, including western ragweed (Hopkins et al. 1948). In Texas, species that seem to thrive up to about three growing seasons after a fire include vine mesquite, tobosagrass, Arizona cottontop, little bluestem, plains bristlegrass, and Texas cupgrass (Wright 1974). Generally, these are the species that accumulate the most mulch and thus would be most adversely affected by such an accumulation (Launchbaugh 1964, 1973). A species's response to fire is also affected by climate. Most grasses tolerate fire during years with normal to above-normal precipitation but are adversely affected during dry years (Hopkins et al. 1948; Wink and Wright 1973; Wright 1974). When subjected to fire in dry years, some species, such as sideoats grama, Texas wintergrass, and little bluestem, have been shown to decrease productivity by as much as 40 to 58 percent, requiring up to three years to recover to preburn standing crop. Yet during wet years, little bluestem increased as much as 81 percent (Wink and Wright 1973). While fire is a natural component of the mixed prairie, burning more frequently than every five to eight years will result in a decline in standing crop of the dominant herbaceous species (Sharrow and Wright 1977; Neuenschwander et al. 1978). The response to burning also depends on species composition. Where annual, cool-season grasses are few, Texas wintergrass standing crop declines. Where cool-season grasses are abundant, fire increases production of this species, although the increase is greater with fall than with spring burning (Whisenant et al. 1984). The standing crop of cacti, an abundant group of plants in the southern mixed prairie, is also reduced by burning (Wright and Bailey 1982).

In the absence of burning and grazing and the concomitant increase of mulch, significant reductions in the Kansas-Oklahoma southern mixed prairie occur for the dominant grasses (e.g., blue grama, buffalo grass, and sideoats grama) while other species (e.g., sedges, smooth brome, and tall dropseed) increase dramatically (Nagel 1994). In the Texas mixed prairie, however, fire is particularly important as a control against the invasion of honey mesquite, juniper, and other woody species (Wink and Wright 1973; Neuenschwander et al. 1978; Steuter and Wright 1983). Presumably because of fire suppression efforts, honey mesquite, for example, is considerably more dense now than is indicated from historical records (Wright et al. 1976). The invasion of these woody species reduces forage, causes a deterioration of the native prairie habitat, and is of sufficient concern that various techniques are used in their control (e.g., Bryant et al. 1983).

Contemporary management has altered the mixed prairie structure, function, and occurrence by reducing or eliminating keystone species, cultivating large areas, redistributing surface water and groundwater, altering fire frequency, developing extensive transportation corridors, introducing exotic species, promoting the development of woodlands, and establishing long-term management unit boundaries.

Extent

As a result of human activities, mixed prairie has been substantially reduced (Samson and Knopf 1994). Klopatek et al. (1979) estimated the reduction of Küchler's potential mixed prairie vegetation based on a set of land-use variables collected by county during the late 1960s (see table 4.2). Their estimates do not account for the expansion of cropland that occurred during the 1970s nor for the conversion back to perennial vegetation that occurred under the Conservation Reserve Program in the 1980s. We compared the Klopatek et al. (1979) local data with a recent analysis of remotely sensed data (see table 4.2) (U.S. Geological Survey 1993). These data were derived from the land-cover characteristics database created at the EROS (Earth Resources Observation System) Data Center (U.S. Geological Survey 1993). The database portrays regions composed of similar land-cover mosaics as defined by a multitemporal, advanced very high resolution radiometer normalized vegetation index obtained from a National Oceanic and Atmospheric Administration satellite during 1990 and attributes such as terrain, climate, and ecoregion (Loveland et al. 1991). The land-cover product has been resampled from a 1.1-km^2 resolution to a 1-km^2 resolution. Although 159 land-cover types are defined in the database, we selected only those identified with native vegetation typical of the five mixed prairie types of Küchler (1985).

Table 4.2

Percent of remaining mixed prairie area

Mixed Prairie Type	Percent Remaining	
	Klopatek et al.	EROS
Northern mixed		
Wheatgrass-bluestem-needlegrass	31	17
Wheatgrass-needlegrass	64	61
Sandhills	94	72
Southern mixed		
Bluestem-grama	35	8
Mesquite-buffalo grass	73	58

Source: Based on estimates by Klopatek et al. 1979 and the Eros Data Center, Sioux Falls, South Dakota 1995.

The lower estimates of extant mixed prairie provided by the EROS data set may be the result of additional mixed prairie loss, although it is also possible that they represent differences due to masking of small tracts and edges when analyzed at the resolution used. This is suggested by the relatively large differences between the estimates in the two most intensively farmed types (wheatgrass-bluestem-needlegrass and bluestem-grama). The EROS data emphasize the larger, less-fragmented tracts of remaining mixed prairie, while the estimates of Klopatek et al. (1979) include large native prairie tracts as well as small isolated tracts surrounded by croplands. These two estimates have significantly different implications for conservation, since ecosystem function within expansive grasslands differs greatly from small grasslands surrounded by croplands (Shafer 1995).

Function and Composition

The percent of land surface remaining in native mixed prairie vegetation is relatively large compared to the tallgrass prairie. But this is a very different mixed prairie ecosystem than the one European settlers took from the Plains Indian cultures. The native vegetation is still a dynamic reflection of the interactions between climate, soils, weather, grazing animals, and fire. But the present grazing and fire regimes are determined by a different set of ecosystem rules. These new rules operate at smaller (individual landowner) and larger (national and international commerce) scales than in the mixed prairie of five hundred years ago. Relatively few species have been extirpated by current management practices. However, major changes in community composition and landscape patterns have resulted from the replacement of bison with cattle and the imposition of croplands, transportation corridors, and urban areas. Although changes that followed European settlement have significantly reduced critical grassland and wetland habitats, they have significantly expanded woodland habitats. Mixed prairie woodlands are expanding due to changes in river flows (Johnson 1994), grazing and fire regimes (Steuter et al. 1990a), and directly from shelter-belt planting. The presence of many large browsing mammals in the pre-Holocene fossil record suggests that woodlands were more common in the mixed prairie region prior to the arrival of humans twelve thousand years ago or more. We expect that mixed prairie will continue to undergo change and that humans will continue to manage the changing ecosystem.

The semiarid climate of the mixed prairie ecosystem places a premium on the linkages between the uplands and the riparian and wetland parts of the landscape. The major functional linkage between uplands and lowlands is water. Highly mobile and sedentary species using the mixed prairie were adapted to the temporal and spatial patterns of available moisture. Surface water storage, drainage for crop production, and flood control have affected plants and altered the landscape pattern too rapidly for many migratory species to adapt. Thus water conservation is central to many issues in conservation.

Conservation

Our understanding of mixed prairie ecology and management suggests a conservation strategy based on land management that acknowledges the ecosystem's adaptations to limited water availability, grazing, and periodic fire. Demands on water resources will continue to exceed supplies in the mixed prairie region. Municipalities and recreation interests will increasingly compete with wildlife conservation interests for water currently allocated to agriculture. Groundwater, as well as surface water supplies, will need to be used more efficiently, with emphasis on water quality. Mixed prairie communities can play a natural role in meeting these water quantity and quality objectives. When properly managed, mixed prairie provides a renewable source of high-quality water, food, and habitat for a wide range of species and uses. Although limited in extent, wetland and riparian areas will continue to be critical to a healthy mixed prairie ecosystem.

Most of the remaining mixed prairie is grazed by cattle, although bison ranching is becoming more common. Large herbivore grazing is a required process for sustaining mixed prairies. Prairie diversity and productivity may be adversely affected by the absence of grazing or when grazing is too intense or occurs at inappropriate times of the year or for too long a period of time. The dominant plants, whether cool- or warm-season grasses, require regular grazed and ungrazed periods during the growing season for optimum future growth. Either continuous heavy grazing or excessive litter accumulation may adversely affect individual species and reduce overall productivity and diversity. By physically disrupting the soil surface, grazing may increase erosion although conditions for plant establishment that enhance long-term diversity may be able to be improved. Thus grazing, whether for livestock production or natural area management, is an appropriate management tool in the mixed prairie.

Fire is also important to mixed prairie conservation, although it is used less often today than it is believed to have occurred in the past. Like grazing, the season, intensity, and frequency of burning are variables that need to be considered when using fire as a management tool. When used appropriately, fire can prevent or slow woody plant invasion and improve forage for grazing animals, although at some loss of standing crop. Fire is also important for establishing seedlings, thereby assisting in the maintenance of long-term plant diversity. Land management using combined fire and grazing is widely applied in the Kansas Flint Hills and has maintained a productive grassland for over one hundred years. The expansion of woodlands, which results from a reduction in fire frequency, may provide opportunities for offsetting habitat declines in other regions (i.e., eastern woodland bird habitat). However, this may lead to hybridization between previously disjunct populations, the cosmopolitanism of the landscape (Knopf 1986), and loss of biological diversity at some scales. These are effects that need to be considered in conservation planning on a regional scale.

The vagaries of climate dictate that a dispersed and redundant system of appropriate upland and wetland habitats be available for both migratory and nonmigratory species. It should be possible to manage for both if these landscapes are appropriately managed by grazing and fire and integrated by the critical water resource (see fig. 4.1). The mixed prairie is predominantly privately owned and managed. Except for the sandhills prairie, cropland acreage is regionally similar to perennial grassland. A successful conservation strategy will need to move land management of croplands, grasslands, and wetlands toward a more conservative use of resources. Several private-public efforts with this objective in mind are underway through the U.S. Fish and Wildlife Service Partners for Wildlife Program and through joint ventures by landowners, state and federal agencies, and private conservation organizations.

Mixed prairie conservation is best approached with a humble and open mind rather than a dogmatic focus on a single point in time. Change is a constant and usually unpredictable characteristic of the ecosystem. It is the species diversity of mixed prairie that gives the ecosystem resilience and can also provide the foundation for a resilient human economy. Conservation of the mixed prairie will be

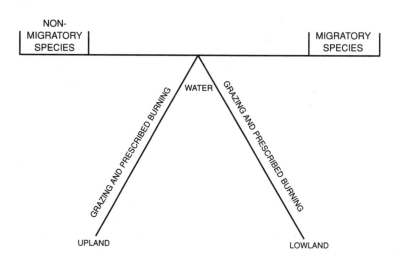

Fig. 4.1 Schematic representation of a conservation program that acknowledges the critical role of water resources in balancing both migratory and nonmigratory species protection.

achieved by monitoring the biological community and being willing to change management techniques to maintain its diversity.

Acknowledgments

We acknowledge the support of our respective institutions in this effort. B. K. Wylie and D. D. DeJong of Augustana College, and the EROS Data Center, Sioux Falls, South Dakota, provided information for the remote-sensed estimates of current mixed prairie extent. S. Winckler and D. Sutherland provided a critical review and technical support for the chapter. Omissions or errors, however, remain solely our responsibility.

CHAPTER 5

Prairie Ecology—The Shortgrass Prairie

T. Weaver
Elizabeth M. Payson
Daniel L. Gustafson

The shortgrass prairie extends eastward from the Rocky Mountains about 320 km to the mixed prairie and reaches south from central Alberta to central Texas. The eastern boundary is an arbitrary point on the east-west precipitation gradient and shifts eastward in dry years and westward in moist years (Küchler 1985). To the north, shortgrass prairie is replaced by fescue grasslands, aspen forests, and tiaga. To the south, prairie fades into shrub communities dependent on water entering deep soil horizons during the wet season.

The shortgrass prairie landscape was once one of relatively treeless stream bottoms and uplands dominated by blue grama and buffalo grass, two warm-season grasses that tolerate intensive grazing. Thread leafed sedge and junegrass are more prevalent northward; and buffalo grass and hairy grama are more common southward. Small forbs, scarlet globe-mallow among others, appear between the grasses and thrive in years when rainfall encourages their growth. Weedy grasses, including sand dropseed, squirreltail, and forbs, for example, curlycup gumweed, also unite the shortgrass prairie. Few plants are endemic to the shortgrass prairie.

As its name implies, the shortgrass prairie is recognized by its low stature (Daubenmire 1978). The low stature is due in part to water stress. Thus, when the natural system is irrigated, production is increased dramatically (Dodd and Lauenroth 1979, Weaver 1983).The strongest expression of this is associated with heavy grazing (Coupland 1961).

Perhaps in no other North American system are impacts of grazing so apparent (Knopf, in press). In the mid-1800s, numbers of large native herbivores—

bison, pronghorn antelope, and elk among others—and predators, such as the grizzly bear and wolf—rivaled or exceeded those now evident in the African Serengeti (Howe 1994b). In plants, evolutionary antigrazing mechanisms including short stature and phytoliths evolved to reduce the impact of grazing. Small native herbivores—prairie dogs, other rodents, birds, insects, mites, and especially plant-parasitic nematodes—may have had an even larger impact on the natural community. Modern-day grazing pressures—cattle to nematodes—may be comparable.

An Environmental Gradient

Climate

The shortgrass prairie environment varies strongly from north to south (see fig. 5.1). The mean average January temperature rises from -15°C in the north to 4°C in the south, and July temperatures increase from 18°C to 26°C. Greater precip-

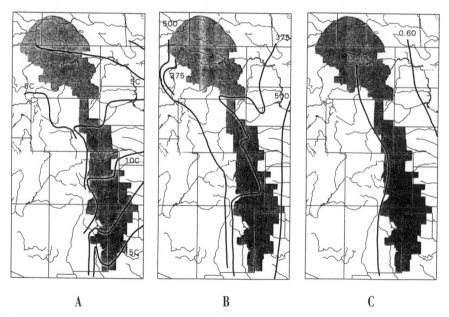

A   B   C

Fig. 5.1 Climate of the shortgrass prairie (shaded region). (A) Average annual temperatures (°C) increase steadily from north to south (5°C to 15°C). (B) Average annual precipitation in the central prairie is 375 mm with rainfall slightly less in the north. (C) Precipitation throughout the shortgrass prairie is approximtaely 20 percent of potential evaporation; the constant precipitation/potential evaporation ratio indicates equal water availability in the shortgrass prairie. The high temperatures in the south increase evaporation and compensate for the higher rainfall.

itation in the south—30 to 56 cm—is counteracted by increases in rates of evaporation associated with the higher temperatures. An index of water availability is the ratio of precipitation to evaporation (p/e) and remains constant (p/e = 0.3) from Texas to Montana (Jenny 1941).

Regardless of latitude, the shortgrass prairie varies little from east to west, since it is a short segment of the long east-west gradient across the prairie. Temperatures are almost invariant because isotherms extend east and west. Precipitation does increase eastward, but the changes are slight. Shortgrass sites have a one- to two-month summer drought; in contrast, no mixed or tallgrass prairie sites have such a drought (Walter 1975). Most of the shortgrass prairie lies on fairly uniform sediments washed east from the Rocky Mountains (Fenneman 1931).

Species

While the tolerances of a few plant species are so wide that they occupy the entire temperature gradient down the shortgrass prairie, most species have more limited distributions (Barkley 1977). Unless plant communities are tightly integrated, southern species should then disappear, not in concert, but at a constant rate as one moves northward. Our tabulation of Barkley's data shows that species do disappear at the rate of 67 ± 44 per degree of latitude as one moves north (see fig. 5.2). Exceptionally high rates of disappearance occur at rivers and mountains—at the Cimarron, Arkansas, and Missouri rivers, and Niobrara–Black Hills mountains. We hypothesize that these exceptions occur because many southern species find their most northerly refuges on south-facing slopes along these features.

Similarly, one predicts the new northern species should appear regularly as one moves northward. An average of 57 ± 45 new species do appear per degree of latitude (see fig. 5.2). Unlike the disappearances of the southern species, the appearance of the northern species is not associated with the steep slopes of the rivers and mountains of the Great Plains. Why are north-facing banks less refugial for northern species than south-facing banks are for southern species? Microclimates may be less strong and less diverse on north-facing than south-facing slopes if south-facing slopes are steeper and more eroded. Similarly, competition against species at the edge of the ranges of northern species may be more in the closed vegetation on north-facing slopes than in the open vegetation of dry south-facing slopes.

The gradual changes in the shortgrass prairie demonstrated in climate, species distribution, and productivity-decomposition have lead some to map the shortgrass prairie as a single unit and others to map the shortgrass prairie as two (Küchler 1964) or even three units (Omernik 1987).

On the north-south temperature gradient, the shortgrass prairie community varies in physiology as well as taxonomic composition. For example, the C_4

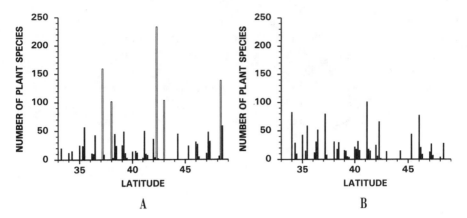

Fig. 5.2 Species disappear in occurrence from south to north (A) while other species appear (B). Exceptional numbers of disappearances (transparent bars in A) appear at major rivers and the Black Hills.

grasses that compete best under warm-dry environmental conditions may comprise 68 percent of the shortgrass prairie community in Texas and only 31 percent in North Dakota (Terri and Stowe 1976).

Productivity and Decomposition

Productivity throughout the shortgrass prairie should be equal if water availability is the limiting factor and if water availability as indexed by the p/e ratio is constant. Information is limited, but dryland wheat yields vary little from north to south—182 ± 40 to 229 gm/m^2 in Montana to 168 ± 27 gm/m^2 in Texas. A similar lack of a north-south trend in wheat yield has been reported in tallgrass prairie.

While production is constant from south to north, annual decomposition of plant material increases from south to north. Aboveground, grazing, fire, and photooxidation consume most production promptly within years. Belowground, primary consumers—earthworms, bacteria, and fungi—are more important in the north than in the south. Thus soil darkness increases as one moves north up the shortgrass prairie with 0.81 percent soil carbon in Texas, 1.26 percent in Colorado, and 1.87 percent in Montana.

Inclusions

Environmentally atypical areas allow organisms unable to survive in normal shortgrass prairie to survive in this region. Examples include igneous intrusions of the northern plains, that is, the Crazy Mountains in central Montana, Devil's Tower in northeast Wyoming, and the Black Hills in South Dakota, or lava flows that created the Black Mesa of the southern plains (King 1967). Species found on inclusions may have arrived by recent travel along a corridor, for example, the river valleys of the Arkansas, Cimarron, and Niobrara rivers (Kaul et al. 1988) or by migration in the more distant past (e.g., white spruce colonized the Black Hills when glacial climates connected the Black Hills to spruce population in the northwest).

Prairie Processes

Any stand of prairie vegetation changes in response to climate and other factors outlined below. As a result, artificial control with modification of natural processes is expected to have significant impacts on the natural biota of the region (Knopf and Scott 1990). There is an urgent need to conserve natural processes on both reserve and privately owned grasslands.

Climate

Ecological processes occur under the influence of the dominant controlling variable, precipitation, which determines trends in vegetation structure and species numbers at any point on the shortgrass prairie (Weaver 1950). Plant death during drought limits the previous community's ability to respond to an increase in moisture. Succeeding increases in water availability may thus support increases in biomass of surviving plants, increase in numbers of existing species, or colonization by new species.

Grazing

Hornaday (1889) as quoted by Larson (1940) reported that bison "at times so completely consumed the herbage on the plains that detachments of the United States Army found it difficult to find sufficient grass for their mules and horses." Domestic cattle exert similar grazing pressures but concentrate more on lowland than upland plant communities (Plumb and Dodd 1994).

Milchunas and Lauenroth (1993) studied the impacts of grazing on species composition, aboveground plant production, and root biomass among grasslands throughout the world. They concluded that species composition between grazed and ungrazed areas is more sensitive to changes in environmental variables—water availability and length of association with grazing animals—than to grazing variables such as intensity. Grazing selects for low growing plants, an avoidance mechanism. In areas with a long history of grazing and low productivity, moderate grazing had no negative effect on plant production. Moreover, grazing had no clear effect on root mass, soil carbon, or soil nitrogen. On this basis, they suggest observations at one level of the ecosystem may not be indicative of another and therefore managers should consider scale. For example, short-term increases in nutrient cycling can be accompanied by unexpected long-term decreases in large nutrient pools. Second, evaluation of grazing on the basis of species presence alone may be misleading in that long-term changes in species composition may be small compared to changes in other ecosystem attributes such as soil nutrients. And, due to slow responses, species changes may not reflect recent changes in land-use, such as the imposition of grazing. Third, geographic location of the grazing unit serves as an index to grazing condition if concerned about the long-term sustainability of a system.

Fire

Like grazing, fire removes plant material and recycles nutrients. Thus, fire may either compete with grazers or favor them by exposing or stimulating new plant growth. Fire simultaneously selects against fire-intolerant species and may also significantly affect microclimate, plant invasion, and nutrient loss as well. Despite its probable importance, our knowledge of fire effects on the shortgrass prairie is slight (Daubenmire 1968, Wright and Bailey 1980). Because only 1 percent of the Montana shortgrass prairie burned in a major fire year, 1988, and current effects of fire seem small. Was fire more important in the past? Wells (1965) suggested that without fire sagebrush, juniper, and ponderosa pine may have been much more important in the northern plains. In the southern plains, fire is reported to have excluded mesquite from the shortgrass prairie (Archer 1989) and, if so, was very important.

Invasion

Many exotic plants evolved with disturbance and colonize open sites where competition by native species is reduced by excessive grazing, excessive trampling, insufficient grazing, fire, or other stresses. Exotics are widespread in the shortgrass prairie but not necessarily dominant. Important exotic species on the northern plains include grasses (crested wheatgrass and Japanese brome) and forbs (yellow sweetclover, spotted knapweed, goat's beard, prickly lettuce, pale alyssum, and false flax) (Weaver et al. 1993).

In the southern plains, shortgrass prairie is more affected by invasive native species. In Texas, much of the prairie is now either farmland or grassland so invaded by honey mesquite as to form a shrubland with an understory of shortgrass species (Sims 1988). One or more of the following factors may account for the invasion of grassland by woody vegetation: change in rainfall patterns during the last one to two centuries, fewer fires, and reduction in grass cover by grazing

(Archer 1989). Once initiated, the changes are self-reinforcing—shrubs and trees capture available water, grasses become patchy and unproductive, frequency of fire is reduced so fire no longer kills young woody seedlings, and mesquite trees proliferate to capture more water.

Sodbreakers

Sod is broken by organisms ranging from native rodents to man. Shortgrass plant community response varies with the size of the disturbance. For example, blue grama, a major shortgrass prairie species, tends to recolonizes patches similar in size to that of an adult plant (Coffin and Lauenroth 1988). In contrast, if the disturbance is large, re-establishment of blue grama is low (Briske and Wilson 1977). Similarly, grama reinvades all but the smallest abandoned fields slowly. At a still larger scale, recolonization is reported to be dependent on topographic position and most often in the uplands and at moderate grazing intensity. Methods to increase rates of natural succession on large disturbed patches are being developed.

Prairie Conservation—Reserve Acquisition and Management

Three issues important to prairie conservation are: locating and evaluating prairie fragments and identifying their part in the reserve system; managing the fragments; and linking the fragments.

Finding fragments in the shortgrass prairie is easier than in the tallgrass and mixed prairies. Approximately 60 percent of the region is in grassland and a large percent of that is unplowed. This is because the regional climate is too dry to invite farming and, except where irrigated, tends to evict established farmers.

An existing network of large shortgrass prairie fragments is healthy except in Texas and New Mexico (see fig. 5.3). Smaller reserves (250 to 25,000 ha) are well distributed in the north (i.e., Montana, Wyoming, and Colorado). The largest fragments mapped (>25,000 ha) are National Grasslands remarkable for their size and distribution. They include areas in Canada (i.e., Sheffield Air Force reserve in Alberta and the public grasslands in Saskatchewan) and the United States (i.e., the U.S. Department of Interior Fish and Wildlife Service Charlie Russell National Wildlife Refuge in Montana; U.S. Department of Agriculture Forest Service National Grasslands in Wyoming, Thunder Basin; Nebraska, Ogallala; Colorado, Pawnee and Comanche; Kansas, Cimarron; New Mexico, Kiowa; and Texas, Santa Rita; and U.S. Department of Interior Bureau of Land Management lands across the West).

Reserves in the shortgrass region vary greatly in size (fig 5.3) and condition but given the dominance of grasses in the matrix and ongoing natural succession,

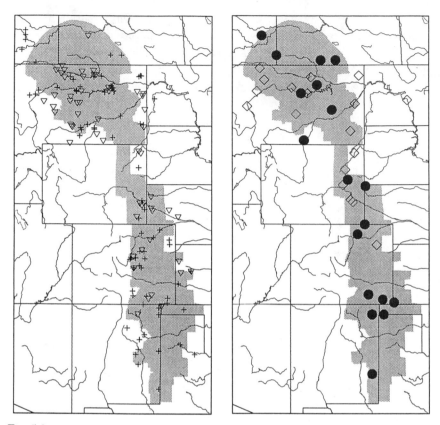

Fig. 5.3 The distribution of small (0 to 250 ha), medium (250 to 2500 ha), large (2500 to 25000 ha), and very large (>25000 ha) reserves in the shortgrass prairie (shaded region). A small preserve is represented by a plus, medium by a triangle, large by a diamond, and very large by a circle. The largest reserves are federal grasslands recommended as the backbone of a shortgrass preserve system.

most should contribute significantly to the system of reserves. With reasonable management, smaller organisms (plants and invertebrates) should survive in small to middle-size reserves. Populations of larger animals could survive on the national grasslands as well as on small reserves if the effective areas of such reserves are increased by animal use of surrounding nonreserve grassland. Lands surrounding reserves could further provide resources and serve as corridors between established reserves. Where such surrounding areas are limited, management should not hesitate to transplant native species, either small or large, among nearby reserves to maintain both community diversity and the genetic health of component populations. Natural processes, as well as species, should be preserved.

We suggest that areas large enough to support genetically healthy populations of natural predators—and perhaps even ruminant herbivores—will support natural processes, i.e., fire, grazing, and trampling. As areas shrink, however, management of these processes must become more intense or more costly and may include fencing to regulate grazing and controlled burning. In our inventory of prairie fragments, we found few statements about the ecological condition of shortgrass prairie fragments large or small.

One might assert that the fragments should be managed to approximate pre-European settlement environmental condition. This may involve appropriate levels of grazing and burning to control standing crops of plants and litter, to select against invading species, and to provide disturbed sites for mobile noncompetitive species. Due to the similarity of their grazing (Plumb and Dodd 1994), managed grazing by cattle can be substituted for bison. Grasslands in poor condition may require reclamation efforts, for example, elimination of exotics or reintroduction of missing species. As farmland is abandoned outside the reserve system, planting of oldfields to native grasses—never exotics—will increase areas of productive, well-adapted grassland. This assignment is easily stated, but will be accomplished with some difficulty. Prairie managers must first establish what pre-European conditions were and must then determine what grazing and/or fire reclamation treatments will reinstate those conditions.

We suggest that shortgrass prairie reserves be united in a loose federation for at least four reasons. First is to facilitate cross-system experiments. Across-reserve discussion and experimentation is needed for application of management goals as described above and the the blend of processes (e.g., grazing and fire) necessary to reach them.

Second is to educate. On the shortgrass prairie, information gained through cross-system experimentation is useful both to private landowners and conservationists and should be extended to them. Moreover, to increase and stabilize the nation's commitment to prairie conservation, the public should be informed of the ecological and economic benefits of good prairie management.

Third is to maintain cross-system conservation standards, for example, private grazing is needed, but it will likely result in pressure to maximize grazing yields at the expense of other processes, a tragedy of the commons. The application of conservation standards will be most successful if managers include representatives from throughout the region.

Fourth is to provide cross-system management. While a single management approach may be less expensive, a federation of separate managements is preferred because it is less likely to institute a single system-wide destructive management process.

CHAPTER 6

Prairie Ecology–Prairie Wetlands

Bruce D. J. Batt

Wetlands of the Great Plains Region

Five major wetland regions of the Great Plains include the prairie pothole region, the Nebraska Sandhills, the Rainwater Basin, Cheyenne Bottoms, and the Playa Lakes. This chapter reviews the geological origin, wetland characteristics, land-use patterns, historical changes, and conservation needs in each region. Importance to water birds is discussed for each region because these ecological components have been fairly well studied and are at the basis of the most promising conservation initiatives. Other flora and fauna are unevenly but generally less well documented. Other wetland systems, including rivers and floodplains, are excluded as are most reservoirs, most lakes, and man-made wetlands, including those constructed for livestock, industrial uses, domestic waste management, and power generation. Nevertheless, these latter wetland types often play vital roles in some locations within the Great Plains.

The life history strategies of wetland plants and animals are adapted to the dynamic fluctuations in wetland numbers and quality. Vegetation dynamics have been particularly well studied and useful models of plant responses to cycles of drought and water abundance have been presented by Weller and Fredrickson (1974) and Weller and Spatcher (1965) who also described the responses of wetland wildlife. Van der Valk and Davis (1978) advanced a model that is now widely embraced as reflecting the patterns that are observed throughout the prairie pothole region. In general, wetlands go through four recognizable phases in a typical wet-dry cycle. During dry periods the marsh bottom is colonized by

Prairie Conservation
Island Press (Washington, DC • Covelo, CA)

plants that emerge from the seed bank in the marsh substrate. Following re-flooding, annual plant seedlings are eliminated and the perennial emergent species mature and establish typical marsh vegetation. After extended flooding, the perennials become senescent, are removed by muskrat or insect herbivory, or are intolerant of extended deep flooding, and eventually the marsh advances to an open marsh stage wherefrom the emergent species will become established again after another dry cycle. Not all wetlands have the same pronounced vege-tative cycles. For example, basins that dry annually are dominated by species that tolerate those conditions (van der Valk and Davis 1979; Galatowitsch and van der Valk 1994).

Prairie Pothole Region

The prairie pothole region extends across approximately 870,000 km^2 of the Great Plains region of Canada (about 80 percent) and the north-central United States (about 20 percent). Very little of the native prairie persists, as most if it has been cultivated. Where cultivation has not occurred, grazing is the dominant land-use. Because this region was recognized early in this century as having a vital role in the size and distribution of migratory waterfowl populations (Batt et al. 1989), it received relatively abundant attention from private and public wa-terfowl conservation interests. This spawned a great deal of research and conser-vation activity, summarized in van der Valk (1989b).

Geological History

The northern portions of the North American continent were impacted by four major glacial periods during the Pleistocene (Pielou 1992). The most recent was the Wisconsin glacial stage, which reached its maximum extent about thirteen thousand years ago (Mayewski et al. 1981). Essentially all the dominant land-scape features of the prairie pothole region are products of that geological event, including the potholes themselves. This region is dominated by poorly drained glacial drift consisting of rock and debris scoured in a slurry of ice as the glacier advanced and retreated (Bird 1961). Glacial drift occurs today in deposits that range from 20 m to 150 m in depth.

The generally rolling surface relief is comprised of glacial features such as es-kers; ground, end, and lateral moraines; and glacial drift deposited as stratified drift and till. The till deposits were mixed extensively with blocks of ice that were covered with outwash sand and gravel. When the ice melted, the depressions that were left, known as kettles, filled with water and are the potholes of mod-ern terminology. This rolling landscape of depressions and higher areas is termed knob and kettle topography. Potholes occur in various densities throughout the region, ranging from a few to forty-seven or more per square kilometer. Toward the southern extension of glacial drift, the till is very shallow, as were the potholes. Further north, glacial effects were more extensive; the till is deeper and the potholes tend to be deeper and more permanent (Kemmis 1991). The depth of potholes and the relief of the terrain is inversely related to the ease and pace at which pothole drainage has taken place since European settlement (Galatowitsch and van der Valk 1994).

Wetland Characteristics

The vast prairie pothole region has marked regional and local differences in climate and pothole characteristics. Topography, soil, and groundwater flow characteristics are also variable and cause considerable micro and macro differences in wetland characteristics (e.g., Lissey 1971). Various schemes have been developed to classify wetland types in this region, but Stewart and Kantrud (1971) devised the most widely used system for whole pothole classification. It recognizes seven classes of wetlands: (1) ephemeral ponds, (2) temporary ponds, (3) seasonal ponds and lakes, (4) semipermanent ponds and lakes, (5) permanent ponds and lakes, (6) alkali ponds and lakes, and (7) fen or alkaline bog ponds. Characteristic vegetation for each type has been described (Stewart and Kantrud 1971; Kantrud et al. 1989a). The more generally applicable system of Cowardin et al. (1979) classifies floristically homogeneous stands of vegetation and thus will have more than one wetland type represented in each basin.

The climate is continental, and precipitation is exceeded by evaporation, more so in the north (e.g., -55 cm in southwestern Saskatchewan to -75 cm in Eastern Montana) than in more southern portions (e.g., -10 cm in Iowa) (Winter 1989). Annual precipitation is highly variable, with corresponding impacts on wetland numbers and quality. Because of the relationship between wetland conditions in the prairie pothole region and waterfowl populations, the federal governments of Canada and the United States conduct coordinated surveys each May and July to count wetland numbers and waterfowl populations. Between the years 1955 and 1985, wetland numbers in the prairie pothole region ranged from a low of about 2 million ponds to a high of over 7 million (Batt et al. 1989).

Water Bird Use

Water bird use of pothole wetlands is correlated with the interspersion of cover and water. Maximum use occurs when the marsh is in a hemi-marsh phase with an approximate 50:50 ratio of cover and water (Weller and Spatcher 1965; Weller and Fredrickson 1974). Waterfowl are the most thoroughly studied of pothole breeding wildlife. Waterfowl use of wetlands is associated with clusters, or complexes, of wetlands, with various types providing different requirements during spring migration, courtship, nesting, brood-rearing, molt, and fall migration (e.g., Swanson and Duebbert 1989). Thus the dynamics within individual wetlands and within regional wetland complexes affects the numbers and kinds of waterfowl settling in the prairie pothole region to breed (e.g., Johnson and Grier 1988).

During the years 1955 to 1985, considerable variation, positively correlated with pothole numbers counted in May, of the mean total number of the twelve main game ducks settling in the prairie pothole region (21.6 ± 4.75 million, CV 22.1 percent) was calculated by Batt et al. (1989). The mean proportion of all ducks estimated to be in the region was also highly variable (51.1 ± 1.4 percent, CV 15.2 percent). This variability is caused primarily by the displacement of birds, usually northward, during drought (e.g., Hansen and McKnight 1964; Smith 1970; Derksen and Eldridge 1980).

Besides wetland numbers and quality, waterfowl breeding success in this region for most species is also intimately linked to the condition of upland nesting cover in the intensively farmed landscape surrounding the wetlands. The relationships are complex, but the pattern is that low nest success now occurs across most of this region in most years (e.g., Greenwood et al. 1995). Low nest success is correlated with the expansion of agriculture, which reduces nesting cover and concentrates nesting birds in remnant patches that are easily searched by predators. Predators are the agents of nest destruction, and numerous studies provide documentation (Higgins 1977; Cowardin et al. 1985; Greenwood 1986; Klett et al. 1988). These relationships are best understood for waterfowl, but similar cause-and-effect relationships are thought to exist for declining grassland-nesting birds in general (e.g., Johnson and Schwartz 1993).

Historical Changes

The wetlands of the prairie pothole region have been extensively drained, filled, and otherwise degraded. The only published tabulation of the degree of wetland loss in the United States was presented by Dahl (1990). However, these data were presented by state, and no separation was made for the various wetland types. Because potholes dominate the wetlands in the glaciated prairie region, it is reasonable to assume that the following loss statistics closely reflect the proportional loss of potholes throughout the region in the United States: Iowa, -89 percent; South Dakota, -35 percent; North Dakota, -49 percent; and Minnesota, -42 percent.

Similar data are not available for Canada (Adams 1988). Lynch-Stewart (1983) attempted to document changes but found the information base to be incomplete and inconsistently recorded. Loss in most areas is likely considerably less than what has occurred in the United States because the Canadian prairies have been settled for a shorter period of time and the infrastructure of government support for activities that degrade the wetlands has been less. Throughout most of prairie Canada, investments in tile drainage are still relatively uncommon, and the depth and relief of glacial till is greater, making drainage more difficult and expensive. Shorter growing seasons and lower land values also make the economics of drainage less feasible. Nevertheless, drainage has been extensive in a few sites. Glooschenko et al. (1993) cited studies that showed 57 percent loss in southwestern Manitoba and 39 percent in the Alberta prairie parkland region. Several reviews have recorded total impacts on wetlands that included drainage or other causes such as overgrazing, partial filling, burning, and cultivation. Glooschenko et al. (1993) reported that 84 percent of wetlands in southwestern Saskatchewan had been negatively impacted in some way by 1979.

Nebraska Sandhills Wetlands

The sandhills region of Nebraska is often considered in conjunction with the prairie pothole region in discussions of northern prairie wetland ecology and management (e.g., van der Valk 1989b; Ducks Unlimited 1994). However, there are important differences in geological history, ecology, conservation, and management issues that will only be summarized here. Most of the information provided is developed in more detail in Novacek (1989).

The sandhills represent the largest stabilized sand dune field in the Western Hemisphere. Entirely vegetated by mixed grass prairie, it covers a 51,800-km² area of fifteen north central counties of the state. The sandhills were formed during several periods of dune activity that began in the Hypsithermal Interval about seven thousand years B.P. (Swincheart 1984; Warren 1976). Prevailing winds moved the dune material during the postglacial period and during several periods of extended drought that weakened the vegetative cover of the dunes. Groundwater-supplied wetlands occur in the interdunal valleys in the form of marshes, lakes, and wet meadows. Other wetlands are perched alkaline mineralized lakes on poorly drained soils and a third type of wetland that is poorly connected to the groundwater reservoir (Wilen and Tiner 1993; Ginsberg 1985). Novacek (1989) concluded that wetlands in the western third of the sandhills are more linked to precipitation than are lakes in the eastern two-thirds, which usually have some connection to the groundwater reservoir. As a result, lakes in the western region are more variable in aerial extent and tend to be more alkaline. The limnological characteristics of sandhills have been described thoroughly by McCarraher (1959, 1971, 1972, 1977 in Novacek 1989).

During the 1970s, sandhills wetlands covered 557,591 ha in the following three major classifications: open water, 51,655 ha; marsh, 26,224 ha; riparian vegetation, 9,272 ha; and subirrigated meadow, 470,440 ha (Novacek 1989). Sandhills wetlands have undergone little drainage and are relatively secure compared to wetlands in other areas because the dominant land-use is cattle ranching, wherein wetlands are used for haying and grazing (Gabig 1985).

Water Bird Use

Sandhills wetlands provide breeding habitat for a similar range of waterfowl species as those found in the prairie pothole region. The most abundant species are the mallard, blue-winged teal, northern pintail, and northern shoveler (Wolfe 1984). The most abundant upland-nesting species is the blue-winged teal, while the ruddy duck is the most common overwater nesting species. The giant Canada goose has also been reintroduced and is now a relatively common species, and the trumpeter swan occurs as a rare nester (Novacek 1989). Wolfe (1984) also recorded about eighty species of aquatic birds. Among the less-common species are the long-billed curlew, upland sandpiper, black tern, and white pelican.

Historical Changes

While sandhills wetlands are relatively secure, they have been impacted by farming practices, primarily center-pivot irrigation (Novacek 1989). The wet-meadow wetlands are the most extensive and most important wetlands of the sandhills for both wildlife and agriculture, but they are also the most threatened by increased irrigation, which lowers water tables and reduces productivity. Between 1969 and 1984, the number of wells increased from about 150 to 2,300 (Dreeszen 1984). Wet meadows in some portions of the sandhills have been almost completely eliminated, and impacts occur throughout the irrigated areas (Novacek 1989). Negative impacts on blue-winged teal and other wet-meadow water bird species are suspected, but further evaluation is required (Lewis and Bockelman 1988).

Rainwater Basin

The Rainwater Basin covers an area of 10,878 km^2 in seventeen counties in south-central Nebraska, south of the Platte River (U.S. Environmental Protection Agency 1986). The area includes portions of both tallgrass and mixed-grass prairie (Weaver and Brunner 1954). It was formed from irregularly deposited, wind-blown silt loam deposits and is known as the loess plains region (Condra 1939). Precipitation runoff has leached and concentrated clay in the subsoils of depression areas, forming impermeable deposits that cause the development of wetlands (Erickson and Leslie 1987). Wetlands of these depressions depend heavily on surface runoff from natural precipitation and are not linked to groundwater tables. They are thus very dynamic, both seasonally and annually.

Water Bird Use

The Rainwater Basin is the most important spring staging area for waterfowl in the midcontinent area (Gersib et al. 1989, 1992). Large concentrations of birds stop there to rest and feed before continuing north to breed in the northern Great Plains and the Arctic. Spring concentrations include 5 to 9 million ducks and several hundred thousand geese, consisting of 90 percent of the midcontinent white-fronted geese; over 1 million lesser snow geese; 50 percent of the continental breeding population of mallards; 30 percent of the pintails; 0.5 million Canada geese; and large numbers of blue-winged teal, green-winged teal, gadwall, northern shoveler, lesser scaup, canvasback, ring-necked duck, redhead, and ruddy duck (Gersib et al. 1992; Windingstad et al. 1984; Ducks Unlimited 1994). Less well recorded are the large numbers of waterfowl and other water birds that stage in the region during the fall and the staging of endangered species such as the whooping crane, piping plover, and bald eagle (Gersib et al. 1992).

Historical Changes

By the turn of the century, nearly all upland areas had been converted to some form of agricultural use (Ducks Unlimited 1994). This was followed by an intensification of drainage and wetland loss that was particularly effective during the 1960s and 1970s (Farrar 1982; Nebraska Game and Parks Commission 1984). Currently, less than 10 percent of the original wetlands remain. Between three survey periods in 1900, 1965, and 1983, wetland numbers dropped from 3,907 to 685 to 384, respectively. For these same periods, wetland area dropped from 38,323 ha to 13,164 to 9,750 ha, respectively. Wetland loss is nearly all associated with agricultural development. Over half of the lost wetlands were drained through roadside ditches, and the remainder were either drained to concentration pits or filled with material from pits or land leveling. Essentially every wetland remaining today has modified physical or hydrological features (Nebraska Game and Parks Commission 1984; Ducks Unlimited 1994).

Since 1975 fowl cholera has killed a varying number of birds, mostly waterfowl, each spring in the Rainwater Basin Region (Windingstad et al. 1984; Smith et al. 1989). The largest die-off occurred in 1980 when between seventy-two thousand to eighty thousand ducks and geese died. The sources and mechanisms of the spread of the disease are not well understood, but there is an inverse statistical relationship with the density of semipermanent wetland basins (Smith and Higgins 1990). This has led to recommendations and programs to restore wetlands in this region as a means of reducing the occurrence of fowl cholera (Smith and Higgins 1990; Gersib et al. 1992).

Playa Lakes

Playa lakes are described as wind-deflated depressions that developed on the surface of the southern high plains of the Great Plains during the Pleistocene (Judson 1950). Playa basins are underlain by nearly impermeable clayey soils quite distinct from surrounding soil (Reeves 1966). Over 25,000 playas exist—19,300 in Texas and about 6,000 shared among the states of Oklahoma, Kansas, Colorado, and New Mexico (Guthery et al. 1981; Guthery and Bryant 1982). Almost all (about 99 percent) of the playas occur on private land (Nelson et al. 1983). The average size of playas in Texas is 6.9 ha with a range of 0.4 ha to 27.5 ha (Guthery et al. 1982). In the five-state area, playa size is distributed as follows: 49.8 percent less than 4 ha; 27.0 percent from 4.1 ha to 8 ha; 10.0 percent from 8.1 ha to 12.1 ha; 5.8 percent from 12.2 ha to 16.1 ha; 3.0 percent from 16.2 ha to 20.2 ha; and 4.4 percent greater than 16.2 ha (Guthery et al. 1981). The southern high plains region has a very dry climate, resulting in less than 5 percent of the playas holding water at any point in time (Bolen et al. 1989a).

Water Bird Use

The Playa Lakes Region provides the most important wintering habitat for waterfowl in the Great Plains (Bolen et al. 1989b). Numbers of wintering ducks range from a high of 2.8 million ducks and 750,000 geese to a low of 500,000 ducks and 100,000 geese (U.S. Fish and Wildlife Service 1989a). The main factor influencing these numbers is water abundance and distribution. Over twenty species of migratory game birds winter there, but four species of dabbling ducks make up the majority (Simpson et al. 1981): green-winged teal, mallard, northern pintail, and American wigeon. The other remarkable species wintering on the playas is about 360,000 sandhill cranes, primarily in the southwestern Texas panhandle (Iverson et al. 1985).

Playas are also used extensively by breeding waterfowl, with variation depending on water abundance (Bolen et al. 1989a). Five species of ducks are common breeders: mallard, blue-winged teal, cinnamon teal, northern pintails, ruddy ducks, and northern shovelers (Rhodes and Garcia 1981; Simpson et al. 1981). Mallard and blue-winged teal make up about 80 percent of the breeding ducks (Rhodes and Garcia 1981), but surveys indicate that mallards alone made up 80 percent of the production (Traweek 1978). In the best year, 13,754 ducklings were produced in a twelve-county region of Texas (Traweek 1978), but other estimates place production at 250,000 birds in exceptional years (U.S. Fish and Wildlife Service 1989a).

Historical Changes

Playas have been altered in many ways, but because of the nature of the soils, the general topography, and runoff patterns, almost all of the original playas still exist. However, all the playas are modified either directly or indirectly by virtue of surrounding land-use changes (U.S. Fish and Wildlife Service 1989a). The prevalent land-use impact is caused by irrigation. By 1977 over seventy thousand wells were tapped into the Ogallala Aquifer (New 1979). The main agricultural crop grown with irrigation is corn, which likely supported the masses of waterfowl wintering in the Playa Lakes Region (Bolen et al. 1989a). Other land-use modifications may have been beneficial, such as increased runoff in some areas as a result of irrigation and more permanence of water for breeding birds because of pit excavation in playas.

The most prevalent modification of playas is the digging of pits within the basins to concentrate irrigation runoff water for reuse in irrigation or to ensure water for cattle. Guthery et al. (1981) found that over 70 percent of playas greater than 4 ha in size had pits dug in them, and 20 percent of all playas had pits dug for livestock use. Although they may increase waterfowl use during the breeding season, pits reduce the surface water available for wintering birds, reduce the littoral zone of the wetlands, and may serve to increase the spread of epizootics such as avian cholera (Friend 1982). Other major agricultural impacts on playas occur as a result of planting (over 16 percent of all playas) and grazing (about 50 percent of all playas). Feed lots are also commonly located on playas where animal waste is impounded and concentrated. During dry winters the only water available for waterfowl might be feedlots, the use of which is also associated with the occurrence of diseases (Friend 1982).

Other wetlands in the Playa Lakes Region are impacted by sludge and spillage pits associated with petroleum production (Flickinger 1981). Mortality has been extremely high. About 225,000 birds have been estimated to have died annually in these pits in New Mexico alone (U.S. Fish and Wildlife Service 1989b). However, conditions have improved greatly during the last five years as oil companies have responded by covering sludge ponds with netting that excludes most water birds. The accumulation of pesticides and nutrients in runoff and irrigation tail water and siltation into the basins also occurs but is not likely to be very significant relative to the above impacts (Ducks Unlimited 1994).

Disease, especially avian botulism and avian cholera, has become a very significant mortality factor for waterfowl wintering in the playas (U.S. Fish and

Wildlife Service 1989b). Precise estimates of the numbers of birds dying each year are not available but are thought to be in the tens of thousands during many winters. Avian cholera may have been introduced into North America from poultry waste discarded near playa lakes in the 1940s (Quortrup et al. 1946). The spread of cholera is believed to be enhanced by the concentration of birds on limited water areas and in cattle feedlots. It may be endemic in resident waterfowl and other birds, and its occurrence may be stress related in the crowded conditions that occur on many playas, especially during cold weather. Botulism, on the other hand, occurs during warm weather and is more likely to affect resident birds and early fall migrants (Ducks Unlimited 1994).

Cheyenne Bottoms

The Cheyenne Bottoms is a 16,000-ha natural land sink located in central Kansas. It is one of only three wetland complexes still remaining in the state of the twelve that were there when the Europeans arrived. In Kansas, nearly 50 percent of the wetlands that existed in 1950 have been lost.

Water supply was historically derived from two intermittent creeks, Blood and Deception, which resulted in the bottoms being dry two or three years out of ten. It has been under management by the state of Kansas for the enhancement of waterfowl habitat since the late 1940s, but an unreliable and diminishing water supply has threatened the integrity of the basin and the wildlife resources that use it.

Migratory Bird Use

The importance of the Cheyenne Bottoms to migratory water birds has been well recognized, most significantly in 1988 as a Wetland of International Importance under the Ramsar Convention of the United Nations and as a Hemispheric Shorebird Reserve by Wetlands for the Americas. Shorebirds may be the water birds most dependent on Cheyenne Bottoms. About 45 percent of the continent's migratory shorebirds stop there during spring migration, including over 90 percent of the known world population of five species: the white-rumped sandpiper, Baird's sandpiper, stilt sandpiper, long-billed dowitcher, and Wilson's phalarope. The area has also been designated as critical habitat for the whooping crane by the U.S. Fish and Wildlife Service and is commonly used by other endangered or threatened species of birds, including peregrine falcon, piping plover, least tern, snowy plover, and long-billed curlew. Cheyenne Bottoms has also been recognized as an important migration and breeding area for waterfowl. Half a million ducks and forty thousand geese have been reported during fall and spring migration, and twelve species of ducks are known to commonly nest, including mallard, canvasback, blue-winged teal, and mottled duck.

Historical Changes

Water was historically replenished by spring and fall rains that entered the basin through Deception and Blood Creeks, both of which have very small watersheds inadequate to sustain high rates of evapotranspiration (Zimmerman 1990). Since records have been kept, the basin has gone through many periods of water abundance and extended drought. A major agricultural industry, which includes extensive irrigation systems, has developed near Cheyenne Bottoms and along all the river tributaries that supply the Bottoms with water. Over the years, numerous attempts have been made to secure a water supply for Cheyenne Bottoms by diverting water from other rivers, streams, and canals and by securing rights for water supplies from these systems.

There have been many controversies as to how the water has been allocated and these can be traced through a complex of legal documents. Regardless of blame, the end result has been that the Bottoms's water supply decreased dramatically during the past three decades and its existence was greatly threatened. By the end of the 1980s the Bottoms was dry for much of the year. During the 1980s a comprehensive twenty-five-year plan for securing, managing, and conserving a water supply for Cheyenne Bottoms was started and several portions of the work plan have been completed. Much remains to be done.

The Conservation of Prairie Wetlands

The roots of historical conflicts between wetland conservation, agriculture, urban development, transportation, and other economic development on the prairies remain. However, there are now organized interests that can speak to other wetland values important to quality of life, such as local and migratory natural resources, flood control, groundwater abundance and quality, and sustainable development. For the wetland systems described above, the most prominent local and national government infrastructures under which these values are advocated are the Prairie Pothole, Prairie Habitat, Rainwater Basin, and the Playa Lakes Joint Ventures of the North American Waterfowl Management Plan. The basic operating mechanisms of the joint ventures are described in chapters 15 and 16 of this volume.

Other key programs include the Conservation Reserve and Wetlands Reserve Programs of the 1985 Food and Security Act in the United States, the North American Wetlands Conservation Act, Canada's Prairie Conservation Action Plan, Wetlands for the Americas, the Rainwater Basin Management Plan, and the Cheyenne Bottoms Renovation Plan.

Most of the easily drained and previously unappreciated seasonal and ephemeral wetlands have been lost and are no longer in the way of development pressures. Protection of remaining wetlands is critical if the integrity of the complete prairie ecosystem is to be maintained. Protection is complex though because the majority are found on private lands yet support wildlife and groundwater resources that are of national and international interest. True sustainable protection, in both the United States and Canada must occur because of real and perceived values of wetlands by individual landowners or by consequence of gravity, topography, or water quality making wetland drainage unfeasible.

Protection of basins must be accompanied by sensitivity to what else in the environment actually makes them into functioning ecosystems. Hydrology is the most obvious linkage (Fredrickson and Reid 1990). In cases where basins are linked to groundwater, such as in the Nebraska Sandhills, the physical protection of the basin is only one phase of sustaining wetlands. Where wetlands are dependent on surface water runoff, maintenance or restoration of regional and local drainage patterns must be considered. Linkages between basins and their wildlife resources that depend on associated uplands are the other most important factor for prairie wetlands. Many waterfowl, for example, cannot be sustained on the prairies if their upland nesting cover is not available.

Because the hydrology of most regions of the continent has been dramatically altered, wetland protection and restoration cannot work under a passive, set-it-aside and leave-it-alone philosophy. Restoration of functioning wetland ecosystems will not occur in most cases. Fortunately, there is a solid base of information for most prairie wetland systems that can assist managers in simulating natural hydrological cycles and in enhancing the likelihood of achieving natural wetland ecosystem function.

One of the wonderful things about prairie wetlands is that they are generally fully restorable in form and function if the hydrological regime that created them in the first place is restored. Time required to full restoration is generally dependent on how long a wetland has been lost and the nature of the alternate land-use to which the basin had been subjected (Galatowitsch and van der Valk 1994). Thousands of wetlands have been restored during the last ten years in the prairie pothole region of the United States by simply plugging the drains or removing the drainage tiles that had converted them to farmland during the last century or so—thus restoring the original hydrology. In most instances, portions of the seed bank of native wetland plant species survive for many decades and germinate when conditions are again suitable. Some species are lost over time but can be reintroduced by natural mechanisms, such as transportation by wind or wildlife, or by man-made mechanisms, such as transplanting plants or transporting soil from more complete seed banks.

Wetlands are as much a part of North America's prairies as once the buffalo were. However, wetland systems remain in place in most regions, have their own intrinsic values, and provide local models for restoration efforts. Major wetland systems often require large expenditures to maintain or restore, but a great deal can be accomplished cheaply and with simple tools on individual smaller basins.

PART 3

Prairie Legacies

C H A P T E R 7

Prairie Legacies—Invertebrates

Cody L. Arenz
Anthony Joern

Invertebrates are the little things that run the world.

E. O. Wilson

As an entomologist of European origin, I was appalled on my first
United States experience at the paucity of handbooks for the identification of insects. . . . It is a deficiency in American culture.

Anonymous grant reviewer

Despite the conspicuous and important contribution that both vegetation and
vertebrates make to characterizing North American grasslands, the most significant contributor to the diversity of the prairie system resides with the tiniest animal residents, the invertebrates. While vegetation fuels the grassland system,
providing energy through photosynthesis as well as physical structure from subsequent growth, invertebrates overwhelm the plants and vertebrates in their contribution to local diversity—both numerically and biologically. Each plant
species supports at least one and usually many more invertebrate herbivores
above- as well as belowground. Diversity continues to build from the herbivore
base such that each plant-feeding invertebrate species supports several invertebrate predators, parasites, or parasitoids, and on again as one moves up the invertebrate food chain. Diversity begets diversity in its most striking form among
the invertebrates.

Prairie Conservation
Island Press (Washington, DC • Covelo, CA)

The nature of invertebrate diversity in grasslands takes many forms. Invertebrates vary greatly in size and morphological structures, modes of life, and life cycles. Sizes span several orders of magnitude, a range unequaled in any other taxonomic group. As might be expected, this wide range in sizes leads to significant ecological opportunities unavailable to the other coexisting taxa. Many species live aboveground and many belowground, each with their own adaptations to different environments and roles in the grassland system. Niches that one might never have believed existed are filled by grassland invertebrates, including dung and carrion beetles, acrobatic aerial predators, and ants tending lepidopteran larvae like nomadic herders. Functionally, ecosystems absolutely depend on the presence of invertebrates to facilitate nutrient cycling.

Aesthetically, the prairie landscape would be incomplete without the striking beauty of the many butterflies slipping from flower to flower, pollinating as they feed. Nor would the acoustic ambiance of the prairie air be satisfying without the persistent buzz of stridulating grasshoppers searching for mates or the chirp of the prairie mole cricket dominating the evening wind. In a shimmering web strung among grass clumps, a large yellow, black, and gold orb spider balances on its laclike architecture like a tightwire acrobat, providing visual repose. Finally, the dramatic pervades when a robber fly perches on the branch of a short forb waiting patiently until it spots an insect prey. An immediate and definitive pounce means certain death for the hunted. Invertebrate diversity of North American grasslands clearly plays many roles to the prairie enthusiast.

The grassland invertebrates are poorly studied. Our chapter reflects this problem. Important groups are only superficially treated, in part because the critical information and synthesis do not exist and because key detailed syntheses require more space than is available here. First, we briefly describe the diversity and functional contributions of key groups to grassland systems. In this section, we separate the arthropod component from the remaining invertebrates for both convenience and in recognition that arthropods, especially insects, contribute greatly to biodiversity and deserve emphasis for this reason alone. Then, we focus specifically on some well-studied insect groups that we hope provide representative patterns typical of the invertebrates in North American prairies. Finally, we address some of the principal conservation issues as they impact invertebrates, with citations to direct the reader to more detailed literature.

Diversity of Grassland Invertebrates

Of the thousand interesting stories to be told, none is more astounding than the remarkable diversity contained within the invertebrates. Of the approximately 8 to 10 million total species anticipated worldwide, insects represent about 90 percent of the terrestrial species. Experts anticipate approximately 5 to 10 million insect species will be identified (Gaston 1991). Fewer than 10 percent of these species presently possess a scientific name. For perspective, there are about 12,500 species worldwide of the relatively small order Orthoptera (grasshoppers, katydids, and crickets), approximately the same as the number of bird species and more than the number of mammal species. In North America, about 600 grasshopper species are recognized, again rivaling the number of bird species. Grasshoppers are particularly well represented in the grasslands of the central and western states (see fig. 7.1). Nebraska or Texas grassland sites contain upward of 60 species annually, clearly many more taxa than all of the vertebrates combined. A similar claim can be made for the butterflies, which represent just a fraction of the number of Lepidoptera species either worldwide or within the Great Plains. Both the grasshoppers and macrolepidoptera fall somewhere in the middle in terms of insect species diversity but far below the diversity seen in the Coleoptera (beetles), Hymenoptera (wasps and bees), or Diptera (flies). And just how many nematodes or spiders are there? Again the numbers demand attention.

The significant but wide-ranging grassland invertebrate fauna includes the protozoans, nematodes, earthworms (annelids), mollusks, mites, spiders, insects, isopods, and millipedes/centipedes (Curry 1994). In fact, some have estimated that body mass of just the family Formicidae (ants) includes half the range of body mass recorded for all animals (10^{-5} to 10^5 g) (Bonner 1988). Species from

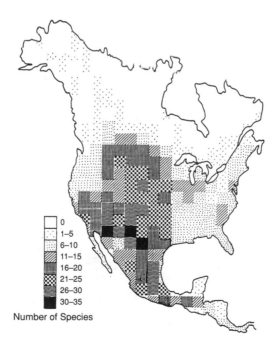

0
1–5
6–10
11–15
16–20
21–25
26–30
30–35

Number of Species

Fig. 7.1 Geographic distribution of the number of gomphocerine and acridine grasshoppers in North America (Otte 1981).

these groups perform remarkably different roles or else perform similar functional roles in very different ways.

The diversity of grassland invertebrates on all fronts precludes detailed descriptions of individual species biology in this chapter, even as representative examples. Moreover, so little is really known about the range of responses practiced by invertebrates in grassland environments that any detailed account is likely to be grossly in error as a generalization. In this section, we briefly sketch the nature of the dominant groups and their significant contributions to grasslands.

The Non-arthropod Component

The non-arthropod invertebrates are not less important than the arthropods, just taxonomically less diverse. Still, the diversity of this group overwhelms when compared to the plants or vertebrate component of the grassland biodiversity.

Protozoans

These widely distributed and numerous single-celled animals exploit almost all available grassland habitats and microhabitats (Curry 1994). The need for sufficient moisture becomes the most definitive unifying theme for understanding the contribution of this group to grassland diversity. However, because of their small size, moisture among soil particles or dew on leaves may be sufficient to support protozoan populations for extended periods. Vegetation-dwelling ciliates and flagellates develop rapidly on leaves when sufficient dew is present and form resistant cysts to overcome inhospitable periods. Soil environments are generally more favorable and support a wider range of taxa, including amoebae, ciliates, flagellates, and slow-growing testate amoebae, that perform significantly more functions and occur at even higher population densities. In semiarid shortgrass prairie of Colorado, for example, soil protozoans ranged from twenty thousand individuals (g^{-1}) in dry soil to over one hundred thousand (g^{-1}) in irrigated and fertilized plots (Elliott and Coleman 1977).

Soil protozoans play important roles in grasslands, often feeding on bacteria while yeasts, algae, and fungal mycelia also contribute significantly to their diets (Curry 1994). Rapid turnover capabilities of soil protozoan populations implicate this group in the transformation of soil organic matter (Elliott and Coleman 1977), such as the mineralization of organic nitrogen or accumulation of organic carbon (Coleman et al. 1977), the genesis of grassland soil fertility.

Nematodes

Occupying all corners of the grassland realm, nematodes are diverse in every sense and play incredibly important roles. Species diversity of grassland nematodes can be great. In Kansas true prairie, for example, Orr and Dickerson (1966) found 228 species from 80 genera. While capable of attacking a range of food sources, including living plants, bacteria, fungi, protozoans, and other animals, root nematodes have received the most attention because of the incredible impact root-feeding habits exert on grassland primary productivity (Smolik 1974; Smolik and Lewis 1982). Very large increases in primary production result (30 to 60 percent or more) when nematodes are suppressed. Like the protozoans, nematodes are tiny and quite sensitive to moisture availability; inappropriate moisture levels may often restrict nematode distributions and activity (Curry 1994). While density estimates for grassland nematodes vary greatly, an average value was about 9 million individuals per m² (range 2.4 to 30 million) for twenty temperate grassland sites (Sohlenius 1980). Nematode densities correlate very strongly with primary production (Yeates 1979). In mixed prairie in Saskatchewan (Matador), the nematodes contributed 98 percent of the total density and 75 percent of the biomass of belowground invertebrates (Willard 1974). Similar results were obtained in South Dakota (Smolik 1974). These individuals may have great impact in grassland function. Nematodes were the single most important invertebrate group in terms of energy consumption (Scott et al. 1979); annual intake was greater in tallgrass prairie than in mixed or shortgrass prairie. Along with the microarthropods, nematodes account for much of the belowground predatory consumption, while primary microbivorous consumption occurs primarily by the protozoans and nematodes in shortgrass steppe (Lauenroth and Milchunas 1992).

Earthworms (Oligochaetes)

Unlike the protozoans and nematodes, earthworms are generally much larger but less diverse and abundant. Earthworms feed primarily on dead organic matter and associated microflora. A major functional role also played by individuals in this group appears to be soil mixing and general contribution to soil fertility. European earthworms are now common in North America, often responding to environmental variation differently than native taxa (James 1988). What is less clear is the impact of exotic species of earthworms on the native fauna.

Arthropods: Spiders and Insects

Arthropods, including both macro- and microarthropods, contribute noticeably to grassland diversity, comprising most of the species, typically representing the largest fraction of the biomass, and contributing significantly to total energy flow (Coupland 1992; Lauenroth and Milchunas 1992). Taxa are unequally distributed between above- and belowground environments depending on grassland type or land-use practices. Macroarthropods primarily include insects and spiders, while mites (Acarina) or small insects such as Collembola typically dominate the microarthropod component. As with the non-arthropod taxa, much remains to be learned before we will understand either the diversity or functional importance of grassland arthropods. In addition, it is clear that prairie type and land-use exert major impacts on the invertebrate fauna in grasslands (Coupland 1992; Kucera 1992; Lauenroth and Milchunas 1992).

Arthropod Diversity in North American Grasslands

In North America, approximately 100,000 species of insects and spiders have been described, representing 34 orders and about 1,250 families (Kosztarab and Schaefer 1990). Species are widely and unevenly distributed among these diverse orders or families within orders (see table 7.1). The orders Coleoptera, Diptera, Hymenoptera, and Lepidoptera clearly dominate the list in terms of numbers of species, each containing a variety of families with an uneven distribution of species. As indicated in table 7.1, many undescribed species are anticipated even in well-populated and well-studied North America; most experts postulate that only about half of North American arthropods have been described.

This type of analysis is not presently available for the more restricted geographic region delineating North American grasslands. However, a sense of the distribution of insect diversity can be seen in figure 7.2 (Danks 1994). While representing only 4 percent of the described species from North America, distribution of these representative species likely describes responses of most insect taxa at higher taxonomic levels. Based on this analysis, the grasslands clearly exhibit relatively well-developed insect faunas. The southern portion of this range clearly includes the most taxa, but this may not be the case in the grasslands. Because of their geographic position, the southern states probably include taxa that are more typically subtropical. Geographic factors probably play major roles in invertebrate diversity in many of the central plains states as well, especially Nebraska and Kansas. In these cases, eastern versus western as well as northern versus southern faunas converge in this grassland region. The enlarged regional species pool most likely results in increased local species diversity within and among sites independent of the vegetation physiognomy or species composition (Ricklefs and Schluter 1993). In addition to uncovering the geographic trends in species diversity, Danks's (1994) assessment of North American insect diversity found that many species are widespread.

Determinants of Arthropod Species Richness

Multiple factors contribute to local biodiversity, some of which are regional in nature while others are local. Ultimately, the regional species pool constrains the total number of species that can be found in an area. Representing the results of multiple speciation events, the evolutionary history in an area must be considered in order to understand biodiversity. Careful systematic and biogeographic

Table 7.1

Species diversity of North American arthropods

Order	Families	Described Species	Undescribed Species
Non-acarine Arachnids	114	4,200	1,100–1,700
Acari	447	5,106	26,000
Protura	3	20	—
Collembola	7	700	250
Diplura	4	64	—
Microryphia	2	35	15
Thysanura	2	30	15
Ephemeroptera	17	555	60
Plecoptera	9	578	0
Odonata	11	415	0
Notoptera (Grylloblattoidea)	1	13	—
Phasmatodea	5	31	5
Orthoptera	20	1,800	600
Mantodea	1	20	0
Blattodea	5	66	10
Isoptera	7	41	0
Dermaptera	6	23	0
Embiidina	4	13	1–2
Zoraptera	1	3	0
Psuoptera	26	257	110
Anoplura	15	76	68
Mallophaga	7	700	655
Siphonoptera	4	258	20
Hemiptera	44	3,834	600
Homoptera	38	6,970	2,650
Megaloptera	2	43	—
Raphidioptera	2	21	—
Neuroptera	17	400	—
Coleoptera	113	23, 640	2,627
Strepsiptera	4	109	—
Mecoptera	5	75	20
Diptera	121	19,562	41,622
Trichoptera	23	1,340	—
Lepidoptera	75	11,300	2,700
Hymenoptera	81	17,429	18,570

Source: Collated from Kosztarab and Schaefer 1990.

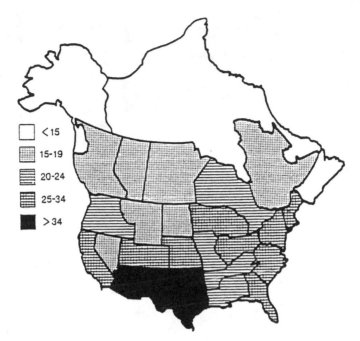

Fig. 7.2 Percentages of insect species throughout North America to illustrate the trends in insect biodiversity based on selected groups from the following insect orders with the families indicated in parentheses: Dictyoptera; Hemiptera (Corixidae, Pentatomidae, Tingidae); Coleptera (Carabidae [part], Curculionidae); Diptera (Culicidae, Blephariceridae, Dolichopodidae, Syrphidae); Lepidoptera (Papilionidae/Pieridae, Nymphalidae, Lycaenidae, Hesperiidae); and Hymenoptera (Pamphilidae, Chrysididae, Bethylidae, Dryinidae, Pompilidae) (Danks 1994).

efforts will continue to provide information on these processes (Kosztarab and Schaefer 1990; Brooks and McClennan 1991).

Other regional, biogeographic rules that affect local and regional species diversity often apply as well. For example, insect herbivore distributions and diversity often track the distributional area of their host plants (Strong et al. 1984; Lawton and MacGarvin 1986). The number of British insect herbivore species that accumulate on these host plants (as a species) can be represented as a logarithmic relationship:

$$S = cA^b$$

where S is the number of species associated with a host plant, A is the habitat area of the plant, and c (y-intercept) and b (slope) are constants. Such relationships have not been carefully delineated for North American grasslands. However, Whitcomb et al. (1984) suggest that the large number of cicadellids (leafhoppers: Homoptera) on dominant grasses and forbs of shortgrass steppe reflect related biogeographic patterns in an analogous process.

Local invertebrate species richness is greatly influenced by additional features of the environment, such as the number and taxonomic identity of plant species, plant productivity in an area, local habitat structure, colonization rates of possible component species relative to disturbance and local extinction, and the quality and timing of essential nutrients (in plants) or hosts for predators or parasitoids. The three-dimensional architecture of the plant assemblage often exerts a dominant influence on patterns of local diversity (Lawton 1983; Strong et al. 1984).

Finally, some landscape-level features can greatly influence local diversity even though regional processes are operating: source-sink relationships (Pulliam and Danielson 1991), fragmentation, and the mass effect (Shmida and Wilson 1985). Most natural and managed communities are open systems, and events in neighboring habitats can influence local patterns of diversity. As such, local and regional diversity must be considered simultaneously. Few examples presently exist for grassland insect assemblages, however.

Functional Roles

In addition to their astounding contribution to diversity, arthropods contribute significantly to the functioning of grassland communities and ecosystems (French 1979; Coupland 1992; Lauenroth and Milchunas 1992; Kucera 1992). The biomass of arthropods regularly exceeds that of vertebrates if domestic livestock are excluded (Lauenroth and Milchunas 1992). In addition, critical differences appear between the above- and belowground components of grassland systems. Greater than 90 percent of the total arthropod community resides in the soil and litter, including the large component of the aboveground taxa that either nest (e.g., ants) or have significant life stages (e.g., scarab beetles) belowground.

Herbivores typically dominate the macroarthropod taxa at almost all grassland sites (French 1979), providing about 85 percent of the total arthropod biomass in shortgrass steppe (Lauenroth and Milchunas 1992). In addition, spiders may dominate as secondary consumers in grassland systems, with little impact from vertebrate predators in terms of total energy flow (Wiens 1977; Schmidt and Kucera 1975). However, bird predation may contribute significantly to the population and community dynamics of dominant insect herbivores such as grasshoppers (Joern 1992).

Grassland Invertebrate Conservation

Conservation and management issues as they affect invertebrates are many and complex. Innumerable sites and untold numbers of invertebrate assemblages challenge our ability to establish firm guidelines or explicit conservation goals. In this section, we aim to spur others to design effective local prairie management plans that promote invertebrate conservation as a goal equal to all others. Figure 7.3 will help in identifying how these issues fit into the flow of a conservation outline. However, it is also clear that our present on-line conservation initiatives for invertebrates are rather basic operations.

Protection of natural-seminatural systems will require combinations of two different approaches: (1) preservation, the cessation of all or most of the detri-. mental anthropogenic activities and environmental influences upon the site to allow the site to remain intact; and (2) conservation, the manipulative actions designed to maintain the integrity and functioning of a system through management (Samways 1994). As described in other chapters, however, the cessation of management may greatly affect the integrity of the natural system through changes in species diversity and habitat structure and actually harm the functioning unit (see Naeem et al. 1994; Tilman and Downing 1994).

Common wisdom typically asserts that protecting habitat for charismatic megavertebrates also conserves invertebrates that require habitat on a smaller scale (Murphy and Wilcox 1986). This assumption may be unfounded. Although posing a popular mammal or bird as a "poster child" often facilitates the protection of lands, subsequent management for that vertebrate will not necessarily afford the habitat requirements for the native invertebrates (Samways 1993a). Many declining invertebrate populations require microhabitat diversity and structure not commonly achieved on preserves without specifically accounting for their needs when determining management techniques.

Threats to Grassland Invertebrates

As with biodiversity in general, invertebrates are most threatened by habitat alteration. Fragmentation of suitable habitat isolates populations, making them more susceptible to extinction through further habitat change, demographic stochasticity, and deleterious genetic effects (Wilcove et al. 1986; Wilcox and Murphy 1985). Basic threats include habitat loss and degradation, competition from edge and nonnative species, and chemical pollution. Pesticides, herbicides, and fertilizers are all threats to invertebrate biodiversity when used indiscriminately. About 320,000 tons of pesticide are used annually in the United States (Pimentel 1986). The impact to invertebrates from such chemical use is not presently known, but it is likely to be significant in regions where grasslands are interspersed with row crop agriculture. Many prairie insects exhibit wildly fluctuating populations (Gaston and McArdle 1993; Dempster 1991) and poor dispersal abilities while encountering fragmented habitats, significant landscape alteration from human land-use, and general apathy toward invertebrates. Each factor increases the vulnerability of these animals to extinction (Panzer 1988). Apathy, in particular, is a significant but largely unrecognized problem to grassland inverte-

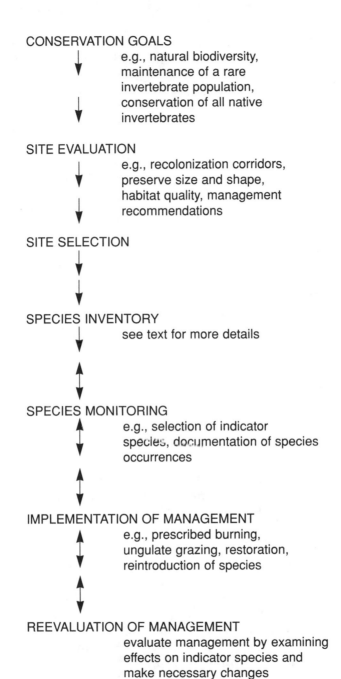

CONSERVATION GOALS
      e.g., natural biodiversity,
      maintenance of a rare
      invertebrate population,
      conservation of all native
      invertebrates

SITE EVALUATION
      e.g., recolonization corridors,
      preserve size and shape,
      habitat quality, management
      recommendations

SITE SELECTION

SPECIES INVENTORY
      see text for more details

SPECIES MONITORING
      e.g., selection of indicator
      species, documentation of species
      occurrences

IMPLEMENTATION OF MANAGEMENT
      e.g., prescribed burning,
      ungulate grazing, restoration,
      reintroduction of species

REEVALUATION OF MANAGEMENT
      evaluate management by examining
      effects on indicator species and
      make necessary changes

Fig. 7.3 Conceptual flow-diagram of the conservation process (see Pearsall et al. 1986; Goldsmith 1991; Goldstein-Golding 1991; Noss 1987, 1990; Usher 1985, 1991; Panzer 1988; Samways 1993a, 1993b, 1994; Shafer 1995).

brate conservation. Many citizens think of the "creepy and crawly" when considering invertebrates, often focusing on those taxa that act as pathogens, pests, or merely undesirable agents of a misunderstood world. Public education about the lives and styles of prairie invertebrates provides the most effective solution to this issue.

Invertebrate Monitoring

Invertebrate conservation requires species inventories that assess a range of habitat types and types of land-use. Land managers must know specifically which taxa are present in order to develop useful, rational recommendations. For the invertebrates in particular, separate sites require individual inventories. Subtle variation among sites can exert profound effects on invertebrate assemblages that will go undetected from cursory inspections. At present, imprecise gross trends of invertebrate distributions will not suffice, thus making species inventories at each site necessary (Opler 1981).

While most vertebrates possess formal scientific names, only about 50 percent of the insects from North American grasslands have been described (Kosztarab and Schaefer 1990). Such limitations hinder our ability to make a full, species-level inventory of the invertebrates let alone have information on critical aspects of the biology needed to develop management plans, including behavior, distribution, ecology, population densities, or vulnerability to anthropogenic change (Opler 1991; Samways 1994). Fortunately, from the standpoint of preserving North American grasslands, this taxonomic impediment (New 1984) is much less pronounced than in the tropics where only 10 to 20 percent of the insect species have been described (Brown, K. S. 1991). Using an indicator assemblage of well-known species to design conservation protocols may minimize the loss or mismanagement of unknown invertebrate taxa in many cases.

Indicator Assemblages

Indicator assemblages include groups of organisms that closely track a particular habitat type or set of changes to a landscape. An indicator species or assemblage is used to *indicate* changes or influences on a landscape that are too difficult or expensive to measure directly (Landres et al. 1988). While vertebrate indicator species have been used in this capacity for some time, their utility is often challenged because large, endothermic vertebrates experience the environment very differently than the much smaller but more diverse invertebrates (Landres et al. 1988; Murphy and Wilcox 1986). Certain invertebrate taxa can be monitored to determine how they respond to management practices or anthropogenic threats, providing information that can then be extrapolated to some or all of the remaining biota (Kremen et al. 1993; Samways 1994).

Land managers cannot monitor all species at a site. Choosing species as members of indicator assemblages will depend on several factors, including the ability to readily sample and identify indicator species (especially by nonexperts), the conservation and management goals, the nature of the threat(s) to the site, the invertebrates present, and the resources available to the land manager. The effect to be monitored is very important when deciding on an indicator assemblage because the main reason for conservation monitoring is to document changes to the habitat and the subsequent effects. An indicator assemblage should include species that can be assayed easily and that are known to respond to the anticipated environmental impact.

In addition, an ideal indicator assemblage should include invertebrates from a variety of taxonomic groups that fill a range of ecological roles (Brown, K. S. 1991; Samways 1993a, 1993b, 1994; Kremen et al. 1993). Such diverse assessment strategies provide a clearer picture about the impact of management techniques and other forces on the invertebrate fauna. It is also crucial that the indicator assemblage be well known ecologically and relatively easy to find, identify, and evaluate (Brown 1991) (see table 7.2).

Terrestrial arthropods make excellent indicator assemblages because they are relatively easy to identify, exhibit rapid population growth, and are sensitive to habitat fragmentation, disturbance, chemical pollution, and other environmental and anthropogenic challenges (Kremen et al. 1993). Although insect population sizes often fluctuate, making point estimates difficult to interpret (Panzer 1988; Dempster 1991; Samways 1994), the rapid population growth rate potential and short time of population turnover of most arthropod species make them eminently more suitable than mammalian groups as indicators of environmental change. Insects rapidly track changes in their environment with changes in abundance or distribution (Erhardt and Thomas 1991).

The Oklahoma Field Office of the Nature Conservancy started a monitoring program using butterflies (Papilionidae) as the indicator assemblage (Arenz, in press), a study patterned closely after Pollard 1982 and Pollard and Yates 1993. As subjects of conservation interest, butterflies are often promoted as indicators for other invertebrates. Butterfly life histories and ecology are often better known than many other insects; they exhibit population characteristics that make them useful bioindicators for assessing local conditions (Erhardt and Thomas 1991); and volunteers can be readily recruited to count butterflies (Pollard 1982; Arenz, in press; Panzer, personal communication 1993). As with other potential indicator assemblages, not all butterflies respond equally to environmental degradation. Certain butterflies (e.g., orange sulphur and sachem) are ubiquitous throughout their range. Such taxa could provide a land manager with valuable information about changes in the surrounding region that will help in interpreting local responses occurring within preserve boundaries. Conversely, some butterfly species require highly specific habitats or have complicated life cycles (e.g.,

Table 7.2

Selected insect groups and their corresponding rankings for selected criteria deemed valuable to indicator groups

Desirable quality for an indicator group in ecology and hazard assessment	Insect Groups[a]				
	Odonata	Orthoptera	H1	H2	Diptera
Taxonomically and ecologically diverse within a system	+	++	+	+++	++
Relatively sedentary	++	++	++	++	+
Species narrowly endemic or, if widespread, well differentiated (local or regional)	+	++	+	+	+
Ease of identification	++	+++	++	++	+
Well studied (genetics, behavior, distribution, ecology)	++	+++	++	++	++
Easy to find in the field (if present)	++	+++	++	++	++
Response to disturbance predictable, rapid, sensitive analysable, and linear	++	++	+	+	++
Functionally important in ecosystem	+	++	+	++	++
Relative specialists or complicate life history	++	+	+	++	+
Responds to wide variety of disturbances (i.e., sensitive indicator)	+	++	+	+	0
Reacts similarly to or is closely associated with other species which this group can indicate	+	++	+	++	+
Total Value of Indicator[b]	17	24	15	20	14

Source: Adapted from K.S. Brown 1991.

Note: All groups represented are hypothesized as having value as indicator groups. + represents value, ++ represents greater value, +++ represents a unique status, and 0 represents lack of information to make a decision.

[a]*Odonata* (dragonflies and damselflies), *Orthoptera* (Acrididae, Grylidae, Tettigonidae), *H1* (Hymenoptera Formicidae: ants), *H2* (Hymenoptera: Apoidea, Ichuemonidea, Mutillidae,

Heteroptera	Homoptera	Coelo1	Coelo2	Lepidop1	Lepidop2
+	++	+	++	++	++
++	++	++	++	++	+
+	+	+	++	++	+
+	+	+	+	+++	++
+	++	+	++	+++	++
+	++	+	+	++	++
+	++	+	+	++	+
+	++	+	++	+	+
++	++	+	+	+++	+
+	++	+	+	++	++
++	++	++	++	++	++
14	20	14	17	24	17

Sphecidae), *Diptera* (Asilidae, Syrphidae, Bombylidae), *Heteroptera* (Pentatomidae, Reduviidae, Coredae), *Homoptera* (Cercepidae, Cicadellidae), *Coleo* 1 (Beetles: Silphidae, Elateridae, Coccinellidae), *Coleo* 2 (Beetles: Meloidae, Scarabaeidae), *Lepidop* 1 (Lepidoptera: butterflies: Papilionidea, Hesperoidea), *Lepidop* 2 (Lepidoptera: moths: Sphingidae, Saturniidae, Arctiidae, Noctuidae).

[b]Total value as indicator is a relative ranking among the groups listed.

regal fritillary), making them more sensitive to habitat alteration than the ubiquitous species. A diverse indicator assemblage comprised mostly of these sensitive species could reveal precise threats to the invertebrate fauna or the biota as a whole.

Grassland Management and Invertebrate Responses

Grazing, mowing, and prescribed burning are the most commonly used management practices presently available in native grasslands for maintaining or altering the diversity of vegetation or for creating a range of successional stages. Management practices can be tailored to match the needs of the invertebrate community once these needs are known. A heterogeneous mixture of species and habitat structure coupled with natural disturbance appears to maximize invertebrate diversity (Samways 1994) as many arthropods are associated with early and midstages of vegetational succession (Usher and Jefferson 1991).

Prescribed Burning

Fire coupled with extreme temperatures and other climatic fluctuations has been an evolutionary force that sculpted the North American grasslands (Vogl 1974; Bragg, in press). Fire is a disturbance that adds impetus to a dynamic habitat mosaic system (Remmert 1991; Pickett and White 1985), and many grasslands are maintained by the action of fire, especially in areas that would otherwise support woody vegetation (Vogl 1974; Bragg, in press).

Fire tends to stimulate production in tallgrass prairie, but the opposite is true in drier grasslands (Bragg 1995). In general, late spring fires tend to favor warm-season plant species over cool-season species. High-frequency burns tend to decrease overall plant species diversity, usually at the expense of the forbs, and decrease the grassland's nitrogen levels through the detrimental indirect effects of fire on the nitrogen-fixing microbial fauna (Seastedt and Ramundo 1990). Each of these vegetation responses favors different invertebrate groups.

Some limitations and guidelines regarding burning exist (Masters et al. 1990; Higgins et al. 1989; Wright and Bailey 1982). Spring burning is a common range management tactic that can severely impact invertebrates just emerging from diapause. For small sites (about 20 ha or less) that have been set aside for the protection of one or several remnant-dependent invertebrate species, the use of fire as a management tool should be evaluated carefully to ensure that its use does not introduce too much disturbance considering the small size of the preserve (Panzer 1988).

It may take tallgrass prairie three to six years for vegetation to return to preburn conditions and longer for drier prairies. Panzer (1988) recommends that controlled burns not exceed 25 to 50 percent of intact plant communities to retain refugia for the native invertebrates. These refugia are very important because total burns (100 percent of a particular plant community) endanger many taxa (Opler 1991). Although fire in general is not considered bad for invertebrates (Samways 1994; Curry 1994), 100 percent burns may result in 100 percent mortality for those invertebrates within the burn unit. While this 100 percent rule is not necessarily true for soil invertebrates (James 1988; but see Ahlgren 1974), fire exerts a profound negative effect on those aboveground individuals that are present at the time of the fire (Curry 1994). The positive effects of an intermediate fire disturbance frequency (Evans 1984) on invertebrates depends upon the recolonization ability of the species and the pattern of burning. Patchy burns with abundant refugia encourage rapid recolonization of suitable habitat. Many fire sensitive taxa (e.g., the plant-boring Papaipema moths) rely heavily upon refugia, not only because of the presence of food plants but also as a site from which to recolonize the burned areas.

Given the sensitivity of invertebrates to fire, why burn at all? Primarily, fire provides a tool to introduce heterogeneity into grassland systems that contributes directly to local invertebrate diversity (Remmert 1991). Burn frequencies of not less than three years for moist tallgrass prairie or somewhat longer duration in drier prairies, will benefit most invertebrates. The common five- to ten-year interburn interval for drier prairies may be too long (Loucks et al. 1985), but burns at too high a frequency result in lowered invertebrate diversity (Curry 1994; Samways 1994). Some invertebrates increase following a fire and others decrease, but given enough time between fire applications, most invertebrates bounce back readily (Anderson et al. 1989). Most burned areas respond with rapid regrowth, which lures herbivorous insects (Curry 1994) as well as native ungulate grazers (Vinton et al. 1993). A land manager contemplating a prescribed burning program needs to read the completed research and realize that more research into the effects of fire upon invertebrates is clearly needed.

Grazing

Grazing affects invertebrates indirectly by altering the floristic composition as well as the structure of vegetation, nutritional quality of host tissues resulting from regrowth, and in response to soil compaction from trampling. While drastic changes in microclimate involving increased soil temperature, reduced soil moisture, and compaction may affect some soil invertebrates, invertebrates that inhabit the vegetation and the litter will experience the most marked impact (Curry 1994).

Care must be taken in designing grazing treatments. Bison grazing, for example, imposed natural pressure from grazing on grassland invertebrates, but most preserves are not able to support this activity in a naturalistic manner. High floristic diversity resulting from low-intensity grazing may decrease if grazing is removed or reduced (Curry 1994; Samways 1994). Therefore, grazing by native ungulates (or domestic cattle if necessary) can be an important disturbance and selective force for promoting invertebrate diversity. Grazing is often not a suitable option for small sites (44 ha or less) because of the need for near-constant monitoring to ensure that the grazing pressure does not become too intense on these sites with their already reduced species richness and genetic diversity. Usually, a light grazing regime can be instituted on large grassland preserves once the proper timing, frequency, and intensity have been established. Exceptions include the identification of a native, remnant-dependent invertebrate species or group that would be negatively impacted by all forms of large herbivore grazing or lack of a method by which to ensure that the ungulates would exert pressure approximately evenly across the preserve as opposed to intense localized grazing pressures.

Mowing

Judiciously timed cutting can sometimes be an effective alternative to grazing and fire for maintaining floral and faunal diversity. However, most invertebrate species are adversely affected if mowing occurs during their most susceptible stages of development. Compared with grazing and fire, mowing is not a "natural" force upon a grassland system, but may efficiently promote invertebrate responses in desired directions.

Fire, grazing, and mowing have various consequences. Each can alter the floristic composition, structure, and height of the vegetation. But mowing is nonselective (Curry 1994); the mower blade cuts everything to the same height, whereas grazing, and often fire, is selective and patchy. As with fire and grazing, the effects of mowing will differ depending on the timing, intensity, and frequency of the treatment. The timing of mowing is important because mowing can favor those plant species that have completed their growth for the season over those that have not and can impact invertebrates aboveground through direct mortality (Erhardt and Thomas 1991) or indirectly through changes in habitat structure.

Summary

While often inconspicuous, the invertebrate fauna is synonymous with diversity in North American prairies. Unfortunately, despite their incredible significance to grassland function and overall biodiversity, invertebrates are poorly understood. The importance of this group clearly warrants much additional research in a number of arenas, including the best use of invertebrate indicators for monitoring local environmental health and the development of suitable land-use management schemes to promote local and regional biodiversity of this group.

Fortunately, it is clear that the invertebrates will readily respond to appropriate management treatments once the specific treatments are carefully delineated. Land managers need to develop specific conservation goals with highly structured management plans. It should also be clear that grassland invertebrates exhibit a myriad of responses and require a diverse array of micro- and macrohabitat components. Heterogeneity among neighboring habitats, presence of specific food plants, diversity in plant species and structure, and the capacity for rapid population responses each contribute to local responses. The use of invertebrate indicator assemblages can greatly facilitate the assessment of management techniques and threats. Each of these important factors can be manipulated; therefore, effective use of options to promote invertebrate diversity merely awaits proper understanding.

CHAPTER 8

Prairie Legacies–Fish and Aquatic Resources

Charles F. Rabeni

The two dozen major stream systems of the prairies are home to an extremely diverse group of freshwater fishes (see fig. 8.1). Some 204 species from twenty-eight taxonomic families are represented. About three-quarters of all species are from five families, which in descending order of abundance are minnows (Cyprinidae), perches and darters (Percidae), suckers (Catastomidae), sunfishes (Centrarchidae), and catfishes (Ictaluridae). Drastic changes in the streams and rivers of the prairie region and their fish fauna have occurred in the last 150 years. Reduction in stream flow for agricultural purposes and the moderation of flows by dams have reduced the ranges of many fish species, while other species have declined because dams have cut off migration routes to historic spawning areas.

Agricultural activities in the tallgrass prairie eliminated prairie grasses, decreased water tables, and allowed excessive erosion to turn small, clear-flowing perennial streams into turbid, intermittent creeks. Turbidity and suspended solids affected normal feeding and reproductive activities of many species. Streams of the Great Plains are naturally more turbid, with high variations in flow and increasing intermittency from east to west. Flow modifications and water withdrawal for irrigation have further stressed fish populations. Species highly adapted to turbid situations were reduced or eliminated following the silt-reducing effects of impoundments (Cross and Moss 1987).

Throughout the prairies the majority of small headwater springs and marshes are gone; midsize streams are typically straightened; and larger streams are leveed to separate the channel from its floodplain. Mean particle size of the substrate is smaller; in-channel habitats are more homogenized; both riparian vegetation

Prairie Conservation
Island Press (Washington, DC • Covelo, CA)

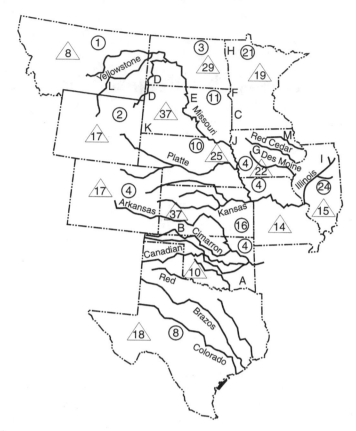

Fig. 8.1 Major stream systems of the prairie. Free-flowing rivers >97 km in length indicated by letters A to M (adapted from Benke 1990, and Stanford and Ward 1979). Numbers in circles represent number of stream segments in each state with high natural qualities as listed by NRI. Numbers in triangles are number of fish species each state considers endangered, threatened, or of special concern. Key to rivers: (A) Blue; (B) Cimarron; (C) Crow; (D) Little Missouri; (E) Little Sioux; (F) Minnesota; (G) Raccoon; (H) Red Lake; (I) Spoon; (J) West Fork, Des Moines; (K) White; (L) Yellowstone; and (M) Wapsipinicon.

and instream aquatic vegetation are much reduced; and barriers to fish migration are much more frequent. Urban domestic and industrial wastes continue to pollute. Introduction of exotics has increased fish diversity and sportfishing opportunities in some drainages, but has often had the negative effect of decreasing the abundance of or eliminating native fishes. All these factors have influenced the native stream fauna, and it is indisputable that, by any biological or physical measure, the aquatic resources of the prairie are highly altered in comparison to their original condition. Surprisingly, no species in the prairie region has become extinct since presettlement, but many are considered endangered or threatened.

Regardless of geographic location, the presettlement prairie stream system possessed a high level of biological integrity (Karr and Dudley 1981) where the fauna had adapted over very long periods to conditions of a particular flow regime, a balance of energy sources, and good water quality—at least no contaminants, adequate physical habitat structure, and a balance of biotic interactions. Alteration of any one of these factors degrades an aquatic system; multiple alterations severely impact ecological functioning and many species will be reduced in abundance or eliminated. The ecological integrity of most prairie streams has been compromised, because every important relation has been affected: flow conditions by dewatering, altered land-use, and disruption of headwaters; energy source balance by the increase of instream primary production with nutrients and less shading; water quality by modern synthetic compounds and organic wastes; physical habitat by channelization and riparian degradation; and the biotic balance by the introduction of fish predators and competitors and elimination of important food sources. Although restoration to presettlement conditions is highly unlikely, there is a growing feeling that the quality of our aquatic resources reflects society's values and that preserving or restoring some part of our heritage is in the best interest of society. This chapter examines the present condition of prairie streams in relation to presettlement condition, with particular reference to fish fauna, and offers some recommendations for preservation and restoration.

Prairie Streams and Fishes

In relation to the rest of North America, prairie streams are of low gradient, highly meandering, with a rich aquatic biota. Within the prairie region, streams are highly diverse and linked to geography and climate.

The Eastern Prairie (Tallgrass) Region

This region includes parts of the upper Mississippi River and its drainages in southwestern Minnesota, most of Iowa, northeastern Missouri, and most of the northern half of Illinois including the Illinois River (Burr and Page 1986). Almost all the streams drain directly into the Mississippi River. Soils in this region tend to be organically rich, and precipitation is relatively abundant. In presettlement times habitat complexity was relatively great and streams are believed to have run clear. This region, which has an extremely rich fish fauna, has undergone some of the most profound changes of any region in North America.

A typical stream system in the presettlement eastern prairie region was likely characterized by numerous upstream prairie swales in headwater areas connected to fewer but larger midreaches (Menzel et al. 1984). The prairie swales were areas of marshy, wetland grasses that acted as sponges that slowly and continuously contributed water, making the whole system perennially flowing—except in the driest years.

These watersheds were periodically burned, and the upstream channels were probably not covered by a canopy cover of large trees. Within this relatively open stream autotrophic processes dominated, and energy to drive the system was supplied principally by algal production and secondarily by riparian grasses. Farther downstream, forested areas may have been more prevalent, and riparian shrubs and trees provided the prime organic energy source. This pattern of energy allocation in prairie streams does not conform to, and is essentially opposite to, the generally accepted river continuum concept (Wiley et al. 1990), which purportedly explains how streams function and which further indicates how little the basic ecology of prairie streams has been studied (Matthews 1988).

Midreaches of presettlement streams meandered considerably. While the flat topography and fine soil often prevented extensive riffle-pool sequence development, the sinuosity contributed to a continuous variety of depths and current velocities that helped provide a diverse habitat. Farther downstream, the stream (now a river) consisted of a main channel and the associated floodplain. Periodic, almost annual, flooding onto the floodplain was essential in maintaining the biological integrity of the system. Flooding rejuvenated the river by providing a major supply of organic matter and nutrients. Periodic flushing flows helped maintain channel configuration and coincided with important life history events of fishes and other biota. This interdependency of channel and floodplain in a properly functioning river is the flood pulse concept (Junk et al. 1989).

Intensive agricultural development began in the late 1840s and produced rapid ecological alteration. Significant ecological damage—probably most of the damage—was done in the first few decades after settlement, and the relative rate of environmental change probably has declined over the last fifty years (Menzel et al. 1984). The elimination of prairie swales and small-stream floodplains by plowing, draining, and installing tiles, the straightening of midreaches and eliminating riparian vegetation, and the decoupling of the river from its floodplain by ditching and levees all contributed to the degradation of habitat and the loss of species.

Today these streams carry greatly elevated sediment loads. Streams believed to originally have had narrow channels, less than a meter deep, presently flow through straightened, wide, heavily eroded channels. Mean particle size has undoubtedly been reduced so that sand and finer particles now dominate. Because of altered stream morphology and changed land-use, storm hydrographs peak rapidly and flooding can be severe, while summer low-water conditions are more extensive.

In many eastern prairie streams, phosphorus and nitrogen are no longer limiting; the sparse shading is gone; and benthic algae constitute the bulk of primary production. Fish communities have changed. Many of the fish existing for so long in an environment of clear water, hard substrates, and abundant vegetation have been extirpated or significantly reduced, and in their place are species more general either in feeding habits, reproductive requirements, or physiological tolerances.

Karr et al. (1985) provided one of the best historic accounts of the changes occurring to a fish fauna in the region concurrent with human development. On the eastern edge of the tallgrass prairie, parts of Illinois that were historically a mosaic of bluestem prairie and oak hickory forest are now mostly cropland. The Illinois River appears to have suffered all possible insults that directly and indirectly affect its biota. The flow has been supplemented by wastewater from the city of Chicago. Floodplains were reduced by drainage and levee construction. Soil erosion became significant and filled floodplain lakes. Barge traffic resuspends sediments, many of them toxic, in the main channel. The combined effects of this environmental degradation have devastated the fish community.

Fish distribution in the Illinois River system has been recorded since about 1850. Out of a total of 140 species of fish, 5 percent are increasing, 23 percent are considered stable, 6 percent have been extirpated, 61 percent are declining, and 5 percent of the species have been introduced (Karr et al. 1985). Two-thirds of the species are declining in abundance or have been eliminated from parts of their historic range. The magnitude of the changes in the fish fauna depended upon the location of the fish community in the river basin as well as each species' ecological role. The trophic balance of the stream fish community was greatly altered. Omnivores, considered generalists in their feeding habits, were less affected than were any other feeding group (i.e., the more specialized carnivores, insectivores, and herbivores).

Invertebrate communities of prairie streams have undoubtedly changed, but no historical evidence is available. However, Menzel et al. (1984) studied a series of Iowa streams on a continuum of degraded to less degraded and made some conclusions regarding the changes in the invertebrate fauna over time. Taxa requiring coarse substrates, such as caddisflies, blackflies, and stoneflies, have undoubtedly decreased in favor of those taxa that burrow or otherwise use soft substrates (many dipterans—especially Chironomids).

The Western Prairie (Great Plains) Region

This region (central and northern Great Plains) includes the Missouri River drainage of parts of Montana, Wyoming, Colorado, and all of North Dakota, South Dakota, Nebraska, Kansas, and Oklahoma. The description of a presettlement stream of the eastern prairie needs to be modified for streams of the western prairie. Rivers and streams of this region have a relatively low gradient, because the area was originally a shallow sea (Cross and Moss 1987). The plains are arid, owing to their midcontinental placement and the rainshadow produced by the Rocky Mountains. Erosion of the sedimentary rocks on the slopes of the mountains eventually covered much of the plains with a blanket of sand and gravel and has an overriding effect on riverine biota in present times. The exceedingly erodible soils and typically brief, intense storms control stream geomorphology and the biota.

The larger rivers with mountain headwaters, such as the Platte and Arkansas, had channels that were wide and shallow. They were extremely turbid at high flows because of erosion and at low flows because of the contribution of sediments from tributaries. The riverbeds had a sandy texture, and quicksand-like conditions were a common occurrence for early settlers. Much of the rivers' lengths were unshaded, and summer suns caused high temperatures and increased salinity from evaporation (Cross and Moss 1987).

Approximately two hundred species of fish presently occur in the central and northern Great Plains region (Cross et al. 1986). Among the native species, minnows (Cyprinidae), perches (Percidae), suckers (Catastomidae), and catfishes (Ictaluridae) comprise about 70 percent of the fauna, and minnows and perches make up about half the species. Over two dozen species have been introduced into the region, most into lakes as either sport fish or as forage for sport fish.

The distribution of fishes appears to be controlled by ecological factors, with drainage boundaries being a secondary factor. Often species will inhabit parts of many drainages, while only a few species will be present throughout a single drainage. Fish communities tend to decrease in species diversity both northward and westward, which is attributed to increases in flow variability, turbidity, and temperature extremes, and a decrease in habitat diversity.

Great Plains fishes have been most affected by the widespread damming of almost all large rivers in the eastern plains and excessive water withdrawals in the west. Several species common thirty years ago are now seriously threatened. In many situations where the fish fauna as a whole has not been nearly extirpated, the newer conditions created by dams (i.e., clearer water and more moderated flows) have produced a fish fauna dominated by sight-feeding planktivores and piscivores.

Water withdrawals, primarily for agriculture, have been especially significant. The best example is the High Plains aquifer, which provides about 30 percent of the water used in irrigation in the region. Large-scale irrigation between 1940 and 1980 caused the average groundwater level to drop some 3 m (Dugan et al. 1994). Since that time, water levels have continued to drop but at a slower rate, probably due to a combination of a reduction in irrigated lands, agricultural conservation practices, and higher than average rainfall. Nevertheless, examples of the effects of water withdrawals such as the completely dry Arkansas River in western Kansas are common.

The turbid river fish community of the central Great Plains (Cross and Collins 1975) has species once common to streams having channels with fluctuating shallow flows and shifting sand beds. Fish of this community are specialized to exist in highly turbid, highly fluctuating water environments. In the very large streams, species include the pallid sturgeon, paddlefish, and blue suckers, all of which have drastically declined in abundance. In the moderately large streams, common species are goldeye, gizzard shad, buffalo fish, gars, carpsucker, sicklefin chub, sturgeon chub, speckled chub, sharpnose shiner, red shiner, Arkansas River shiner, silverband shiner, chub shiner, smalleye shiner, western silvery minnow, and plains minnow. All the minnows in this group have declined in abundance in some part of their historic range.

The central and northern plains, especially glaciated areas, were dotted with a habitat type composed of clear brooks, marshes, springs, and seeps that contained relict populations from the remains of glacial climates (Cross and Collins 1975). Few of these fish are endemic to the region; most also occur widely dispersed in eastern North America. Many of the species have low tolerance to variable flows, high temperatures, and turbidities and were greatly diminished in numbers because of habitat alterations shortly after settlement. Especially detrimental was the disturbance of prairie sod, which was a main factor in maintaining clear-running water. Prominent members of this community include the creek chub, southern redbelly dace, hornyhead chub, common shiner, blacknose shiner, Topeka shiner, plains topminnow, Arkansas darter, and the orangethroat darter.

A third type of stream habitat and associated community were described by Cross et al. (1986) as the residual pools of highly intermittent streams. These are smaller streams dependent upon runoff that is not supplemented by springs. The existing members of this community are widespread and highly tolerant of harsh conditions. Actual community composition depends upon the duration of intermittency. Isolated pools are a hazardous habitat, and a successful species has to endure high temperatures, freezing, dissolved oxygen fluctuations, and an increased vulnerability to predation by fish, birds, and mammals. Four species dominate in these situations: red shiner, flathead minnow, black bullhead, and green sunfish.

Introductions and Major Changes in Distributions

Introduced fishes and range extensions have been most noticeable into the rivers west of the Missouri-Iowa western border (Cross et al. 1986). At least twenty-seven species have been introduced, and forty species have extended westward to where they were not originally found. Many were already present in the prairie but restricted to the eastern region. Other species were from nearby ecoregions, such as the Mississippi Valley, which supplied walleye, yellow perch, white bass, northern pike, and mosquito fish (Cross and Collins 1975). A few species were truly exotic—carp, trout, and striped bass. Many of the introductions were deliberate, due to state agencies stocking sport fishes and their potential prey in newly created lentic situations. Many introductions occurred shortly after settlement by the pioneers. In 1883 the territorial legislature provided for the appointment of a fish commissioner who stocked fry received from the U.S. commissioner of fisheries. Stocking of carp began two years later (Bailey and Allum 1962). Many fish have established populations and also moved into the connecting rivers. Many introductions and range extensions were accidental bait-fish releases, or were because of habitat changes—lowering of turbidity below reservoirs increased sightfeeders, which likely outcompeted native turbid river species. Introductions, for whatever reason or cause, have increased overall fish diversity, while their predation effects have undoubtedly reduced populations of many of the small-stream fishes.

The Coastal Plain

The several large streams that cross the southern prairie and end at the gulf coast of Texas have unique fish faunas because of the high incidence (76 out of 195 species) of fish that occur at some time in their life cycles in marine waters (either euryhaline or diadromous) (Conner and Suttkus 1986). Of the forty-seven families and 195 species, 15 species have been introduced by humans. The fauna is dominated by the same families as elsewhere in the prairie, but Centrarchidae are disproportionately greater and Percidae are much less dominant. The representation of livebearers, killifish, is exceptionally high.

The most important anthropogenic impacts affecting the fish fauna are not unique to this region; they include physical degradation of channels and watersheds and alteration of the hydrograph by dewatering the channel or destabilizing flows.

The Missouri River

The Missouri River is the longest river in the United States and in presettlement times was highly dynamic, with characteristics of extensive migrations of the stream channel and high levels of turbidity (Galat et al., in press). One-third of the river has been channelized, one-third impounded, and the hydrograph of the whole river severely altered. Much of the ecological damage done to the Missouri River was initiated by separating the mainstem from its floodplain by eliminating wetlands, channelizing river stretches, and constructing levees, dams, and impoundments. Even during flooding, except in 1993, less than 10 percent of the original floodplain is now inundated. A properly functioning river depends upon the periodic inundation of its floodplain for energy (detrital organic matter and primary production) on which much of the fish community depends and for appropriate physical habitat for spawning, feeding, and as nursery areas for many species. The flood pulse concept (Junk et al. 1989) is becoming a central focus of many restoration plans. Besides the decoupling of the river from its floodplain, agriculture, urbanization, and industry have all had significant impacts by increasing the traditional factors of pollution.

The Missouri River was aptly nicknamed the Big Muddy and in presettlement times was one of the most turbid systems on the continent. The river travels through highly erodible soils and even in normal flow times had a channel characterized by continuous bank erosion, a highly braided configuration, and extensive sandbars and islands. Floodplains averaged over 8 km wide and ranged up to 27 km wide (Hesse et al. 1989; Galat et al., in press).

Constraining and impounding the river has significantly reduced the sediment load to 60 to 99 percent less than in earlier times. Six mainstem storage reservoirs, along with about thirteen hundred smaller impoundments, have been constructed on the river and its tributaries. Sediments have been retained above dams, and therefore, water discharged from the dams has considerably more power available to erode the downstream channel.

The presettlement river experienced annual or near-annual flushing flows, which helped maintain channel configuration and coincided with important life history events of fishes. Today water is controlled for navigation at essentially constant levels from April through November.

An examination of the changes in the fish fauna of the lower Missouri River (channelized but not impounded) from 1940 to 1983 showed an increase in the number of species and large changes in the relative abundance of species in the community (Pflieger and Grace 1987). New or more abundant species included those that were pelagic planktivores and sight-feeding carnivores such as skipjack herring, gizzard shad, white bass, bluegill, white crappie, emerald shiner, and red shiner. Ecological reasons given for these increases relate to the effects of decreased turbidity.

Declines were seen in common carp, river carpsucker, bigmouth buffalo, and two species uniquely adapted for life in the presettlement Missouri River, pallid sturgeon and flathead chub. Increased competition for ever-scarcer habitat with shovelnose sturgeon may be responsible for pallid sturgeon's decline, and predation may have reduced the flathead chub, because sight-feeding carnivores were originally absent. Two common species that favor silty backwater habitats, the western silvery minnow and plains minnow, have declined. Silty backwaters have become less prevalent as channelization increased and the suspended load decreased. The modified river has actually become favorable for some species. Speckled chub, sturgeon chub, and sicklefin chub prefer open channels with swift current and firm substrate.

Some new species were accidentally (grass carp, silver carp, bighead carp) or intentionally (white bass and rainbow smelt in upstream reservoirs) introduced. An increased frequency of some tributary fishes was noted—spotted bass, longear sunfish, Ozark minnow.

In the more modified, that is, impounded, Nebraska section of the Missouri River, drastic declines have been noted in several dominant species, including paddlefish, sauger, flathead catfish, and blue sucker (Hesse et al. 1989). Minnow species were also greatly impacted, most notably five chub species (flathead, speckled, sturgeon, sicklefin, and silver) and the Plains minnow and the western silvery minnow (Hesse 1994). These species have been reduced in abundance 70 to 80 percent, while several have been extirpated above Gavins Point Dam. These apparently more severe effects to fishes in the Nebraska portion of the Missouri River waters are probably a result of fish being influenced by the more stressful hydraulic and temperature conditions of upstream impoundments.

Preservation and Restoration

Effects of Hydropower

Perhaps because of the relatively low topographic relief and low average stream gradient, development of hydropower and associated dams has been less extensive in the prairie region than in other ecoregions—but still substantial. Benke (1990) examined Federal Energy Regulatory Commission documents and concluded that in the prairie region about a hundred hydropower dam sites represented about one-third of all potential dam sites in the region. These developed sites also represented about half the potential generating capacity available in the region. The largest and most productive sites have been developed, and the remaining sites would accommodate only smaller plants. Small plants add little to overall generating capacity. Although plants less than thirty megawatts account for over 80 percent of all power plants in the United States, they provide less than 15 percent of total electrical output. Obviously, new hydropower projects on small rivers of high natural value would be difficult to justify because of the relatively small return at the cost of forever altering a stretch of river.

Spawning

Many species have either declined in number or been extirpated from entire stream systems because the habitat required for spawning has been destroyed. Many species are adapted to spread eggs in specific habitats, for example, clean gravel or aquatic vegetation. When gravel is embedded with silt or vegetation is eliminated, these specialized species disappear. Other species, adapted to enter the floodplain during annual overflow events, have been eliminated by the construction of levees. Some big river plains fish spawn in the open river during the first flood event of the year. Others, such as the paddlefish, are required to travel considerable distances to historic spawning streams that are now blocked by dams. Even when hydraulic and habitat conditions are favorable, many impoundments release water hypolimnetically, and the historic coupling of flow with temperature, which together served as cues to spawning, are no longer available. The significance of changes in faunal composition because of impaired spawning is that our communities are becoming composed of ever more generalist species. Since generalists tend to be widespread and rather common, we are losing a special element of our fauna and reducing overall biodiversity.

Endangered Species

No species in the prairie region is known to have become extinct in the last hundred years (Miller et al. 1989). This is no doubt due to the widespread ranges of most species, their ability to disperse along river systems, and their general high tolerances to environmental insults. Numerous species have seen their range decrease; some have seen their range widen. Two prairie species are on the federal endangered species list: the pallid sturgeon and the fountain darter. The Neosho madtom is federally listed as threatened. The Arkansas river shiner is proposed for listing as an endangered species. Four species, the Arkansas darter, sicklefin chub, sturgeon chub, and Topeka shiner have federal C-1 status, which allows the Fish and Wildlife Service to develop proposals for listing as threatened or endangered. A category recently eliminated by the FWS is Category 2 (C-2), given when there was evidence of a species decline but further work was needed to ascertain distribution and abundance. Nine species were listed as C-2: the blue sucker, flathead chub, Guadelupe bass, lake sturgeon, Ozark minnow, plains minnow, plains topminnow, smalleye shiner, and western silvery minnow. Of course, the elimination of this category does not affect the plight of these species, except perhaps to give the impression that there is less urgency to protect and restore their populations. The American Fisheries Society classifies twenty-five prairie fishes as endangered, threatened, or of special concern (Williams et al. 1989).

Individual states have endangered species programs and together list 109 species in need of protection (fig 8.1). Twenty-seven species are endangered, and twenty-nine are considered threatened. The others are considered in need of conservation, officially unlisted but concern warranted, or on a watch list. Thus, over one-third of all fish species in the prairie region are believed to be in at least potential trouble in terms of their distribution or abundance. The benefits accorded state-listed species vary, but are always much less than the protections and actions specified for federally listed species under the Endangered Species Act. While state designation provides a higher visibility for a troubled species, often with some regulations on capture and killing, overall protections are limited, and recovery plans and designations of critical habitats for listed species are rare.

The variety of listings for the states indicate how a species may decline in one part of its range while remaining reasonably healthy elsewhere. The reason given for the vast majority of the listings is "present or threatened destruction, modification, or curtailment of habitat or range" (Williams et al. 1989).

Streams of Special Concern

Given this rather bleak picture concerning the state of waterways and fishes in the prairie region, the question becomes whether we actually have much to save. The Nationwide Rivers Inventory (NRI), a database now maintained by the National Park Service, identifies streams or stream segments with high natural quality. An analysis of every entry in the NRI indicates a total of 111 river sections within the prairie region that have values related to scenic, recreational, geologic, fish and wildlife, historic, cultural, or other conditions. The number of stream segments listed in the NRI and the total miles of stream for prairies in each state are Colorado 4 (360 km), Illinois 24 (2240 km), Iowa 4 (712 km), Kansas 16 (1270 km), Minnesota 21 (1940 km), Missouri 4 (408 km), Montana 1 (13 km), Nebraska 10 (668 km), North Dakota 3 (644 km), Oklahoma 4 (430 km), South Dakota 11 (1625 km), Texas 8 (787 km), and Wyoming 2 (80 km). Segment length varies widely but averages 103 km. The listed rivers are not necessarily pristine but, in many situations these rivers would be considered the best of what remains.

The NRI was analyzed by Benke (1990) to determine free-flowing streams and rivers of high quality. His criteria for a quality stream was one that was essentially free flowing for more than 200 km, in a relatively undeveloped corridor, and that possessed outstanding natural or cultural values. Nationwide, only 1.9 percent of all stream miles met these criteria. The percentage is even less in the prairie region. A similar attempt to identify streams of particular value was made by Stanford and Ward (1979) using a 97-km uninterrupted stream length as one of their criteria. A combination of both studies results in only fourteen rivers in the entire prairie region that are free flowing for at least 97 km and possess a moderately high degree of biological integrity (see fig. 8.1). Few of the rivers listed receive any sort of special protection from the state or federal government.

A more detailed examination at a smaller spatial scale for stream legacies was conducted by the conservation organization American Rivers (Huntington and Echeverria 1990). The American Rivers Outstanding Rivers List compiles information from a variety of state, federal, and nongovernmental conservation organizations—all with different selection criteria. The listing includes shorter streams and stretches, and designates several hundred stream stretches in the prairie region. While each report differs somewhat in its selection of streams in need of preservation, it is evident that only a small percentage of our original heritage remains.

Protection

From a natural resource perspective, the history of humans' relation to our prairie streams and rivers has been one of development with little regard for biological or ecological considerations. Laws and regulations have historically favored uses related to economic development. It has only been within the last thirty years that uses of water related to biological, recreational, and aesthetic goals—rather than power generation, navigation, and waste disposal—have been asserted. But the situation varies widely among states, and there is often a wide discrepancy between the language of a law and its application.

An essential element in the protection of fisheries and other aquatic resources of the prairie region will be the maintenance of water in the rivers at times and at levels sufficient to protect aquatic life. The term "instream flow" has been given to this process, and several prairie states have enacted instream flow legislation. Colorado, Iowa, Kansas, Montana, Nebraska, and Wyoming are considered to have the most effective systems. Illinois, Minnesota, Oklahoma, and Texas have regulations that are either untested or inadequate. Missouri has yet to enact regulations (Berton Lamb, National Biological Service, personal communication). In the Canadian prairies, the provinces have been delegated authority to carry out the Canada Fisheries Act, which designates a sufficient flow to protect fishery resources. The enactment of legislation or regulations for instream flow does not guarantee preservation of fish and other aquatic biota but is a necessary first step.

In addition to adequate flows, water quality and the integrity of the riparian corridor must be maintained. The most straightforward approach is to designate streams for total protection of existing conditions. The most comprehensive protection for a stream is its designation within the federal Wild and Scenic Rivers Program. Unfortunately, only the Middle Fork of the Vermilion River in Illinois and short sections of the Missouri and Niobrara Rivers in Montana, Nebraska, South Dakota, and Wyoming are so designated. Some state programs afford significant protections. Iowa has five streams designated as State Protected Waters. Illinois, North Dakota, Minnesota, and Oklahoma have state wild and scenic river programs but only Minnesota lists more than one prairie river. Two states, Texas and Nebraska, have a list of streams for a proposed wild and scenic river system.

Attempts by individual states to publicize, designate, and protect individual streams vary widely. All prairie states have published a listing of outstanding rivers. The natural heritage programs of North Dakota and Wyoming have listings of streams deserving protection. Most states attempt to promote the value of streams by publishing listings of outstanding streams, outstanding fishing waters, designated canoe trails, or exceptional boating waters. These efforts provide little actual protection but do publicize the value of particular waterways.

Summary

While the protection of particular stream segments or particular species is important, more comprehensive management is needed. A segment of a stream cannot logically be isolated from the rest of the watershed. Streams are the ecologically unifying element of the landscape and reflect all activities of a highly populated, highly technological society. A simple but important question emerges in our study of water-use in the prairie. Can we manage our resources for the benefit of fisheries and wildlife, for quality recreation, and for essential economic purposes? A solution is easy in principle. Regard a stream as an interconnected system with component parts of channel, riparian, headwaters, and floodplain and strive to achieve a measure of conservation of all these elements.

Acknowledgments

I thank Jennifer Brady for research assistance. The following people supplied information on the status of fish and streams in various states: T. Nesler, C. Taylor, J. Fleckinstein, J. Horak, H. Drewes, D. Figg, C. Hunter, G. Zuerlein, G. Par, M. Howery, E. Dowd-Stukel, G. Garrett, and L. Bergstedt. B. Lamb assisted with information on stream flow. This is a contribution from the Missouri Cooperative Fish and Wildlife Research Unit (National Biological Service, Missouri Department of Conservation, University of Missouri, and Wildlife Management Institute cooperating).

CHAPTER 9

Prairie Legacies–Amphibians and Reptiles

Paul S. Corn
Charles R. Peterson

Amphibians and reptiles are an important part of the prairie for a number of reasons. First, they are significant elements of biodiversity with at least 124 species occurring in prairie landscapes. Second, several species currently have federal or state status as sensitive, threatened, or endangered species. Third, amphibians and reptiles are functionally important components of ecosystems. They prey on a wide variety of animals, particularly invertebrates. Amphibians may determine community structure in ecosystems such as fishless ponds. Amphibians and reptiles are prey for many other vertebrates (including each other) and larger invertebrates, providing important links in food webs that might otherwise be lacking. Because of their high efficiency in converting food to tissue and their often high numbers, they can form a significant portion of the biomass in a system and serve to buffer energy flow in the ecosystem (Pough 1983). They may also play an important role in transporting nutrients, especially between aquatic and terrestrial systems (e.g., aquatic larvae and terrestrial adults of many amphibians). Fourth, changes in some amphibian and reptile populations may be early indicators of general environmental problems, such as pollution from pesticides or the effects of increased ultraviolet radiation (Beiswenger 1986; Blaustein 1994).

Prairie Herpetofauna

Characterization of the prairie herpetofauna involved several steps. To determine which species of amphibians and reptiles occur in the prairie, the distribution

Prairie Conservation
Island Press (Washington, DC • Covelo, CA)

maps from recent field guides (Stebbins 1985; Conant and Collins 1991) were overlaid on a map (Küchler 1985) depicting the geographic extent of the tall-grass, mixed-grass, and shortgrass prairies of North America. Those species peripheral to the prairie area were not included (e.g., the blackneck garter snake). A few of the species have been described since the publication of the field guides (e.g., the dunes sagebrush lizard). The overall range of each species was visually estimated and assigned to one of six percentage categories (<5, 5–25, 25–50, 50–75, 75–95, or >95 percent). The field guide maps were then used to determine the general pattern of geographic distribution (continental, west, southwest, Great Plains, north-central, south-central, southeast, northeast, or east). Species with greater than 75 percent of their range within the prairie region were considered endemic. Using the habitat descriptions from the field guides, supplemented by personal field experience with some of the species, we assigned each species to one of four general habitat types—general, prairie-desert, forest, or aquatic. (A complete listing of the species and characterizations of their patterns of distribution and habitat associations is available from the authors.)

Distribution Patterns

Ninety species of reptiles and thirty-four species of amphibians occur in the prairie region. The majority are snakes (48 species) and lizards (23 species). Nineteen species of turtles complete the reptiles. Most of the amphibians (29 species) are anurans. Salamanders (5 species) are underrepresented in prairies relative to the other herpetofauna.

Two strong decreasing gradients in species richness, south to north and east to west, were identified, echoing Kiester 1971. For example, 106 species were recorded in Texas (80 reptiles and 26 amphibians), 19 species in Manitoba (12 reptiles and 7 amphibians), and 73 species in Illinois (52 reptiles and 21 amphibians) compared to 31 species in Wyoming (21 reptiles and 10 amphibians). These gradients are determined by continental climatic and associated vegetation gradients. Simplified, there are fewer reptiles where it is cold and fewer amphibians where it is dry. These trends converge in the northwest corner of the prairie in Alberta, which has just eight reptiles and seven amphibians.

Similar to birds and mammals, true prairie herpetofauna is a minor component of total richness. Most of the herpetofaunal diversity of tallgrass prairie landscapes is contributed by species with primarily eastern and southern distributions, and the diversity of shortgrass prairies is dominated by species with southwestern affinities (see fig. 9.1). Only twelve reptiles and three amphibians have distributions centered on the central North American prairie.

Of these fifteen species, only eight reptiles (Texas map turtle, ornate box turtle, prairie skink, dunes sagebrush lizard, Brazos water snake, Concho water snake, plains garter snake, and lined snake) and two amphibians (plains leopard frog and plains spadefoot) are considered endemic to the prairie. However, as

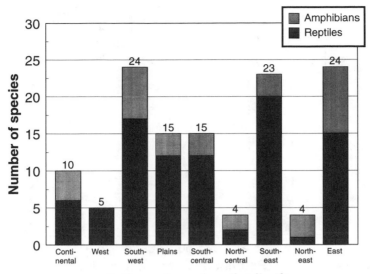

Pattern of species distribution

Fig. 9.1 Species richness of prairie amphibians and reptiles arranged by distributional affinities. Only 15 of 124 species (12 percent) are distributed primarily in the prairie. Numbers above the bars are the total number of species in each distribution category.

biochemical systematic techniques are applied to more taxa of amphibians and reptiles, a splitting of species previously described on the basis of morphology (Frost and Hillis 1990) is anticipated. Consequently, the number of endemic species will most likely increase. This trend in herpetological systematics should be considered relative to strategy design to conserve biodiversity in the prairie.

Much of the diversity of the prairie herpetofauna is contributed by peripheral species, particularly those with eastern and southern affinities. Less than 5 percent of the range of twenty of seventy species with eastern, northern, or southern distributions occurs in the prairie, and the prairie captures less than 25 percent of the ranges of another twenty-eight of these seventy species (see fig. 9.2). Most of these eastern and southern species occur in the prairie in aquatic habitats or in riparian forests. They can be important components of the prairie herpetofauna, although the distribution of a species in the prairie may not seem to be an important component of that species' total range.

Habitat Associations

Of the 124 species of amphibians and reptiles occurring in the prairies of central North America, 42 species are associated with grassland or desert habitats (e.g.,

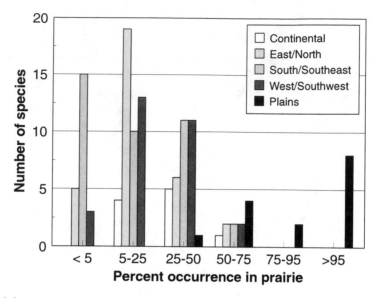

Fig. 9.2 Numbers of species in broad distribution categories and the percent of their ranges that occur in the prairie.

spadefoots, ornate box turtle, earless lizards, western hognose snake, bullsnake); 38 species are primarily aquatic or require permanent water at some stage in their life cycle (e.g., leopard frogs, most turtles except box turtles, water snakes); 28 species use forests or other woody vegetation (e.g., gray treefrog, ground skink, common kingsnake, earth snakes); and 16 species are generalists with no readily definable habitat association (e.g., tiger salamander, Woodhouse's toad, racer) (see table 9.1).

Broadly defined prairie landscapes (including other habitats such as ponds, rivers, riparian forests, and woodlots) have different herpetofaunas. This largely reflects increasing aridity from east to west. For example, tallgrass prairie landscapes include more species associated with forests than species associated with grassland or desert habitats, while shortgrass landscapes have more than twice as many grassland and desert species than forest species (see fig. 9.3). Mixed-grass landscapes fall between these extremes.

Amphibians and reptiles, however, are often associated more closely with either zoogeographic patterns or specific habitat features than with broad categories of vegetation (Bury et al. 1991). While prairie herpetofauna is largely a mixture of eastern and southwestern species with few unique elements, a majority of species also depend on specific habitats. The aquatic species require per-

Table 9.1

Amphibian and reptile species in broadly defined habitat associations for general zoogeographic characterizations of species' distributions

Pattern of Distribution	General Habitat			
	Prairie/Desert	Forest/Woody	Aquatic	General
Prairie	9	0	5	1
North/East/South	9	26	29	6
West/Southwest	24	0	2	3
Continental	0	2	2	6

manent water in some form, and species associated with woody vegetation occur mainly in riparian areas or at the eastern margins of the prairie.

Many species associated with grasslands and deserts also require specific habitat features. Amphibians such as spadefoots or the Great Plains toad breed in playas or pools that temporarily fill with water following summer thunderstorms (Collins 1982; Hammerson 1986). Winter hibernacula, or dens, are important for many species of snakes, particularly at higher latitudes. Several species may den communally, sometimes in large numbers (Gregory 1982). Deep crevices in south-facing sandstone or limestone outcrops are typical den sites for prairie rattlesnakes, wandering garter snakes, and bullsnakes (Klauber 1972; Duvall et al. 1985), while limestone sinks in southern Manitoba harbor thousands of red-sided garter snakes (Gregory 1982). A rookery may be located within a few hundred meters of a den. This is a location with abundant cover—large table rocks—where pregnant prairie rattlesnakes spend the summer before giving birth (Duvall et al. 1985; Graves and Duvall 1993). Male and nonpregnant female snakes may disperse several hundred meters to several kilometers from dens to summer foraging areas (Gregory 1982; King and Duvall 1990). Dens may be a limiting resource in some landscapes (Parker and Brown 1974; Gregory 1982) and are often foci for human persecution of rattlesnakes.

Prairie dog colonies are prominent features of shortgrass prairies that provide habitat for many species of vertebrates. Prairie rattlesnakes may use prairie dog burrows as dens (Klauber 1972). Clark et al. (1982) recorded six species of reptiles on colonies of the black-tailed prairie dog in New Mexico, and Sharps and Uresk (1990) listed four amphibians and six reptiles associated with black-tailed prairie dog colonies in South Dakota. All of the species recorded on prairie dog colonies are common in other habitats, so although populations might be enhanced, prairie dog colonies apparently are not required habitats for any herpetofauna.

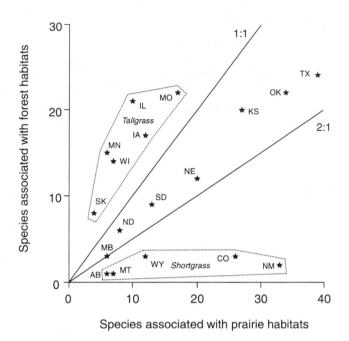

Fig. 9.3 Richness by state or province of herpetofauna associated with grassland or desert habitats versus species associated with forest or woody vegetation. Solid lines describe ratios of 1:1 and 2:1 grassland to forest species. Polygons, drawn by eye, enclose broad prairie landscapes.

Conservation of Prairie Herpetofauna

Status

Protection of prairie herpetofauna has not received much attention from a legal perspective. Few of the federally listed amphibians and reptiles are associated with grassland habitats; most are aquatic species. The endangered Wyoming toad is a glacial relict restricted to the aquatic habitats near Laramie, Wyoming. Although close to extinction, the cause of its decline is not known (Corn 1994). The single threatened species, the Concho water snake, is endemic to a single river drainage in central Texas.

A larger number of species are listed as endangered or threatened by the states. However, the majority of these are rare and are listed because they are just marginally distributed in the states (see table 9.2). Inclusion on a state list of endangered or threatened species may not reflect a threat to the survival of the species as a whole, although rarity and, therefore, vulnerability of amphibians and rep-

Table 9.2

Prairie amphibians and reptiles listed as threatened or endangered by states

State	Amphibians	Reptiles	Number of species with marginal distribution in state (percent)
Colorado	0	0	0 (-)
Illinois	1	9	4 (40)
Iowa	4	15	15 (79)
Kansas	7	14	19 (90)
Minnesota	0	3	2 (67)
Missouri	0	5	5 (100)
Montana	0	0	0 (-)
Nebraska	0	1	1 (100)
New Mexico	2	4	5 (83)
North Dakota[a]	—	—	
Oklahoma	0	0	0 (-)
South Dakota	0	6	6 (100)
Texas	0	7	3 (43)
Wisconsin	1	6	1 (14)
Wyoming[a]	—	—	

Source: Levell 1995.

[a]No state list of threatened or endangered species.

tiles does not necessarily correlate well to legal status (McCoy and Mushinsky 1992).

Recent Trends

In habitats that have not been altered significantly by human activities, the herpetofauna probably has not changed radically. Valentine and Fort Niobrara National Wildlife Refuges in the sandhills of north-central Nebraska have undergone some changes since European settlement, including tree plantings, aggressive fire suppression, and introduction of game fish to previously fishless lakes (Bogan et al. 1995), but have not been cultivated. In a two-year survey of amphibians and reptiles, twenty-one of twenty-five species previously recorded for Cherry County were captured and the spiny softshell turtle was added to the county's fauna (Corn et al. 1995). The four species not captured were rare or secretive and were not considered to have been lost from the area.

Conversely, Lanoo et al. (1994) surveyed the amphibians of Dickinson County in northwestern Iowa and compared their results to a survey by Frank Blanchard in 1920 (Blanchard 1923). The latter survey found two species not found by Blanchard, but failed to find two species recorded by Blanchard and estimated that the most abundant species, the northern leopard frog, had declined from perhaps 20 million to 50,000 frogs (Lanoo et al. 1994). This decline was attributed largely to the decline in available wetlands as a result of historic patterns of wetland destruction. Abundance of amphibians, however, may fluctuate widely (Bragg 1960; Pechmann et al. 1991), so conclusions from a two-point comparison should be drawn cautiously.

Because more than 90 percent of the presettlement wetlands have been lost from the tallgrass prairie (Lant et al. 1995), abundance of aquatic species of herpetofauna must generally be just a fraction of their former amounts. Farther west, where less of the landscape has been cultivated, aquatic species may not have declined as much from presettlement levels.

One of the recent concerns in conservation of amphibians and reptiles, the apparent decline of amphibian species throughout the world (Yoffe 1992; Blaustein 1994; Corn 1994) has a longer history in the prairie than in most other places. Northern leopard frogs virtually disappeared from tallgrass prairies in Wisconsin and Minnesota in the 1960s and early 1970s (Gibbs et al. 1971; Hine et al. 1981). In Manitoba, 49,907 kg of leopard frogs (more than 1.5 million frogs) were harvested for scientific and educational use in 1972. The amount dropped to 5,900 kg in 1974, and no leopard frogs were harvested from 1975 to 1982 (Koonz 1992). The causes of the decline of northern leopard frogs are not known with certainty, although commercial harvest (Gibbs et al. 1971) and contamination from agricultural chemicals (Hine et al. 1976) may have contributed. Koonz (1992) reports anecdotes of mass overwinter mortality of leopard frogs in Manitoba in the 1970s. Populations of northern leopard frogs have recovered somewhat from the lows of twenty years ago (Koonz 1992; Seburn 1992; Vial and Saylor 1994). In contrast to areas farther west in North America (Corn 1994), few other species of amphibians in the prairie have undergone notable declines (Vial and Saylor 1994). Cricket frogs have become scarce in the northern portions of their range, especially in southern Wisconsin, Minnesota, and northern Iowa (Lanoo et al. 1994; Vial and Saylor 1994). This small aquatic frog, whose typical habitat is a farm pond, may be particularly susceptible to runoff of agricultural pollutants.

Threats

Historically, draining wetlands and plowing the sod destroyed the tallgrass prairie habitat, resulting in fragmented distributions and reduced abundances of the herpetofauna. Although conversion of prairies to agricultural use is less frequent, a current victim may be the dunes sagebrush lizard. This endemic species from the southwestern corner of the prairie in Texas and New Mexico occurs on sand dunes vegetated with shinnery oak.

Pothole ponds in northern prairies are important habitats for many amphibians. Abundance of northern leopard frogs in such habitats at times has been sufficient to support commercial exploitation (Gibbs et al. 1971; Koonz 1992; Lanoo et al. 1994). This is due in large part to absence of fish from many pothole ponds, because these habitats are isolated from running water and water may be available only seasonally. Although draining and cultivating wetlands is no longer a dominant threat, 28 percent of a sample of tallgrass prairie farmers replied that, if allowed, they would drain and cultivate wetlands currently protected by federal legislation (Lant et al. 1995). Another threat is conversion of potholes to permanent water combined with stocking of game fish (M. Lanoo, personal communication). The native amphibians, particularly tiger salamanders and leopard frogs are generally unable to coexist with predatory fish (Merrell 1977). Conversion of temporary ponds to permanent water may also increase the population of terrestrial mammalian predators (Scott, in press).

Persecution of rattlesnakes as dangerous vermin has a long history in North America (Klauber 1972). The rattlesnake roundup, a relatively recent organized and sanctioned form of this persecution, may pose a localized threat to snake populations in the southern Great Plains. These events harvest and kill large numbers of western diamondback and prairie rattlesnakes (Campbell et al. 1989; Weir 1992) and have broader ecological effects. For example, rattlesnakes are often collected by spraying gasoline into dens, killing or injuring nontargeted snakes and other wildlife (Campbell et al. 1989). However, the ecological consequences of rattlesnake roundups have not been studied under controlled conditions. Although roundups are usually sponsored by civic organizations and may benefit local charities, Weir (1992) doubted this justified the environmental cost and suggested conservation measures, including banning the use of gasoline to capture snakes and instituting monitoring of snake populations.

In the Great Basin and arid Southwest, livestock grazing alters habitat structure, allowing abundance of some species of lizards to increase, but diversity and abundance of reptiles is usually lower in grazed areas (Busack and Bury 1974; Reynolds 1979; Jones 1981; Bock et al. 1990). Similarly, fire has become a concern in southwestern deserts, because expansion of nonnative annual grasses has allowed large fires in landscapes that have not experienced intense fires (Brown and Minnich 1986). Prairie fires do kill amphibians and reptiles (Erwin and Stasiak 1979; Seigel 1986). However, fire and intense grazing by bison and prairie dogs were regular features of the shortgrass and mixed-grass prairies of central North America. Fire suppression and light-intensity grazing result in increased vegetative cover, which may indicate lower quality habitat for shortgrass prairie reptiles.

In landscapes with frequent natural fires, burning may increase diversity and abundance of the herpetofauna (Mushinsky 1985). Ballinger and Jones (1985) observed that cessation of grazing at Arapaho Prairie, Nebraska, resulted in lower lizard populations. The survey of amphibians and reptiles in the Nebraska Sandhills by Corn et al. (1995) found the lowest diversity and abundance at a site that had been protected from fire and grazing for several decades.

Summary

The herpetofauna of the prairie is composed of a complex mix of eastern and western species, with relatively few species that evolved in or are restricted to the central part of North America. Most of the species diversity of prairie amphibians and reptiles is associated with nongrassland habitats, such as permanent water or riparian forests. Presence of specific habitats is a better predictor of species diversity than broad categories of vegetation.

Relatively few current downward trends in abundance are known for prairie amphibians and reptiles, and because there are few endemic species, few are listed as threatened or endangered. This is the case even in tallgrass prairies, which essentially do not exist in presettlement form. However, we do not conclude that the lack of listed species means that the herpetofauna of the prairies is relatively free of conservation concerns. The extreme fragmentation of remaining tallgrass prairies suggests the possibility of future extinctions, and relatively little is known about the abundance and population trends of most species, particularly snakes and lizards.

There is considerable potential to manage for enhanced diversity and abundance of the herpetofauna. Specific actions include protection of wetlands, regulation of exploitative activities such as commercial harvest of leopard frogs and rattlesnake roundups, and use of fire and grazing, especially in shortgrass and mixed-grass landscapes.

Acknowledgments

We thank Fred Samson and Fritz Knopf for their encouragement and patience. Michael Dorcas reviewed the list of habitat associations. Shara Howie and Jeffrey Lerner provided us with information from the Network of Natural Heritage Programs and Conservation Data Centers and the Nature Conservancy. Jim Duncan provided information on the legal status of herpetofauna in Manitoba.

CHAPTER 10

Prairie Legacies—Birds

Fritz L. Knopf

To many naturalists, native birds of prairie landscapes are the drabbest, most boring, and ecologically least significant of the North American avifauna. The generally small, mostly nondescript species forage in hidden places of nearly featureless landscapes much as small mammals do, rather than as their colorful brethren that flit through lush forests or display decorative plumes as they stride erectly through shallow wetlands. Yet the sun rises on calm spring mornings and males burst from their herbaceous hidings in a flight that exposes brightly colored underparts and melodious songs, often with accentuated wing movements that more resemble insects than birds. And off on a barren rise, chickenlike fowl of browns-with-barring puff gigantic orange-red neck sacks to boom an antediluvian drumming across still grasses. Wind muffles sounds on the prairie. But when wind stills, birds provide the sounds of the prairie. The consonance of bird songs speaks to the health of the grasslands that once stretched, unbroken, to every horizon.

The Endemic Prairie Avifauna

Prairie, or steppe, landscapes cover 7 ha x 10^9 ha of the world—over half of its terrestrial surface (Imboden 1988). Almost 1.5 ha x 10^9 ha of prairie occur in North America, representing about 17 percent of the continent. Despite the extensiveness of North American prairies, however, they have not resulted in the radiation of a diverse avifauna. Only 5.3 percent of North American bird species

Prairie Conservation
Island Press (Washington, DC • Covelo, CA)

evolved on prairies (Udvardy 1958; Mengel 1970). Excluding wetland and sage-brush associates, nine species are what might be considered narrow endemics (Biddy et al. 1992) of the Great Plains, and an additional twenty species are more widespread but have strong affinities to the Great Plains. These avian representatives of the native prairie vertebrates are equally divided between passerine and nonpasserine forms (see table 10.1).

Avian assemblages on grasslands are locally simplistic, often dominated by only a few species. Four species (common yellowthroat, red-winged blackbird, common grackle, and eastern meadowlark) collectively accounted for 63 percent of all individuals recorded during surveys from 1991 to 1994 on the Goose Lake Prairie area in Illinois (J. R. Herkert, personal communication). Another four species (horned lark, western meadowlark, lark bunting, and chestnut-collared longspur) accounted for 69 percent of all individuals recorded at many sites across the northern Great Plains (Kantrud and Kologiski 1982). Only three species (horned lark, McCown's longspur, and lark bunting) accounted for 913 of 1,047 (87 percent) of all birds that I recorded on 112 point surveys of the

Table 10.1
North American prairie avifauna of the Great Plains

Nonpasserines	Passerines
Primary (Endemic) Species	
Ferruginous hawk	
Mountain plover	Spraque's pipit
Long-billed curlew	Cassin's sparrow
	Baird's sparrow
	Lark bunting
	McCown's longspur
	Chestnut-collared longspur
Secondary (More Widespread) Species	
Mississippi kite	Horned lark
Swainson's hawk	Eastern meadowlark
Northern harrier	Western meadowlark
Prairie falcon	Dickcissel
Greater prairie chicken	Savannah sparrow
Lesser prairie chicken	Grasshopper sparrow
Sharp-tailed grouse	Henslow's sparrow
Upland sandpiper	Vesper sparrow
Burrowing owl	Lark sparrow
Short-eared owl	Clay-colored sparrow

Source: Modified from Mengel 1970.

Note: Excludes wetlands associates and species with stronger ecological associations with sagebrush (*Artemisia* spp.) landscapes of the Great Basin.

Pawnee National Grasslands, Weld County, Colorado in 1990. The total number of native species, excluding wetland associates and exotics, recorded in those three studies were 38, 29, and 14, respectively.

Distribution of Prairie Birds

As defined in earlier chapters, the North American prairie is usually viewed as comprising eastern tallgrass, central mixed-grass, and western shortgrass components. The mixed-grass and tallgrass components extend westward at more northerly latitudes. By definition, mixed-grass landscapes are ecotones between tallgrass and shortgrass associations, but do have specific floral elements (Risser 1985).

Native birds of prairies can be viewed relative to their distribution across the grassland associations of the Great Plains. Of the endemic prairie birds, shortgrass and short–mixed-grass associates include six of the nine species, two species appear to be primarily mixed-grass associates, and one species, the Cassin's sparrow, occurs along the prairie-shrub ecotone of the southwestern United States (see fig. 10.1). Evolutionarily, these species radiated from the short–mixed-grass prairies of north-central Montana (see fig. 10.2). None of the endemic birds of North American grasslands are associates of tallgrass prairies.

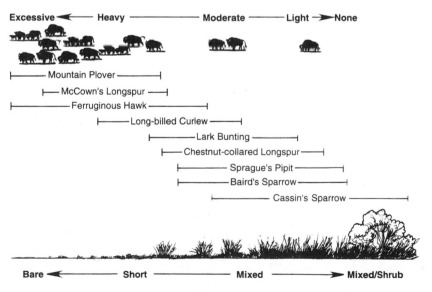

Fig. 10.1 Distributions of endemic birds of prairie uplands relative to grassland type and historical grazing pressure across the western landscapes of the Great Plains.

Fig. 10.2 Collective distributions of endemic birds of North American prairie (including three wetland species) illustrating the importance of shortgrass prairie as the biogeographic region from which these species radiated. Isoclines delineate the number of species nesting within a geographic area (after Mengel 1970). The center of radiation is approximately north-central Montana.

The secondary prairie species are more widespread than the primary endemics (see table 10.2). Six are shortgrass or short–mixed-grass associates, three primarily mixed-grass, and six tallgrass or tall–mixed-grass. The horned lark is the only shortgrass-specific species, and the dickcissel and Henslow's sparrow are the only tallgrass-specific species. The dickcissel will use areas of shrub invasion into tallgrass (Zimmerman 1992), but primarily responds to the grass component locally (Zimmerman, personal communication).

Five additional widespread species characteristically use ecotonal areas of shrub incursion into prairie or forest savannah. These ecotone species include Mississippi kite and lesser prairie chicken on the south, and sharp-tailed grouse and clay-colored sparrow on the north and west. The lark sparrow occurs across prairie types and selects habitats based more on local brush associations than characteristics of the grasses.

Table 10.2
Relative ecological associations of grassland birds in North America

	Grass				
	Short	Short/Mixed	Mixed/tall	Tall	Ecotone[a]
Primary (Endemic) Species					
Mountain plover*	X				
Long-billed curlew	X				
McCown's longspur	X				
Ferruginous hawk	X	X			
Chestnut-collared longspur	X	X			
Lark bunting	X	X			
Baird's sparrow		X	X		
Sprague's pipit*			X		X
Secondary (More Widespread) Species					
Horned lark*	X				
Swainson's hawk	X	X			
Prairie flacon	X	X			
Burrowing owl	X	X			
Vesper sparrow	X	X			
Savannah sparrow		X	X		
Short-eared owl		X	X		
Western meadowlark		X	X		
Upland sandpiper			X	X	
Northern harrier			X	X	
Greater prairie chicken			X	X	
Grasshopper sparrow			X	X	
Eastern meadowlark*			X	X	
Henslow's sparrow*				X	
Dickcissel*					X
Mississippi kite					X
Lesser prairie chicken					X
Sharp-tailed grouse					X
Clay-colored sparrow*					X
Lark sparrow*					X

[a]Areas of significant brush associations or forest savannas.

*Species declining at $P \leq 0.05$ (see Table 10.3).

Historical Ecology of Endemic Birds

The native landscape of the shortgrass prairie was a mosaic ranging from areas of excessive disturbance to areas of infrequent or no grazing. The primary herbivores included prairie dogs, pronghorn, and bison (see fig. 10.3). Bison and pronghorn preferentially graze on prairie dog towns (Coppock et al. 1983b; Krueger 1986), thus intensifying grazing and trampling pressure locally. In addition, areas at the headwaters of the Platte River held large herds of bison for prolonged periods. Large numbers of bison wallows are still discernable (see fig. 10.4), and many remain unvegetated in the 1990s even though bison were extirpated locally in 1867 (Hornaday 1889). Throughout the western Great Plains, prairie dog numbers have been reduced 98 percent (Summers and Linder 1978), and domestic cattle have been substituted for the bison.

Fig. 10.3 Historical distribution of bison in North America. The dashed line represents the approximate widest distribution (Hornaday 1889). The solid line identifies the shortgrass prairie where the large herds, and most animals, were located. The solid line also approximates the distribution of white-tailed and black-tailed prairie dogs.

Fig. 10.4 A buffalo wallow on the Pawnee National Grassland. Wallows are readily discernible although the buffalo was extirpated locally over 125 years ago. Wallows are often up to 50 m in diameter and 3 m deep.

All of the endemic birds of prairies evolved within this grazed mosaic of the grassland landscape (see fig. 10.1). The mountain plover and McCown's longspur occur at sites of heavy grazing pressure to the point of excessive surface disturbance (Knopf and Miller 1994; Warner 1994). The plover will actually nest on plowed ground (Shackford 1991) and spends its winters in California's Central Valley, which was historically clipped by extensive kangaroo rat populations and grazed by an estimated six hundred thousand Tule elk (McCullough 1971). The ferruginous hawk also prefers moderate to heavily grazed sites as taller grasses reduce the detectability of their small-mammal prey (Wakeley 1978).

A second suite of species appear more adapted to moderate grazing intensities. The long-billed curlew, lark bunting, and Sprague's pipit prefer areas of taller grasses interspersed in a shortgrass landscape. All three species use the taller grass tufts to conceal nests (Bicak et al. 1982; Finch et al. 1987; Kantrud 1981), and the longspur may also sing from elevated perches.

The Baird's sparrow and chestnut-collared longspur appear at sites across a broad spectrum of grazing intensities (Kantrud 1981; Kantrud and Kologiski 1982). Both of these species also nest in hayfields of native grasses on the northern plains, that is, sites that are not grazed.

The Cassin's sparrow occurs in areas of at least 6 percent shrub cover that is generally lightly (Bock and Webb 1984) to moderately (W. H. Howe, personal communication) grazed. Breeding habitats of this sparrow probably were outside the distribution of the historically large herds of bison on the plains. The shrub component of its habitat also implies that fire did not play as major a role in maintaining grass vigor as it did in other geographic regions of the Great Plains.

The universal use of grazed landscapes by endemic grassland birds points to the historical impact of grazing as an ecological force on the western Great Plains. Grazing played a significant role in breaking sod. Whereas the ferruginous hawk forages on small mammals, the remaining endemic grassland birds eat insects. Two primary food items, grasshoppers and beetles, require >10 percent of an area to be bare ground in order to lay eggs. Native mammals disturbed the grass sod locally, enabling these insectivorous herbivores to reproduce.

Recent Trends in Populations

Data on changes in bird populations are among the best available for any taxonomic group. The Breeding Bird Survey (BBS) is a coordinated annual inventory of birds in the United States and Canada (Robbins et al. 1986; Sauer and Geissler 1990). The survey presently includes >200,000 point counts at >4,000 locations (Sam Droege, personal communication). Survey results indicate that continental populations of many native birds of the North American prairies were changing rapidly from 1966 to 1991 (Knopf 1994). As a group, grassland birds in general (Askins 1993) and endemic grassland birds specifically (Knopf 1994) have shown steeper, more consistent, and more geographically widespread declines than any other behavioral or ecological guild of North American species.

An update of the population trend information through the most recent analyses (1993) confirms the patterns of decline I reported earlier among the grassland species (see table 10.3). Six of the nine endemic species are declining, and half of those at a statistically significant ($P < 0.05$) rate. The major significant change from the earlier reporting is that the decline in the continental population of Sprague's pipit is now statistically supported.

BBS data are inadequate to determine trends in lesser prairie chickens, but fourteen of the nineteen remaining widespread species were declining, and again half at a significant rate. Only three of the collective twenty-eight species were increasing significantly: the ferruginous hawk, upland sandpiper, and McCown's longspur. Thus, two-thirds of native prairie birds are declining, and half of those declines are supported statistically.

A second major historical inventory of North American birds has been the annual Christmas Bird Count (CBC). Christmas Bird Counts are inventories of all species and numbers of birds recorded in a one-day period within a count area with a radius of 11.2 km. The survey routes tend to focus efforts on more diverse landscapes (wetlands, riparian areas, forest edges) along environmental gradients to maximize the total number of species seen in a day, and thereby probably do

Table 10.3

Annual rates of change in continental populations of grassland bird species 1966–1993

Species	No. of Routes	Birds/ Route	Population Trend[a]
Endemics			
Ferruginous hawk	276	0.20	+1.64**
Mountain plover	46	0.39	−3.69***
Long-billed curlew	253	1.37	−1.67
Sprague's pipit	142	1.43	−3.63***
Cassin's sparrow	190	13.64	−2.54***
Baird's sparrow	138	1.62	−1.75
Lark bunting	390	27.31	−2.13*
McCown's longspur	72	4.68	+7.30***
Chestnut-collared longspur	151	9.87	+0.44
Secondary (More Widespread) Species			
Mississippi kite	188	0.71	+0.88
Swainson's hawk	678	0.82	+1.37*
Northern harrier	1,155	0.43	−0.36
Prairie falcon	296	0.10	+0.33
Greater prairie chicken	54	0.55	−6.85
Lesser prairie chicken	N/A[b]		
Sharp-tailed grouse	187	0.49	+1.05
Upland sandpiper	709	1.90	+2.67***
Burrowing owl	379	0.58	−0.18
Short-eared owl	285	0.12	−0.57
Horned lark	1,830	24.65	−0.70**
Eastern meadowlark	1,807	18.02	−2.25***
Western meadowlark	1,424	42.19	−0.52*
Dickcissel	817	12.44	−1.63***
Savannah sparrow	1,502	7.39	−0.53
Grasshopper sparrow	1,540	3.43	−4.11***
Henslow's sparrow	253	0.11	−4.96**
Vesper sparrow	1,548	7.18	−0.29
Lark sparrow	1,013	4.02	−3.45***
Clay-colored sparrow	464	6.61	−1.20***

Source: Breeding Bird Survey.

[a]Annual rate (expressed as a percent) of change in population numbers;
   * = $P < 0.10$ ** = $P < 0.05$ *** = $P < 0.01$.

[b]Inadequate data.

not represent the true numbers of prairie birds as well as the BBS routes, which are more vegetation-specific, standardized in effort, and conducted when birds are actively displaying on territories. Nevertheless, the data are meaningful for those species for which a substantial proportion of the continental population winters in the United States. The Migratory Bird Management Office of the National Biological Service first entered CBC data from twenty-four hundred local counts for the years 1959 to 1988 into electronic form in 1994 (Sam Droege, personal communication).

For the endemic prairie birds, the CBC data corroborate the annual declines in lark bunting and increases in McCown's longspur populations as statistically valid (see table 10.4). The CBC further imparted some statistical significance to declines of the Baird's sparrow and increases of the ferruginous hawk. For the widespread species, declines of the eastern meadowlark and lark sparrow are corroborated, and the nonsignificant BBS trends for four other species (sharp-tailed grouse, short-eared owl, western meadowlark, and vesper sparrow) gained statistical support. These data all must be viewed with caution, however, pending knowledge of what proportion of each species population winters north of Mexico.

Conservation of Prairie Birds

The native prairie landscape was once viewed as home to a meager avifauna that only needed trees to attract birds and become a lavish showcase of color and song. Studies of avifaunas at forest edges (Kendeigh 1941) and within shelterbelts (Martin 1981; Yahner 1983) corroborate this potential. Three percent of the Great Plains is now forested (Knopf and Samson 1995), much of it with exotic tree species such as Siberian elm and Russian-olive. More generally, however, the number of bird species within a local assemblage increases with any human development of a landscape (Grinnell 1922; Knopf and Scott 1990). Such increases in the number of alien and exotic species in native landscapes often result in population declines, or extirpations, of narrow-endemic species going unnoticed (Knopf 1992).

Endangered Species, Endangered Ecosystem

The present proposal activity to list the mountain plover under the Endangered Species Act may be viewed as a precursor to similar action for other prairie bird species as continental populations of one-third of the species are currently declining at statistically significant rates. These declines support the view that the prairies of the Great Plains are the most endangered ecosystem in North America (Samson and Knopf 1994).

Table 10.4

Annual rates of change in continental populations of endemic grassland bird species

Species	BBS	CBC
Endemics		
Ferruginous hawk	+1.64***	+3.9***
Baird's sparrow	–1.75	–1.4***
Lark bunting	–2.13*	–3.7*
McCown's longspur	+7.30***	+2.13**
Secondary Species		
Sharp-tailed grouse	+1.05	+2.2**
Short-eared owl	–0.57	–1.8**
Eastern meadowlark	–2.25***	–2.6**
Western meadowlark	–0.52*	–1.4**
Vesper sparrow	–0.29	–1.7*
Lark sparrow	–3.45***	–2.0**

Note: Numbers in the CBC column reflect bird species that were adequately sampled on Christmas Bird Counts from 1959 to 1988. An adequate statistical sample does not equate to confirmed trends; the proportion of the continental population wintering in the United States is unknown for most of these species. Numbers in the BBS column are from the Breeding Bird Survey. Numbers are the annual rate (expressed as percent) of change in population numbers.

* = $P < 0.10$ ** = $P < 0.05$ *** = $P < 0.01$

Conservation Priorities for Prairie Birds

Densities of prairie bird populations are highly variable from one year to the next and between locales (Robbins and Van Velzen 1969; McNicholl 1988). Reproductive success also varies annually, with some years exhibiting very poor productivity (George et al. 1992; Knopf and Rupert 1996). Such variability is particularly obvious in seasonally unpredictable climates such as the Great Plains, where drought often affects food supplies over broad areas of landscape. Birds have been argued as relatively poor indicators of local environmental degradation (Morrison 1986). Annual fluctuations in populations led Temple and Wiens (1989) to conclude that the duration of a trend is more important than the magnitude of change. The availability of the BBS data (especially where it can be cross-validated with CBC data) far exceeds the scope and detail of information for any other taxonomic group.

Conservation of shortgrass and mixed-grass prairies appears most critical given that sizable areas of these associations remain intact. Declines of the two shortgrass species, the mountain plover and horned lark, appear relatively universal across biogeographic provinces. The narrowly distributed mountain plover is declining in all regions where it occurs. The widely distributed horned lark is stable at many locations but areas of major decline include such widely separated regions as the Central Valley of California, southern Rocky Mountains, Pinon-Juniper Woodlands, Aspen Parklands, Ozark-Ouachita Region, Ohio Hills, Allegheny Plateau, and Southern Piedmont.

Whereas the endemic species evolved within a mosaic of grazed landscapes, the relatively standardized approach to rangeland management using allotments and applying the same grazing intensity across broad landscapes is contradictory to the historical ecology of these species (Knopf 1996). Alternatively, factors driving widespread declines of prairie birds may not even be in the breeding habitat of the species. Patterns of declines of Cassin's and clay-colored sparrows point to problems at nonbreeding locales (Knopf 1994).

Questions about Widespread Species

Targeting shortgrass and mixed-grass prairies for conservation priority is based on the habitats that they provide for the narrow-endemic species. Unlike the shortgrass prairie, which is still more than half intact, however, the tallgrass prairie is 98 percent tilled (Samson and Knopf 1994), and prairie remnants are isolated and often too small to support some species (Samson 1980; Herkert 1994a). Populations of tallgrass species such as the dickcissel and Henslow's sparrow have historically declined significantly due to the overwhelming fragmentation of the prairie. In contrast to land-management emphasis in the conservation of grassland endemics, conservation of tallgrass species is much more dependent on a series of prairie preserves throughout the region. Management of those preserves also needs to emphasize the wise use of fire, the major natural disturbance critical to maintaining the ecology of tallgrass prairie (Collins and Wallace 1990). Preserves alone may be inadequate, however, and conservation of tallgrass birds will need to address agricultural land-use in many areas.

Population changes among the more widespread species permit some generalizations that can ultimately be posed as hypotheses. For example, declines in eastern meadowlark populations are dominated by trends in eastern states with the species stable and actually increasing at locales on the southern Great Plains. Declines of this meadowlark represent a decline in agricultural pastures and hay meadows due to agricultural intensification in the Midwest (Warner 1994) and succession to forests in eastern states (Askins 1993). Simultaneously, the BBS data indicate that declines in Cassin's and lark sparrows are occurring especially in that region (the Edward's Plateau of Texas) where eastern meadowlark populations are increasing rapidly. Together, these trends point toward ecological phenomena (perhaps brush control) as driving population changes of all three species. A similar ecological scenario may be driving clay-colored sparrow declines, which are primarily occurring in the aspen parklands of Canada.

Assuming that the more widespread species of prairie edges increased historically with fire control and subsequent woody invasion of native prairies, conservation focus for these species strongly depends on the historical reference time that one chooses for comparison. Population distributions of many species in the 1990s may be reverting more toward 1850 patterns. Further diagnostic inquiry into patterns of population change are required to define the long-term significance of declines among the prairie-edge species.

Summary

The endemic avifauna of the Great Plains prairies includes only nine species of upland birds. An additional twenty species apparently evolved on the prairies and currently range broadly into other biogeographic areas. The endemic species are primarily shortgrass or short–mixed-grass associates. Historically, they evolved with the large native herbivore assemblage of the western Great Plains. The landscape was a mosaic of differentially grazed sites. The individual species evolved to specific site characteristics based on local soils, drought cycles, and grazing pressure.

As a group, grassland birds have shown steeper, more consistent, and more geographically widespread declines than any other behavioral or ecological grouping of North American species. Six of the nine endemic and fourteen of nineteen widespread species are declining, and half of all declines are supported statistically. The mountain plover is currently listed as a C-1 candidate species under the Endangered Species Act. The rates of decline in continental populations of the endemic Sprague's pipit, Cassin's sparrow, and lark bunting are cause for immediate concern. Of the more widespread species, declines in the eastern meadowlark, dickcissel, grasshopper sparrow, Henslow's sparrow, lark sparrow, and clay-colored sparrow are also statistically valid. These collective declines point to major landscape-ecosystem changes that warrant immediate ecological inquiry.

Conservation priority for endemic grassland birds is on the historically grazed western Great Plains. In addition, a network of prairie preserves are critical to sustain core populations of tallgrass prairie birds due to the extent of fragmentation of the eastern Great Plains. Unlike other taxonomic groups addressed in this volume, native prairie birds are equally dependent on the quality of breeding, migration, and wintering habitats. Conservation actions require a more cosmopolitan view that extends well beyond the geographical boundaries of the Great Plains.

Acknowledgments

I thank Bill Iko for technical assistance. James R. Herkert, Bill Iko, Stephanie L. Jones, and John L. Zimmerman provided many helpful comments on working drafts.

CHAPTER 11

Prairie Legacies—Mammals

Russell A. Benedict
Patricia W. Freeman
Hugh H. Genoways

Few North American ecosystems have been as dramatically altered by humans as the prairies of the Great Plains. Occupying the immense interior of North America, these deceiving grassland oceans hid their complexity and diversity from many early travelers who saw this area merely as an obstacle to overcome in their westward journeys. But for the careful observer, prairies hold a tremendous quantity of life, arranged in a diverse mosaic of patches ranging in scale from minute anthills to the vastness of the Nebraska Sandhills or Kansas Flint Hills. Not only is a given ridgetop subdivided into a number of areas in varying stages of succession, but this ridge varies from the hilltop adjacent to it and from the valley separating the two. On a broader scale, prairies change substantially as one progresses west and north because of variation in soil characteristics, the rainshadow effect of the Rocky Mountains, and increasing continentality of climate to the north.

But as fascinating and complex as these prairies are, they have a characteristic that may lead to their extinction: they perfectly meet the agricultural demands of a species that does not understand the value of conservation. Thousands of years of decomposing plant matter have created some of the richest soils on earth, and the living prairie plants fulfill the nutritional needs of livestock. Today, no piece of prairie exists that has not been impacted by humans in one way or another. The plant and animal communities that have occupied the Great Plains for thousands of years have been completely restructured by humans in the last two centuries. They have been impacted by such a variety of factors both intentional and unintentional that we will never understand them all.

Prairie Conservation
Island Press (Washington, DC • Covelo, CA)

Prairie Mammals

Defining a community of prairie mammals is difficult because the species comprising it have changed dramatically during the last several hundred thousand years for several reasons. First, although landscapes dominated by grasses have been present in North America during the last 20 million years, prairie as we know it today appears to be quite young (Axelrod 1985; Risser et al. 1981). Taking modern form only in the last several million years, much of the flora and fauna of North American prairies was borrowed from ecosystems surrounding the Great Plains. For instance, many of the grass species that now dominate the plains originally evolved in the forest openings of the East, the meadows of the Rocky Mountains, or the deserts of the Southwest (Risser et al. 1981). Similarly, the community of prairie mammals is formed largely of species whose resource requirements were broad enough to allow them to move into the grasslands from other ecosystems. Additional evidence of the prairie's youth and turbulent history is the paucity of endemic species. Of the mammals found on the central and northern plains for instance, only 11.6 percent are considered true grassland species (Armstrong et al. 1986).

A second factor causing instability in the prairie mammal community was dramatic changes in climate that occurred during the Ice Age. These fluctuations caused rapid distributional shifts for many North American organisms and led to some degree of mixing of ecosystems (Graham 1986; Lundelius et al. 1983). As recently as ten thousand years ago, prairie mammals in the central and southern plains mixed with species typical of tundra (mammoth, lemmings, heather voles, caribou), northern boreal forests (mastodon, red-backed voles, red squirrel), and eastern deciduous forests (fox squirrel) to form a community unlike any found on earth today (Rhodes and Semken 1986; Voorhies and Corner 1985).

Third, near the end of the Ice Age, climate change, hunting pressure by early humans, or both caused an extinction event on the plains that affected large grazing mammals and their predators as well as other groups (Marshall 1984; Martin and Neuner 1978; Martin 1984). The diverse grazing community of horses, antelopes, camels, rhinos, bison, elephants, tapirs, and others was reduced to the modern assemblage dominated by only two or three species. Finally, moderate climatic shifts during the last ten thousand years (Bryson et al. 1970) have caused a continual reshuffling of the prairie mammal community that continues to this day (Frey 1992). The combined effect of the above factors created a temporally and spatially unstable mammal community on the plains and prevented the extensive evolution of endemic species.

The prairie mammal community is dominated by species that colonized the grasslands from surrounding ecosystems. This topic has been examined most recently by Jones et al. (1983) and Armstrong et al. (1986) who analyzed affinities of mammals found in all habitats of the north-central prairie states. Of 138 species found in this region, only 16 (11.6 percent) are geographically centered on the Great Plains and likely evolved there (see table 11.1). The remaining 88.4 percent of the mammals apparently originated in other ecosystems and later colonized the plains. These mammals include 28 species from the coniferous forests of the mountains and northern North America, 25 species from the desert southwest, 18 that are widespread across the continent, 15 from the eastern deciduous forests, 12 from the southeastern United States, 5 from the Great Basin, and 5 of Neotropical origins. If we examine those 138 species listed by Armstrong et al. (1986) and omit mammals found in wetlands and forest habitats, we are left with 57 mammals found primarily in grasslands. This modification, however, still does not eliminate the bias toward nonendemics; the majority of these grassland mammals (41) still evolved in other ecosystems and later colonized the prairies. The present community of prairie mammals therefore represents a unique melting pot where grassland specialists interact and evolve with animals that originated in other regions of North America.

One more point that can be drawn from the data of Jones et al. (1983) and Armstrong et al. (1986) relates to the distribution of the grassland specialists, the

Table 11.1

Mammals whose distributions are centered on the grasslands of the Great Plains

Species	Habitat Affinity
White-tailed jackrabbit	Short–mixed-grass prairie
Franklin's ground squirrel	Tallgrass prairie
Richardson's ground squirrel	Short–mixed-grass prairie
Thirteen-lined ground squirrel	Widespread
Black-tailed prairie dog	Short–mixed-grass prairie
Plains pocket gopher	Widespread
Olive-backed pocket mouse	Short–mixed-grass prairie
Plains pocket mouse	Short–mixed-grass prairie
Hispid pocket mouse	Short–mixed-grass prairie
Plains harvest mouse	Short–mixed-grass prairie
Northern grasshopper mouse	Short–mixed-grass prairie
Prairie vole	Tallgrass prairie
Swift fox	Short–mixed-grass prairie
Black-footed ferret	Short–mixed-grass prairie
Spotted skunk	Widespread
Pronghorn	Short–mixed-grass prairie

Source: Modified from Armstrong et al. 1986, and Jones et al. 1983.

group most vulnerable to the prairie's decline. Of these sixteen endemic mammals, eleven are associated primarily with mixed-grass and shortgrass prairies of the central and western plains (see table 11.1). Only two species, Franklin's ground squirrel and the prairie vole, inhabit tallgrass prairies primarily, whereas the other three are fairly widespread, namely the spotted skunk, thirteen-lined ground squirrel, and plains pocket gopher.

Another important aspect of the prairie ecosystem is the impact of disturbance on the mammal community and the role of mammals in creating this disturbance. Historically, prairies were characterized by frequent perturbations, including fire, drought, grazing, storms, and local factors such as the digging activities of animals (Kaufman et al. 1988; Tomanek and Hulett 1970). The combined impact of these factors created a mosaic environment in which the microhabitat features at a given location could change dramatically in a few days (Collins and Barber 1985; Plumb and Dodd 1993). Prairie mammals were adapted to tolerate these conditions; many large mammals were somewhat migratory, whereas the reproductive capability and rapid dispersal of small mammals allowed them to quickly colonize and populate new patches of suitable habitat (Grant et al. 1982; Risser et al. 1981; Vinton et al. 1993).

Large-scale disturbances undoubtedly played a significant role in determining mammal use of an area. Fire, for example, has positive effects on some species (Coppock and Detling 1986; Kaufman et al. 1983; Vinton et al. 1993) and a negative impact on others (Kaufman et al. 1983; Vacanti and Geluso 1985). Likewise, grazing benefits some species by providing open habitat or by encouraging fresh growth of vegetation and reducing standing dead litter (Coppock et al. 1983b; Hansen and Gold 1977; Miller et al. 1994; Reading et al. 1989). In fact, Whicker and Detling (1988) cite previous research suggesting that female bison could potentially gain 25 percent more weight by preferentially feeding on prairie dog towns versus feeding in mixed-grass prairie. Alternatively, some species, especially those requiring dense litter, are negatively impacted by grazing (Birney et al. 1976; Grant et al. 1982). Drought also appears to have differential effects on various species, with some negatively impacted, others unaffected, and some responding favorably (Tomanek and Hulett 1970). On a landscape scale, the prairie mammal community was probably spatially and temporally diverse, with grazing and burning creating a continually shifting mosaic of habitats and the whole regional flora and fauna slowly recovering from past droughts.

But prairie mammals do more than passively respond to disturbance; they create disturbance and thus greatly impact the diversity of the whole prairie ecosystem. Mammals affect vegetative structure and composition by feeding on plants and by disturbing the soil. Several authors (Anderson 1982; Axelrod 1985; Plumb and Dodd 1993; Risser et al. 1981) have suggested that grazing played a substantial role in the development of North America's prairies by decreasing woody vegetation and favoring the evolution of grazing-tolerant plants. Modern grasslands appear to represent an extreme in the level of herbivory, with rates of removal of annual aboveground net productivity ranging between 50 to 80 percent (Coppock et al. 1983a; McNaughton et al. 1988). Mammalian grazing affects plant structure and species composition and represents a major link in nutrient cycling pathways. By feeding selectively on certain plants, grazers reduce the dominance of these species and allow subdominant plants to become a more important component of the vegetation (Risser et al. 1981). Bison feed heavily on grasses and are thought to have been numerous enough to have significantly impacted the prairie vegetation at a local scale (England and DeVos 1969; Peden et al. 1974; Plumb and Dodd 1993). In addition, other grazers, such as pronghorn, prairie dogs, wapiti, voles, and pocket gophers, were abundant and increased the impact on the prairie vegetation. To attempt to manage today's prairies without some form of grazing ignores the importance of this process in maintenance of the ecosystem and may lead to overdominance by a few species (Howe 1994b).

Mammals also affect vegetative composition and structure by disturbing the soil. Wallowing by bison and digging by badgers, pocket gophers, prairie dogs, and other mammals provide unique microhabitats, affect soil conditions, and break the dominance of perennial grasses to provide habitat for annual forbs and grasses (Collins and Barker 1985; Huntly and Inouye 1988; Munn 1993; Platt and Weiss 1985; Whicker and Detling 1988). The abundance of these disturbances on the prairies of the past undoubtedly led to a substantial increase in vegetative diversity and further enhanced the mosaic nature of grasslands. Unfortunately, of the three most important groups of mammals involved in soil disturbance (pocket gophers, prairie dogs, and bison), the latter two have been drastically reduced in number.

A final aspect of herbivory that potentially has a significant impact on prairie vegetation is the activity of seed-eating mammals and other animals. Although this topic has received little attention in prairies, research conducted in a transitional area between desert scrub habitat and grassland in Arizona has found that kangaroo rats exert such an important force on the vegetation that their removal results in a dramatic change in habitat (Heske et al. 1993). Through seed predation and soil disturbance, these rodents decrease grass cover by nearly threefold and appear to have greater local impact than grazing by cattle.

Present Community

The mammal community that inhabited the prairies of the Great Plains three hundred years ago has been radically restructured by recent human activities. Keystone species have been eliminated or drastically reduced; other species have declined, while still others have increased. Additionally, the prairie ecosystem that these mammals inhabited, the evolutionary landscape in which they evolved and continue to evolve, has been converted, fragmented, and otherwise altered.

Humans have caused the decline or disappearance of a large number of mammals, both directly through overhunting and extermination, and indirectly through habitat modifications (see table 11.2). The removal of several of these animals has resulted in changes in the whole prairie ecosystem because of the im-

Table 11.2
Great Plains mammals extirpated, declining, or extinct

Species	Status
White-tailed jackrabbit[a]	D
Eastern chipmunk	D
Franklin's ground squirrel[a]	D
Black-tailed prairie dog[a]	D
Gray squirrel	D
Southern flying squirrel	D?
Plains pocket mouse[a]	D - L
Plains harvest mouse[a]	D - L
Prairie vole[a]	D? - L
Woodland vole	D?
Gray wolf[a]	Extinct
Swift fox[a]	D
Black bear	Ex
Grizzly bear[a]	Ex
Black-footed ferret[a]	Ex
Wolverine	Ex
Badger[a]	D? - L
Eastern spotted skunk[a]	D?
River otter	D (reintroduced)
Mountain lion	D
Lynx	D
Wapiti[a]	Ex (reintroduced)
Pronghorn[a]	D (now increasing)
Bison[a]	Ex (reintroduced)
Mountain sheep[a]	Extinct

Source: Includes information from Bowles 1981, and Jones et al. 1983.

Note: D = declining; D? = status somewhat unclear, but appears to be declining; D - L = declining in local portions of ranges; Ex = extirpated.

[a]Primarily inhabiting grasslands.

portant role these species played in regulating communities and modifying vegetation.

The near elimination of bison has had a substantial impact on prairies. Grazing activities of bison created patches of open habitat that differed vegetatively from surrounding ungrazed prairie. Further, the wallowing, trampling, rubbing, and excretion of waste of 30 to 60 million bison created a habitat that was highly variable both spatially and temporally (Axelrod 1985; England and DeVos 1969; Plumb and Dodd 1993; Risser et al. 1981). Several species of birds and small mammals (Risser et al. 1981) were apparently adapted to utilize these temporary open patches created by bison activities. Although bison have been replaced by cattle and other livestock in most regions of the prairie, the impact is not the same (Noss et al. 1995). Bison eat different plants than cattle (Peden et al. 1974; Plumb and Dodd 1993), and the confinement of cattle creates an environment that is not as spatially or temporally diverse (Howe 1994a; Knopf 1994). The role that free-ranging bison played in altering vegetative and faunal communities is poorly understood, especially on a landscape scale, and needs further research if genuine attempts are to be made to restore large prairie reserves.

A second keystone species that has been enormously impacted by the activities of humans is prairie dogs. Historically occupying roughly 400,000 km², or 20 percent of the available shortgrass and mixed-grass prairies, prairie dogs alter vegetation, create open habitat, modify soil conditions, affect energy and nutrient cycles, and create burrows that are used by a host of other animals (Munn 1993; Whicker and Detling 1993 and citations therein). In fact, nearly 170 vertebrates have been recorded using prairie dog towns (Miller et al. 1994), although this number is somewhat inflated since it includes birds flying over and not really using the town's unique habitat. Depending on the species, the degree of use may include foraging on the town, use of abandoned burrows as hibernacula, consuming prairie dogs or animals attracted to their colonies, or even total dependence upon prairie dogs. Feeding and clipping activities of prairie dogs stimulate fresh vegetative growth and reduce standing dead biomass, which provides nutritious forage (Coppock et al. 1983a). As a result bison, wapiti, and pronghorn prefer to feed in prairie dog towns (Coppock et al. 1983b; Miller et al. 1994). Although the typical prairie dog colony of today is fairly small and isolated (Miller et al. 1994), towns during presettlement times often stretched for miles. One prairie dog town in Texas covered roughly 40,234 km² and contained an estimated 400 million prairie dogs (Bailey 1905).

Because of early reports of competition between prairie dogs and cattle and the possible role of these rodents in transmitting diseases such as plague to humans, eradication programs were instituted in the late 1800s and continue to this day. During four years in the early 1980s, over 185,000 ha of prairie dog colonies were poisoned, at a cost of $6.2 million. Two agencies of the federal government and numerous state offices are responsible for the control of prairie dogs on an estimated 80,000 ha annually (Miller et al. 1994 and citations therein). In reality, cattle prefer to graze in prairie dog colonies (Miller et al. 1994) and their weight gain is not significantly different from cattle grazing away from towns (O'Meilia et al. 1982; Uresk 1993). In addition, direct transmission of plague from prairie dogs to humans accounts for only 3 percent of the two to twenty human cases per year (Barnes 1993). Although local control of prairie dogs might be justified, the mass elimination of this group of species is not. Populations of prairie dogs have been reduced by roughly 98 percent according to current estimates (Whicker and Detling 1993). Today, local populations are in serious danger of elimination from further poisonings and outbreaks of sylvatic plague, and colonies are so isolated that repopulation through immigration is becoming less and less likely (Miller et al. 1994). Evidence suggests that fragmentation of prairie dog populations has already impacted the genetic and population structure of colonies (Pizzimenti 1981). Given the significance of prairie dogs in creating large patches of habitat that differ from the surrounding prairie and their importance in providing habitat for many other animals, the program of widespread eradication of prairie dogs must be reconsidered.

The effects of removing a keystone species is evident in the plight of the black-footed ferret. Totally dependent on prairie dogs as a food source and to provide burrows, this North American endemic was reduced to dangerously low levels in the mid-1980s when the remaining eighteen wild animals were captured for breeding (Forrest et al. 1988; Thorne and Williams 1988). Because ferrets require a large number of prairie dogs within a small geographic area, decline in populations of ferrets is attributable primarily to the decimation of prairie dog colonies. Further, canine distemper has eliminated local populations because of the ferret's lack of immunity. Assuming canine distemper is a native disease, pre-settlement ferret populations probably remained stable because of immigration from large surrounding populations that recolonized areas decimated by the disease. The highly fragmented and isolated conditions of recent populations of ferrets made local extinction more probable and recolonization unlikely (Forrest et al. 1988). Although the captive ferret population is reproducing, reintroduction programs are hampered by the continued presence of distemper, sylvatic plague outbreaks that affect populations of prairie dogs, and the continuing eradication of prairie dogs (Oldemeyer et al. 1993).

Another group of animals decimated by human activities on the plains was the midsize grazers, including white-tailed and mule deer, wapiti, mountain sheep, and pronghorn. All members of this group decreased because of overhunting or the introduction of nonnative diseases (Genoways 1986; Genoways et al. 1979), but modification of habitat as the human population on the plains increased further hurt their numbers. Wapiti were extirpated from most of the Great Plains; the subspecies that was native to the plains is now extinct (Jones et al. 1983). Similarly, the native prairie subspecies of mountain sheep, which was probably fairly common in the western prairies prior to settlement, is now extinct (Jones et al. 1983). Pronghorn, mule deer, and white-tailed deer decreased to dangerously low levels near the beginning of the 1900s but have rebounded to varying extent due to regulated hunting. Mule deer will likely not return to their previous population levels due to habitat modification and the resulting increase in white-tailed deer. The impact of the loss of wapiti and the decrease of the western midsize grazers on the prairie community is difficult to assess. Wapiti were once quite common in the eastern prairies (England and DeVos 1969; Jones et al. 1983) and apparently have some impact on the vegetation in areas where they are still extant (Frank and McNaughton 1993).

Another group missing from today's prairies are the large predators, including gray wolf, grizzly bear, black bear, and mountain lion. All these animals were apparently somewhat common on the Great Plains but were quickly eliminated as human settlement increased (Jones et al. 1983). The subspecies of gray wolf that inhabited the plains is now extinct. Results of these eliminations may be substantial since mountain lions and wolves are top predators. The recent increase in number of white-tailed deer, raccoons, opossum, coyotes, and red foxes may be attributable at least in part to the removal of these predators-competitors (Jones et al. 1983). Wolves, grizzly bears, and mountain lions have been shown to impact numbers of prey in regions where they are still extant (Bianchet et al. 1994; Messier 1994).

While most of the above reductions or extirpations have been fairly well documented, other prairie mammals have undergone declines quietly (see table 11.2). Documenting declines in population or range for less visible, often small and nocturnal mammals is difficult because intensive trapping or periods of observation are necessary to prove an animal is no longer present or is reduced in number. As a result, the status of many secretive or infrequently trapped species remains uncertain.

Although locally abundant, most of the inconspicuous species that rely on grasslands have decreased throughout their ranges because of wide-scale conversion of prairie to agricultural lands (Armstrong et al. 1986; Bowles 1981; Lovell et al. 1985). Tallgrass prairies of the eastern plains have been eliminated more than other grasslands (Noss et al. 1995; Samson and Knopf 1994), and the landscape of this region bears little resemblance to that which occurred there two centuries ago. Prairie remnants are small and widely isolated, and the only common grassland habitats remaining are roadside right-of-ways, fencerows, streamcourses, and areas too steep or rocky to plow. Without fire, however, the few somewhat natural areas remaining are filling in with woody vegetation. Most of the larger mammals requiring grasslands are either gone or greatly reduced, including white-tailed jackrabbit, black-tailed prairie dog, badger, spotted skunk, mule deer, and wapiti. Of the small mammals inhabiting the tallgrass prairies, many are unspecialized enough to utilize the remaining strips of grasses and may be locally abundant. Perhaps the only mammal that is somewhat restricted to tallgrass prairies is the Franklin's ground squirrel, a species that is, not surprisingly, declining (Bowles 1981; Jones et al. 1983). The prairie vole, a species that likely evolved in the tallgrass prairies, has apparently declined to some extent over much of its original range but now has expanded east to occupy grasslands created as humans cleared eastern forests (Bowles 1981; Jones et al. 1983).

Population and community changes that have occurred in less conspicuous mammals of mixed- and shortgrass prairies are poorly known. In areas where these prairies have been heavily converted, results are likely similar to that discussed for tallgrass prairies. Populations of grassland species have probably declined overall with roadside right-of-ways and fencerows serving as the only somewhat appropriate habitat remaining. Since mixed-grass and shortgrass prairies contain more grassland specialists than eastern prairies, the impact of habitat reduction and fragmentation may be greater. In areas where prairies remain intact, some species are still abundant, while others may be substantially reduced because of the decrease in habitat diversity caused by the control of fire, the near elimination of native grazers, and the control of prairie dogs. So little is known of the community of small mammals in these habitats prior to settlement that more specific statements are not possible. Attempts should be made to quantify the remaining community of mammals in these prairies for use in conservation efforts and to establish data to compare with future surveys.

Although many prairie mammals have undergone declines in recent decades, few of them are reduced to threatened or endangered status. This fact may lead to the false assumption that prairie mammals are not negatively affected by humans. However the negative impact is real because of the decrease in genetic diversity that occurs when populations are reduced or destroyed in large portions of the range of a species (Risser 1988). Like most organisms, mammals, especially smaller and less mobile species, show considerable genetic and morphological variation across their ranges, probably attributable to selection to local environmental conditions. Current methods of measuring biodiversity do not take into account the importance of this genetic diversity to the survival and evolutionary potential of a species. If the next hundred thousand years are as climatically unstable as the last hundred thousand, the loss of genetic diversity will seriously decrease the viability and threaten the existence of many prairie species.

Although this book deals with prairie conservation, nonprairie mammals need mentioning. Forests on the Great Plains have varied considerably in the past. Prior to settlement, strips of true eastern deciduous forest penetrated the plains on steep bluffs along major rivers mostly in the tallgrass prairie region. With the arrival of European settlers, these forests, and the true deciduous forest mammals that require them, were negatively impacted. Unlike white-tailed deer and fox squirrels, species such as woodland voles, gray squirrels, eastern chipmunks, and southern flying squirrels require fairly mature oak-hickory forests for survival.

All of these eastern deciduous forest species have decreased on the Great Plains, probably because of logging, grazing, cultivation, and other modifications (Bowles 1981; Jones et al. 1983). Populations of these mammals are important to conserve since they represent species living on the very edge of their tolerable habitat. As such, these populations might represent evolutionary hot spots where small populations and unique environments lead to genetic diversity and possibly speciation (Frey 1993; Mayr and Ashlock 1991). Similar declines have been found in mammals of the boreal forests that penetrate the northern prairies (Armstrong et al. 1986; Jones et al. 1983). Additional attention must be paid to relict populations of mammals such as the eastern woodrat isolated along the central Niobrara River in Nebraska. Such isolates also represent potential for speciation should selection or random processes lead to sufficient divergence.

The beaver is another nonprairie mammal that requires brief mention. Occurring in riverine habitats in the prairies, this keystone species was decimated by early fur trappers leading to extirpation throughout much of its original range (Bowles 1981; Genoways 1986; Jones 1964). Although this animal has recovered well on the plains, its status requires future monitoring because of the great importance of its damming activities to many other organisms.

The increasing human presence on the prairies has not led to the decline of all species (see table 11.3). Many generalists and successional forest-edge species have increased and expanded their ranges into the plains. Other species have increased as a result of climatic change. Although many of these species are not mammals of grasslands, they are important members of the Great Plains fauna and often have an impact on prairies and prairie mammals.

Habitat and dietary generalists that have increased in the last century are raccoon, opossum, red fox, and coyote (Bowles 1981; Jones et al. 1983). Probable reasons for these increases are numerous and include climatic change, increased habitat, and the ability to live in close contact with humans and feed opportunistically on the abundant food we provide. The impact of these mammals on others in the Great Plains may be substantial because these animals act as predators, competitors, or both, of other, more specialized mammals. Although little attempt has been made to quantify the impact of these increasing mammalian generalists, similar increases in generalist species of birds have been linked to the decline of more specialized birds (Robinson et al. 1995).

Another group expanding its population and geographic range are those species associated with successional forests and forest edges, including white-tailed deer, white-footed mouse, fox squirrel, eastern cottontail, woodchuck, and probably some relatively unspecialized bats (Bowles 1981; Jones et al. 1983). Although oak-hickory forests and their associated fauna have decreased, fast-growing woody plants, both native and nonnative, have increased in abundance on the Great Plains because of the control of prairie fire, decreased flooding, and intentional planting (Johnson 1994). This increase has been most dramatic along

Table 11.3

Great Plains mammals that have increased in populations, range, or both since settlement

Species	Status
Opossum	RI/PI
Masked shrew	RI
Big brown bat	PI?
Red bat	PI?
Hoary bat	PI?
Eastern cottontail	RI/PI?
Black-tailed jackrabbit	RI
Woodchuck	RI/PI?
Fox squirrel	RI/PI
White-footed mouse	RI/PI
Hispid cotton rat	RI
Meadow vole	RI/PI?
House mouse	Introduced
Norway rat	Introduced
Black rat	Introduced, cities only
Meadow jumping mouse	RI
Domestic dog	Introduced, some feral
Coyote	PI
Red fox	RI/PI
Gray fox	RI?
Domestic cat	Introduced, some feral
Raccoon	PI
Least weasel	RI
White-tailed deer	RI/PI

Source: Includes information from Bowles 1981, and Jones et al. 1983.

Note: PI = increase in populations; RI = increase in range; ? = status is uncertain but species appears to be increasing.

rivers, where forested corridors of cottonwood, ash, hackberry, and Russian olive now connect eastern and western woodlands. The impact of this connection on mammals is not at all clear, but in birds, the increased woody vegetation has resulted in the meeting, and occasional hybridization, of faunas that were previously isolated, a conservation issue termed faunal mixing (Knopf 1994). Although mammals have spread west more slowly than birds, the next few decades will see the meeting of the rapidly expanding eastern mammals (white-tailed deer, fox squirrel, white-footed mouse, eastern cottontail) with mammals of western forests, if this is not already occurring. Whether any western mammals will spread east is unknown, but two species of birds, the black-billed magpie and house finch, have apparently done just that. Additionally, the impact of these forest mammals on animals of surrounding prairies is unclear. At least one of these species, the white-tailed deer, probably has a negative impact on its prairie counterpart, the mule deer, through hybridization (Carr et al. 1986) and possibly competition.

Interestingly, of the increasing populations discussed above, none currently require any special conservation efforts. But these are the very species that most often benefit from many Midwest wildlife preserves, game enhancement plantings, and tree planting programs. In addition, the increasing numbers of these often conspicuous species distract the public's attention away from those more secretive species whose populations have been negatively impacted.

Another group of mammals that has appeared recently on the Great Plains includes several introduced species—the house mouse, Norway rat, black rat, domestic cat, and domestic dog. Although the two rats are associated primarily with human dwellings, the house mouse is becoming increasingly common in grassland habitats in the Midwest, especially in the east. It is possible that this species will become an important competitor of native rodents. Domestic dogs and cats also represent a threat to native species through predation and hybridization.

A final example of species that have increased in numbers and geographic range on the plains recently is a group of grassland generalists that probably are reacting to changes in climate. The masked shrew, meadow vole, meadow jumping mouse, and least weasel have northern origins and appear to be spreading south because of an overall cooling trend in the Great Plains since the mid-1960s (Frey 1992; Jones 1964). Apparently responding to climatic warming on the plains during the last several thousand years, species of southern origin including the nine-banded armadillo, opossum, and hispid cotton rat are expanding (Genoways and Schlitter 1967; Jones et al. 1983). The seemingly contradictory nature of the above expansions is probably attributable to the differing time scales of the climatic changes to which the mammals are responding.

Conservation

The most obvious factor impacting mammals of the prairie is the widespread conversion of native grasslands into agricultural fields. As discussed elsewhere in this volume, the loss of prairie habitat ranges from 20 to >99 percent depending on the region. The habitat that has replaced the prairie is primarily monocultural row crops. Although utilization of agricultural fields by large mammals has been mostly unstudied, use of this habitat by small mammals has received some research effort. In the east, use of crop fields by small mammals appears to be limited largely to three species—the deer mouse, white-footed mouse, and house mouse (Houtcooper 1977, 1978; Whitaker 1966); all are generalist species typical of disturbed or successional habitats. Only the deer mouse uses crop fields once they have been plowed. Research by Navo and Fleharty (1983) in western Kansas found that several species used wheat and sorghum fields temporarily, but only the deer mouse used these habitats extensively. They also cited previous research that found that plowed fields had very low abundance of small mammals. Finally, work in western Kansas by Fleharty and Navo (1983) and Reed and Choate (1986) found that irrigated crop fields offered good habitat for several species during some of the year, but the surrounding fencerows were important because they housed permanent populations that could colonize the crop fields when conditions were adequate. Most of the species that used these fields were either generalists or highly mobile species, including northern grasshopper mice, hispid pocket mice, plains pocket mice, western harvest mice, deer mice, white-footed mice, and Ord's kangaroo rats, which are adapted for utilizing temporary booms in resources. Additionally, Fleharty and Navo (1983) suggested that these rodents were not pests, but rather were useful since they consumed waste grain and insect crop pests and benefited the soil through burrowing activities. From these studies it appears that crop fields are used during certain times of year by some species that are generalists or adapted to utilize disturbances but are avoided by more specialized species.

Another habitat feature common in some areas is fields that have been replanted with nonnative pasture grasses or, more recently, native grasses. Many of these fields are enrolled in the Conservation Reserve Program (CRP) instituted by the federal government in 1985 and renewed in 1995. In some areas these fields are quite common, making up over 5 percent of the cropland in shortgrass prairie states and 3.4 percent in mixed-grass–transition states (Knopf 1994). The importance of these plantings to mammals is poorly understood. Generally, the fields are planted with one to four species of grasses and are quite poor in forb diversity. Based on this information it is likely that planted fields are used by mammal species requiring dense cover but lack the vegetative diversity to support a diverse mammal community on a landscape scale. Research in Texas found that although the CRP fields planted in nonnative grasses supported a community with similar values of diversity as local native prairies, the species that were present differed (Hall and Willig 1994). The authors suggested that one of the important differences between prairies and CRP fields was the lack of disturbance in the latter. Further research to understand the importance of CRP fields and nonnative pastures to prairie mammals is badly needed.

Even in areas where prairie vegetation is still intact, overall vegetative structure and diversity are probably quite different from presettlement conditions because of the decrease in perturbations (fire and prairie dogs) and the altered grazing regime. The foraging pattern of cattle is different from that of bison because cattle eat more forbs and shrubs. Confined cattle also create an environment less spatially or temporally diverse than that likely created by free-ranging bison (Howe 1994b; Knopf 1994). The combination of these factors leads to a landscape with different vegetative composition, one that does not contain the diversity of habitats, and one that is not as frequently affected by random disturbance. The impact of this altered vegetative environment on mammals is unknown, although research has shown that grazing by cattle has an impact on small mammals of the prairie, especially in tallgrass or mesic prairies (Birney et al. 1976; Grant et al. 1982; Moulton et al. 1981). From a conservation standpoint, the most important issue is to determine how the current, cattle-dominated landscape differs from the presettlement ecosystem modified by bison, prairie dogs, and fire.

Fragmentation of habitat and isolation of populations are important, but unstudied, issues in conservation of prairie mammals. With increasing loss of habitat, fragmentation gives rise to isolated populations of organisms more likely to become extinct, less likely to be recolonized, more genetically isolated from surrounding populations, and more likely to suffer the negative impacts of inbreeding and genetic drift (Andren 1994; Franklin 1986; Wilcox 1986). In addition, small patches of habitat typically contain fewer species than expected because of a lack of microhabitat diversity and the absence of rare or patchily distributed organisms. These trends become more important as body size of the organism increases (usually lower population density) and vagility decreases (usually lower colonization and immigration rate). To understand the importance of these factors to a given species or ecosystem, additional information, including the life history of the organisms involved and the degree of connectedness of the habitat patches, must be known. To our knowledge, little effort has been made to determine the importance of fragmentation and isolation to the prairie mammal community (but see Robinson et al. 1992). In regions where fragmentation is less intense, flow of animals from patch to patch may be substantial, given the mobility of larger prairie species and the ability of many smaller mammals to use connecting corridors such as fence lines and roadside right-of-ways. In more intensely fragmented areas, those prairie species that are unable to utilize other habitats at least for dispersal may be experiencing the problems associated with small, isolated populations. Finally, the role of interstate highways and channelized rivers in isolating or, in some cases, connecting populations needs to be determined (Genoways 1985).

Simply reading a roster of mammals listed as threatened or endangered by the U.S. Fish and Wildlife Service, one might come to the conclusion that the prairie mammal community has been little affected by activities of humans. Nothing could be further from the truth. The wide-scale destruction of habitat, the alteration of remaining habitat, the extermination or decimation of keystone mammals and top predators, the increase in generalist and introduced animals, and the reduction in genetic diversity within individual species have all contributed to the complete restructuring of the community of prairie mammals. Three characteristics of this community—the relative lack of endemics, the broad resource requirements of many of its species, and the adaptations of many mammals of the prairie to tolerate frequent disturbance—have proved fortuitous from a conservation standpoint. Had the prairies been heavily populated by specialists with narrow geographic ranges and ecological niches, the biological disaster that we are currently battling would have been much worse. Even with this advantage however, prairie mammals need protection to preserve dwindling species, maintain genetic diversity, and retain what remains of the original mammal community structure. The importance of conserving the mammal community of the prairie should be evident to all prairie ecologists and conservationists. For among all the animals that inhabit the prairies, no other group plays as important a role in creating the habitat mosaic that characterizes the prairies of North America.

Action Plan

To protect mammals and other organisms that inhabit the prairies of North America, several courses of action need to be implemented soon. Basic information to make intelligent conservation decisions for managing preserves, public lands, and private holdings is lacking. One of the most important issues to resolve is determining how the current cattle-dominated prairie landscape differs in vegetative and faunal composition from the ecosystem modified by the accumulating effects of bison, prairie dogs, and fire. Although most remaining herds of bison are fairly small and somewhat confined, study of these areas would still be valuable, especially if the area also contains prairie dogs and is managed with fire. We know little about temporal and spatial patterns of grazing by presettlement bison, and some attempt to mimic original conditions could be worthwhile. Especially valuable would be research conducted on preserves containing herds of differing sizes, densities, and degrees of confinement. This research could be compared to similar studies on adjacent cattle-grazed landscapes and would provide a much better understanding of the prairie ecosystem. In addition, this information would be valuable in managing prairie preserves and could lead to means of minimizing the impact of nonnative grazers on the prairie flora and fauna. Research analyzing some of these questions is underway at several locations throughout the Great Plains, and preliminary results can be found in the literature (Pfeiffer and Steuter 1994; Steuter et al. 1995).

Other needed research includes the importance and characteristics of linear grassland remnants (roadside right-of-ways, fencerows, etc.) and planted pastures (including CRP fields) to mammal communities. Because these habitats represent some of the only remaining grasslands in intensively cultivated areas, their use by mammals as corridors and permanent habitat needs clarification.

Many of these areas are heavily managed by grazing, mowing, and planting, and the impact of this management is almost unknown (but see Grimm and Yahner 1987). Additionally, the effects of habitat fragmentation and isolation have been nearly unexplored in prairie mammals. Grassland specialists, such as the black-tailed prairie dog, plains harvest mouse, and plains pocket mouse, in the heavily impacted eastern and central prairies might be suffering from problems characteristic of small, isolated populations that result in declines greater than expected based on the quantity of remaining habitat. Additionally, further surveys to quantify the populations and distributions of prairie mammals, especially those that specialize in grasslands, will provide important information for conservation and supply baseline information against which future changes can be measured. In addition to research, the following actions are recommended:

1. Preservation of large tracts of remaining prairies is crucial to preserving the prairie mammal community. Protecting entire ecosystems is more efficient than conservation measures that work on a species-by-species basis (Noss et al. 1995). Remaining tracts of fairly undisturbed prairies exist, especially in the central and western plains.

2. Existing and future prairie preserves should be managed with the same forces that have impacted this ecosystem for millennia, namely temporally and spatially unpredictable disturbances caused by grazing and digging animals and by fire (Howe 1994a). The results of these management tools will require careful monitoring to determine their impact on other management objectives, such as the control of nonnative, cool-season grasses. Current management techniques, which often prevent grazing and use early-season fires year after year, likely will lead to long-term loss of vegetative diversity and the subsequent loss of faunal diversity.

3. By working with landowners and state road departments, managers could use roadside right-of-ways and fencerows to provide useful habitat and corridors that allow immigration of some mammals between existing patches of prairie.

4. The current trend in using bison to gain income from grazing should be encouraged, especially if research can show how to use bison to restore more native conditions.

5. The senseless eradication of prairie dogs is economically costly (Collins et al. 1984), agriculturally unnecessary, and ecologically detrimental. Government subsidizing of this program should be stopped; prairie dog control and hunting on public lands should be prevented; and educational efforts to improve the image of the prairie dog must be implemented.

6. When it becomes necessary to reintroduce extirpated species, every attempt must be made to use reintroduction stock that is as close to native as possible. Although the release of nonnative genetic stock may be acceptable when the na-

tive gene pool is extinct, the introduction of foreign forms in regions where native individuals are still present contaminates the gene pools that evolved in that area for thousands of years. The reintroduction of river otters in several states for instance, involved subspecies from Louisiana and northwestern North America, even though in some cases native individuals were still present in the region (Genoways 1986).

PART 4

Prairie Conservation

CHAPTER 12

The Great Plains Partnership

Jo S. Clark

For many years the Great Plains has been considered "flyover country"—the area to fly over when trying to get to the mountains or the coasts—or as the bread-basket—the area that produces most of the country's corn and wheat. Children growing up on the Plains often know a great deal more about rain forests than they do about the natural areas outside their own door.

Plains ecosystems reflect this low level of public awareness and are now show-ing some of the sharpest declines in species in the nation. Prairie fish, wildlife that depend on large prairie dog complexes, grassland nesting birds, large river aquatic systems, and the tallgrass prairie are all recognized as systems of concern. Fortunately, agencies of jurisdiction have recognized the need and are conduct-ing research and establishing protection or restoration efforts. States, tribes, non-governmental organizations like The Nature Conservancy, numerous federal agencies, and others all have active programs underway. A major challenge fac-ing all these efforts is learning how to achieve desired goals compatible with and supported by local residents and private landowners. Plains' residents often have deep roots in and strong love for their communities and the surrounding envi-rons. They know the land, they make their living from it, and they own it.

The Great Plains Partnership (GPP), formerly the Great Plains Initiative, is an experimental program to learn how to turn around declining trends of species by working with and through private landowners. It involves taking a broad view of the Plains and the various elements that need to be part of the solution, devel-oping ways to move forward in concert with Plains residents, coordinating among various public and private efforts to increase efficiency and effectiveness, and figuring out how to remove obstacles that make it difficult to get things done. With over 95 percent of the land in private ownership, the possibility of sharp conflicts over threatened and endangered species protection is high. The resulting economic disruption, resentment toward government, and divisiveness would cause major social damage.

Because most of the species in decline have not been listed yet, there is an opportunity to avoid conflict through cooperative, proactive conservation strategies designed to meet the needs of people and wildlife alike. A major challenge will be to develop actions that foster a healthy environment and a healthy economy. At a minimum, conservation strategies should do nothing to exacerbate existing economic trends. Strategies designed to achieve sustainability appear to offer an approach for accommodating both priorities.

Why the Great Plains Partnership?

A former governor of Kansas reflected on the fact that the State of Kansas had spent almost $20 million to restore Cheyenne Bottoms wetlands without knowing what other states and countries were doing for the same migratory birds. He asked the Western Governors' Association (WGA) to pull together other entities to assess whether protection was adequate for the reach of the flyway.

It didn't take long to determine that there were a number of concerns regarding both migratory and nonmigratory species and that protecting birds involved protecting entire ecosystems they depend on. Other chapters in this book set forth those concerns in more detail. In addition, a number of concerns in both science and governance needed to be addressed if effective protection was to be achieved. In addition to WGA and its international counterparts, the initial core of partners shaping GPP included The Nature Conservancy, the International Association of Fish and Wildlife Agencies, the U.S. Fish and Wildlife Service, the Environmental Protection Agency, the province of Manitoba, and the State of Minnesota. It quickly expanded to bring in other government and nongovernmental entities. A number of agricultural organizations were also consulted and have designated representatives.

Vision and Mission

The GPP council has included in its business plan a vision statement that calls for "health, prosperity and generational sustainability . . . perpetuated through economic, biological, social and cultural diversity." The mission goes on to state that "our Mission is to catalyze and empower the people of the Great Plains to define and create their own generationally sustainable future. Our focus is to strengthen and improve biological diversity and ecosystem health in ways that also strengthen and improve the economic, social and cultural foundations of the region."

Science and GPP

When GPP tried to pull together science and data regarding the Plains, it ran into a variety of problems. Federal databases, with the exception of GAP analysis, which is still incomplete, were almost useless for doing a broadscale assessment. They were either too large and complex or too specific; they were not cataloged well; and they used different scales, definitions, classifications, hardware, and software. GPP turned instead to state heritage systems, which although not directly comparable, at least used the same scales, definitions, and classifications.

Under the leadership of The Nature Conservancy, the Environmental Protection Agency, and the Province of Manitoba, GPP created the Great Plains International Data Network (GPIDN) to work on issues related to data exchange. GPIDN includes federal and state agencies. Using the heritage system, the midwest regional office of The Nature Conservancy, working with state and federal agencies as well as others, took the lead in identifying more than eighty areas of high biological concern.

Complicating the practical problems of managing information has been the move toward ecosystem management. Most state and federal resource management agencies are using or considering some form of ecosystem management, whether defined as watersheds, bioregions, integrated resource management, or some other term. Information needs to become much more complex and comprehensive as managers struggle to understand interrelationships.

Governance

In figuring out how to work with private landowners and local residents, it was clear that while some agencies such as the Natural Resources Conservation Service and some individual land managers had a long history of working at the local level, most organizations were not very good at it, even though they tried. Some of that is because decision makers do not understand the public's views and values. Much of it, however, comes from agencies not knowing how to have bottom-up initiatives meet top-down planning and regulations.

GPP has convened a series of focus groups throughout the Plains to better understand public fears and values. What is apparent is that the public has deep concerns about the future of their communities and particularly about loss of control over their future. Individuals have strongly felt conflicting values: they defend private property rights even if land abuses are occurring and simultaneously say that landowners have an obligation to their neighbors, the community, and the future to care for the land. The concept of stewardship resonates strongly; ecosystem management does not.

Results from the focus groups support a bottom-up partnership if government agencies can learn to work on a bottom-up basis. That is the challenge that the GPP has set for itself.

There are other governance challenges. One is figuring out how to connect the dots. There are few surprises in terms of what areas are biologically important. Most are well known. Many have multiple projects underway by multiple agencies, frequently with little connection between them. They emanate from different levels of government, different agencies at the same level, different regions of the same agencies, or different parties altogether—some government and some nongovernment. The result is that good things may happen but perhaps not as good as if the problems were addressed comprehensively and cooperatively.

Another important challenge rests with the ability to be proactive. From the very first days, GPP has adopted the adage "an ounce of prevention is worth a pound of cure." Yet many of the laws, regulations, and government programs get in the way of prevention. Sometimes programs continue to promote harmful practices, even when recognized as harmful, because that is the way they operate. Some regulations or programs are designed to solve one problem even if they create others. Other times the law may address only crisis situations, diverting or precluding resources from being used for prevention.

Just as we need to find a way to mesh and simplify data exchange, there is a need to mesh and simplify resource protection—to treat an area on the ground as a comprehensible unit with integrated programs, shared resources, respect for those who live there, and flexibility to test management ideas that promise improvement and to set aside requirements that don't make sense.

GPP in the Future

The GPP Council is chaired by Nebraska governor Ben Nelson and The Nature Conservancy president John Sawhill. It is made up of a mix of state, federal, and tribal representatives together with agricultural landowners and others in order to provide a common vision for the future of the Plains and to help government work better. It is moving to fulfill the promise made from the beginning for GPP of being based on good science and good sense.

Demonstration Projects

GPP plans to recognize and assist exemplary projects that demonstrate that local action, with assistance if needed, can get the job done. In addition to recognition,

GPP will respond to requests for assistance in problem solving, providing assistance with coordination, technical assistance, incentives, regulatory flexibility, and other positive approaches. The projects will be nominated by states and others with the goal of showing stewardship, protection for biologically important areas, local initiative and control, and commonsense adaptability.

Local Involvement

GPP will encourage not just institutional cooperation but institutional learning as well. To counter the government and scientific mind-set that thinks it knows what is best, GPP will expand on its work with the focus groups to develop meaningful ongoing ways for local input. The intent will be to help agencies value local input more highly, be more open to what they are hearing from locals, and build local involvement throughout agency processes, not just when the Endangered Species Act or some other law requires it. GPP believes that all parties will benefit from knowing each other, trusting each other, and working together regularly.

Good Science

GPP will also encourage development of the best science possible. Recognizing that science will never be able to answer all the questions, the best science available is still needed for many purposes: to guide management choices, to assess trade-offs, to make risks clear, to understand interrelationships, to know when to stop, to correct or modify strategies. These purposes assume that decisions will have to be made in the face of uncertainty, that it is possible they will be the wrong decisions and need adjustment, that science is a guide, not the answer. In these situations, good science may be as important for trust as it is for knowledge.

Summary

There is no question that conserving the biological and social resources of the Plains is a large challenge. Habitat, species, communities, ways of life, quality of life, science, resources, institutions, communications, economies, governance—those are a lot of ingredients to mix together over a thousand-mile range and boundaries that are not just geographic but jurisdictional and discipline-based as well as social. But to quote Governor Nelson, "You can't make a pie a piece at a time." The GPP is one of many experiments underway on the Plains to try to find what works.

CHAPTER 13

Canada's Prairie Conservation Action Plan

Ian W. Dyson

Strategic attempts to address Canadian prairie conservation issues on a regional basis began in the mid-1980s. World Wildlife Fund Canada championed the cause of prairie conservation through its 1986–1988 Wild West Program—a three-year program to "keep the wild in the west." That program supported over ninety demonstration projects for the recovery of endangered species and helped produce the Prairie Conservation Action Plan (PCAP) of 1989–1994 (World Wildlife Fund Canada 1988). The PCAP arose in response to the single most critical issue for wildlife on the Canadian prairies—the loss of habitat.

The prairie and parkland natural region is located within Canada's ecumene— that relatively narrow belt of lands paralleling the Canada–United States border where the majority of the population of the country resides. As such, the region has been radically transformed by settlement and land-use. Infrastructure development, urbanization, petroleum and natural gas exploration and development, recreation development, rural subdivisions, and industrial parks have all had incremental impacts. On a landscape basis, however, agricultural development has had by far the greatest spatial influence.

Background

Today's prairie landscape is predominantly agricultural. This dominance began as national policy in the 1870s—western expansion of both the railway and settlement was intended to counter U.S. expansionism and to open new markets in

Prairie Conservation
Island Press (Washington, DC • Covelo, CA)

the West. The pattern of settlement and land-use bore no resemblance to the bucolic preindustrial agrarian landscapes of Europe. Rather, industrial export-oriented agriculture predominated. The quadrilateral system of survey partitioned the landscape; land was allocated in blocks; towns grew up at grain shipping points; and market-oriented commercial agriculture was able to dominate prairie land-use for a century (Rowe 1990). Until very recently, agricultural policies rewarded the expansion of cropland and made conversion of land-use difficult on lands already dedicated to crops (Girt 1990).

Trends in agricultural land-use have been largely unidirectional—toward more efficient inputs, intensified production techniques, energy-intensive technology, larger farms, and declining rural populations. These trends have produced well-documented environmental consequences: loss of soil and organic matter (Sparrow 1984), fragmentation of habitat (Wildlife Habitat Canada 1991), the draining and disappearance of wetlands (Prairie Habitat Joint Venture Advisory Board 1990), the rise in the number of endangered species on the prairies (Committee on the Status of Endangered Wildlife in Canada 1994), and deterioration of water quality (Alberta Environmental Protection 1995).

Increasingly, the social and economic sustainability of modern agricultural practices has been questioned (Rosaasen and Lokken 1996). Recently, the environmental sustainability of contemporary agricultural practices has received serious attention at national and provincial levels through programs such as the Prairie Farm Rehabilitation Administration's Permanent Cover Program (Agriculture Canada and Alberta Agriculture, Food, and Rural Development 1994) and the Canada-Alberta Environmentally Sustainable Agriculture Agreement. Analysis of the extent to which government policies can work for or against sustainability has been undertaken (Tyrchniewicz and Wilson 1994), and more fundamental measures have been proposed, including directing a portion of Canada's agricultural subsidies to compensate private landowners for delivering ecological services (Sopuck 1993).

Prairie Conservation Action Plan

Global concern over the economic and environmental sustainability of unlimited consumption and growth prompted development of the World Conservation Strategy (WCS) in 1980 (World Commission on Environment and Development 1987). The WCS identified the Canadian prairies as an area of international significance for conservation. The World Wildlife Fund (WWF) accepted the challenge and in 1986 created a steering committee representing diverse agencies and organizations in the West. A prospectus was published and circulated in 1987 to more than five thousand groups and individuals as to what should be included in the plan. His Royal Highness, Prince Philip, president of the World Wide Fund for Nature championed the cause of prairie conservation with federal and provincial ministers, and in 1988 the PCAP was released jointly by WWF and the governments of Manitoba, Saskatchewan, and Alberta.

The PCAP was established to conserve the biological diversity found on the Canadian prairies. The plan includes ten goals (see table 13.1), each supported by various recommendations for actions. These goals and actions provide a blueprint for prairie-wide conservation and management efforts over the five-year period covered by the plan. Land-use planning is recognized as a means to achieve sustainable development objectives (Richardson 1989), but it is only one of numerous tools available to influence land-use practices. The influence of agricultural and taxation policies (Girt 1990), nongovernmental organization projects, and even agency management programs are often viewed as more pervasive and more relevant than planning efforts. This attitude arises because plans have a reputation of not being followed. To maximize the chances of success, Alberta responded by establishing a large, multiparty Prairie Conservation Coordinating Committee (PCCC) to encourage effective implementation of the plan and cooperation among the various stakeholders involved. This committe went on to establish the Prairie Conservation Forum (see table 13.2).

Table 13.1
Goals of the Prairie Conservation Action Plan of 1989–1994

1. Identify the remaining native prairie and parkland.

2. Protect at least one large, representative area in each of the four major prairie ecoregions.

3. Establish across the three prairie provinces a system of protected native prairie ecosystems, and where possible, connecting corridors. This system should include representative samples of each habitat subregion.

4. Protect threatened ecosystems and habitats by preparing and implementing habitat management and restoration plans.

5. Protect and enhance the populations of prairie species designated nationally or provincially as vulnerable, threatened, endangered, or extirpated, by implementing recovery and management plans.

6. Ensure that no additional species become threatened, endangered, or extirpated.

7. Encourage governments to more explicitly incorporate conservation of native prairie in their programs.

8. Encourage balanced use of private lands that allows sustained use of the land while maintaining and enhancing the native biological diversity of the prairies.

9. Promote public awareness of the values and importance of prairie wildlife and wild places.

10. Promote research relevant to prairie conservation.

Table 13.2

Prairie Conservation Forum

Agriculture and Agri-Food Canada
Alberta Agriculture, Food and Rural Development
Alberta Association of MDs and Counties
Alberta Cattle Commission
Alberta Community Development
Alberta Economic Development and Tourism
Alberta Energy
Alberta Energy and Utilities Board
Alberta Environmental Protection
   Corporate Management Service
   Environmental Regulatory Service
   Natural Resources Services (Parks, Fish and Wildlife,
     and Water Management)
Alberta Fish and Game Association
Alberta Municipal Affairs
Alberta Native Plant Council
Alberta Transportation and Utilities
Alberta Wilderness Association
Canadian Forces Base Suffield
Canadian Parks and Wilderness Association
Canadian Wildlife Service, Environment Canada
CN Rail
Coal Association
Ducks Unlimited Canada
Eastern Irrigation District
Federation of Alberta Naturalists
Grasslands Naturalists
Natural Resources Conservation Board
Nature Conservancy of Canada
Palliser Regional Municipal Services
Parkland Community Services
Parks Canada, Canadian Heritage Department
Society for Range Management
Special Areas Advisory Council
Special Areas Board
Sport, Recreation, Parks and Wildlife Foundation
University of Calgary
World Wildlife Fund Canada

The committee developed its own terms of reference and agreed to exist of its own volition with the Department of Environmental Protection providing secretarial support. A chairperson and steering committee are elected annually, and each member organization funds the activities of its representative. Meetings take place about three times a year, in rotating centers throughout prairie and parkland Alberta and usually combining an applied field tour. A number of significant conservation accomplishments have stemmed directly from the activities of the committee. Among these accomplishments was the signing in March 1992 of a memorandum of understanding between the Department of National Defense and Environment Canada to establish a National Wildlife Area (NWA) on Canadian Forces Base Suffield in southeastern Alberta. The NWA will encompass 420 km^2, almost 15 percent of the Suffield Range. This unique prairie ecosystem includes the middle sandhills, a native prairie representative of the dry mixedgrassland ecoregion and riparian areas along the South Saskatchewan River.

The PCCC also organized a training course entitled "Retaining Native Prairie and Wildlife Habitat in an Agricultural Landscape" to expose district agriculturalists, habitat biologists, resource agrologists, range specialists, soil conservation specialists, and agricultural fieldmen to the value of preserving native prairie habitat. Regular educational awareness initiatives include nominations for the Alberta Prairie Conservation Awards (up to three per year) and an Occasional Papers Series designed to provide a forum for nontechnical essays on prairie conservation issues. Perhaps most significantly, the PCCC has published annually a compendium of all member agency activities that have contributed significantly to accomplishing the goals of the PCAP.

Entering the final year of its mandate, the PCCC undertook an assessment of the committee's strengths and weaknesses. In discussing strengths, members stressed the shared experiences, contacts, relationships, and networking stimulated by the PCCC. A major weakness of the committee in its early days was lack of familiarity or support for the PCAP. These early tensions led to a decision to focus PCCC implementation efforts on those areas of the PCAP where there was consensus. Another more pervasive and growing problem has been the limited time and resources PCCC member organizations could bring to the work of the committee. Government and industry have downsized, and nongovernmental organizations are under financial stress. In this situation, organizations without dedicated resources are particularly susceptible since they rely on volunteers and innovation.

Taking Stock and Moving On

Based on Alberta's positive experience with the PCAP and PCCC, the minister of Alberta Environmental Protection supported maintaining the PCAP for another five-year term. In response, the ministers of the Western Accord on Environmental Cooperation in Saskatoon in November 1993 established a working group comprised of representatives from each prairie province, Environment Canada, and WWF to develop more detailed recommendations. Their first step was to develop an assessment of accomplishments of the PCAP in each province.

Specific accomplishments and recommendations highlighted by the assessment compiled by the PCCC are discussed below (Prairie Conservation Coordinating Committee 1994).

Identifying Remaining Prairie and Parkland

The majority of inventory was through environmentally significant area (ESA) inventories, North American Waterfowl Management Plan (NAWMP), and Ducks Unlimited Canada projects, and a reconnaissance inventory of native grass prairie in southern Alberta. Gaps remain for ESAs in some parkland municipalities and the Special Areas in east central Albertan. A reconnaissance-level inventory of native prairie and parkland in central Alberta remains to be undertaken.

Protecting Remaining Prairie and Parkland Ecosystems

In addition to the National Wildlife Area on Canadian Forces Base Suffield, ecological reserves have been designated and management plans prepared for sites in the aspen parkland (Rumsey and Wainright Dunes), fescue prairie (Hand Hills), and mixed-grass (Kennedy Coulee) natural regions. Additional candidate sites are being considered. The World Heritage Site at Dinosaur Provincial Park was expanded in size, and numerous sites have been protected through cooperative habitat programs. On a landscape basis, however, the amount of land under protection designation is small, and opportunities for the establishment of new areally extensive reserves is limited. Approximately 80 percent of the land base in prairie and parkland Alberta is deeded, and remaining Crown lands are virtually all under some form of existing—predominantly grazing—leases.

In March 1995 the government of Alberta announced Special Places 2000, a strategy to complete a network of landscapes representing the environmental diversity of each of the province's six natural regions. Following initial controversy as to whether the policy would allow industrial development on protected sites (the policy includes a tourism–economic development component), most major environmental groups are now participating in the initiative. Any Albertan can nominate sites, and a process to screen nominees and move candidate sites forward for designation has been established. Each natural region is being addressed in turn. Newspaper advertisements soliciting public nominations for the Grasslands Natural Region appeared in June 1996. The province also adopted conservation easement legislation in the 1996 Spring session of the provincial legislature to expand opportunities for protection on private lands.

Protecting Threatened and Endangered Species

Threatened and endangered species recovery plans are in place for the burrowing owl, ferruginous hawk, Baird's sparrow, peregrine falcon, loggerhead shrike, and whooping crane. Plans are under development for the piping plover and swift fox, and mechanisms exist for identifying species at risk. The provinces and the federal government are working toward a harmonized, legislative approach to protecting endangered species and their habitats, but this is a sensitive area because of the well-documented controversies associated with the United States' Endangered Species Act. The federal government released in August 1995 a legislative proposal, the Canadian Endangered Species Protection Act. The proposed legislation limits itself to species that are a federal responsibility and to habitats on federal Crown lands and waters. In the spring of 1996 the Alberta legislature approved modifications to the provincial Wildlife Act that:

- broadens the definition of wildlife to include all plants and animals;

- establishes an independent scientific committee responsible for the assessment and designation of provincial, threatened, and endangered species; and

- creates an Endangered Species Conservation Committee to coordinate the process of protecting endangered species and producing recovery plans.

Creating an Appropriate Policy and Regulatory Environment

Governments at all levels have expressed strong support for conservation objectives, but little progress is evident in converting intent into action. Amendments to agricultural policies detrimental to prairie conservation have been driven by fiscal restraint and economic factors, rather than conservation concerns. Regional and local planning authorities are slow to build conservation objectives into their planning procedures and priorities.

Integrating Conservation into Land-Use Decisions across the Whole Prairie Landscape

Some specific activities and programs have been initiated with individual producers, and a small number of extension and education programs have been initiated for producers, especially programs dealing with range management. However, the general farm community has not become actively involved in the PCAP program and has not agreed to overall objectives. This lack of support is a major shortcoming in the prairie conservation program and, unless it is addressed, will hamper achievement of this goal and other goals that involve land-use and management.

Promoting Public Awareness about the Importance of Protecting Prairie Ecosystems

A number of school programs are in place; interpretative programs and similar activities have been initiated through government agencies and special interest groups; and a number of municipalities have become actively involved in promoting conservation and in providing wildlife-viewing opportunities in urban settings. The challenge is to convert public interest into governmental and societal action.

Promoting Research into Prairie Conservation

There has been discussion but little or no action on these proposals. No prairie research committee has been struck. The concept of a center for research in Alberta has been considered, but no action has been taken. Although range research is going on in a number of centers, this research continues to be oriented primarily toward forage production rather than ecosystem protection.

Similar assessments of the PCAP have also been compiled by Saskatchewan and Manitoba. A summary document highlighting selected accomplishments in all three jurisdictions was compiled by Environment Canada for the Fourth Prairie Conservation and Endangered Species Workshop in Lethbridge in February 1995 (Canadian Wildlife Service 1995).

A New Prairie Conservation Action Plan

Process

To ensure grassroots involvement in developing a new plan, the PCCC held a two-day workshop in Lethbridge in June 1994 to expose a range of stakeholders to the results of the assessment and to generate ideas for a revised plan. Representatives from agricultural, environmental, industry, fish and game, consulting and academic sectors, as well as federal, provincial, and local governments attended. There was consensus on two key elements of a prairie conservation vision for the future: a healthy, sustaining prairie ecosystem where biological diversity is maintained and is secure for the future, and a mainstream societal conservation ethic that is applied to all activities and decisions on the prairies. A PCCC steering committee then drafted an outline for a revised draft plan over the summer and early fall, which was reviewed at a PCCC meeting and by all member agencies and workshop attendees. A revised draft was tabled for review at the Fourth Prairie Conservation and Endangered Species Workshop (Lethbridge, Alberta, February 1995). A general public overview of the PCAP was held in the fall of 1995 and the finalized plan was approved by the Prairie Conservation Forum in March 1996.

Constancy and Change

Based on experience with the first plan and the PCCC, as well as on the results arising from the assessment and the workshop, the following constants were identified:

1. Continued focus on the conservation of native prairie species, communities, and habitats

2. Commitment to a prairie-wide vision

3. Encouragement of multiparty partnerships, networking, and cooperative approaches

The following changes are desirable:

1. Build a plan from the bottom up, that is, grassroots implementation of a prairie plan as opposed to a select group crafting the prairie plan.

2. Broaden the base of support for the new plan, especially in the agricultural community.

3. Emphasize ecosystem and landscape approaches, rather than a species-by-species approach to conservation.

Areas where significant changes have occurred include new awareness of the importance of microfauna, emergence of new information technologies, and major changes in the role of government as facilitator, with an emphasis on cooperative partnerships and multiparty commitment.

Content

The draft Alberta PCAP seeks to conserve the biological diversity of native prairie ecosystems for the benefit of current and future generations. It includes four guiding principles:

1. The main focus of effort during the period 1996 to 2000 will be directed toward conserving Alberta's prairie ecosystems. This principle reflects consensus on the need to move toward ecosystem management and landscape approaches to prairie conservation, rather than focusing exclusively on sites and species.

2. Adopting management practices to more closely approximate disturbance regimes and the range of natural variability holds the promise of preserving biodiversity and ensuring an ecologically and economically sustainable flow of benefits (Bradley and Wallis 1996).

3. A conservation ethic will be applied to all activities and management deci-

sions on the prairies. This principle reflects the idea that human attitudes and values toward the land dictate whether we can accept the role of stewards or whether we will continue to transform native prairie habitats until all are gone.

4. All stakeholders will be involved in the process of achieving the prairie conservation vision. Stakeholders will work cooperatively and form partnerships to achieve prairie conservation objectives. Substantive change will only occur through the efforts of all parties with significant interests, mandates, or influence. Stakeholders will be empowered at a local community level. This principle reflects the belief that most effective conservation action occurs where there is an applied focus at the community level.

The Alberta draft plan has just four goals dealing with information and research, the legislative and policy climate, ecosystem management on a landscape basis, and awareness and education. Goals and objectives are listed in table 13.3. Each objective is also supported by recommendations for actions that are specific, measurable, and include a timetable where appropriate. The new plan retains the substantive content from the original PCAP.

The Future

Alberta is in a relatively good position to retain significant components of native prairie ecosystems. Although very little prairie is protected by legislation, much is protected by the stewardship of ranchers in the Palliser Triangle and in east-central Alberta and by the Canadian military on Canadian Forces Base Suffield. In North America, only Texas and North Dakota retain a larger native prairie land base than Alberta (Samson and Knopf 1994). In southern Alberta alone, there are some 23,968 km^2 dominated by native prairie with provincial Crown lands under grazing disposition at their core. Interest in prairie conservation has grown dramatically in the last five years, and the number and variety of applied conservation projects that have been undertaken is phenomenal (Prairie Conservation Coordinating Committee 1994). Moreover, for the first time in a long time, the external agricultural policy climate is looking more benign. The cessation of the Gross Revenue Insurance Plan in Alberta and the demise of the Crow Rate in the 1995 federal budget will both have the effect of making it even less attractive to bring marginal land into production.

At the same time, the incremental intensification of land-use pressures continues on numerous fronts. The year 1994 was a boom year for oil and gas activity in Alberta, with approximately $1 billion in sales of oil and gas rights and some $2.3 billion in provincial oil and gas royalties (Alberta Energy, personal communication). There is interest in land-use ranging from golf courses to highways to cottage subdivisions on riparian lands adjacent to prairie rivers. There continues to be some wetland drainage and plowing of native prairie. Intensifi-

Table 13.3
Draft Alberta PCAP Goals and Objectives

Goal 1. Advance the identification, understanding, and use of information about Alberta's prairie ecosystems.

Objectives:

Complete identification of remaining native prairie by 1998.

Improve the accessibility and use of available information.

Promote research relevant to prairie conservation and encourage the integration of research and inventory efforts.

Ensure that research and inventory results are applied to ecosystem management in the prairies.

Goal 2. Ensure that governments at all levels have in place policies, programs, and regulations that favor the conservation of Alberta's native prairie ecosystems.

Objectives:

Ensure that management policies applied to Crown land under grazing disposition are compatible with the retention of the existing extensive landscapes of native prairie range.

Amend or remove existing policies and legislation detrimental to the conservation of Alberta's native prairie.

Ensure that all new policies and the legislation, regulations, and activities arising from them (instituted by any level of government) are in accordance with direction in the PCAP.

Encourage the development and implementation of new government policies and programs at all levels that promote the conservation of prairie ecosystems.

Goal 3. Adopt land-use practices, management practices, and protection strategies across the whole prairie landscape that sustain diverse ecosystems.

Objectives:

Encourage the adoption of ecosystem management practices to sustain and conserve all prairie landscapes.

Determine the biotic and abiotic requirements of native prairie species and communities and incorporate habitat considerations in management practices to sustain these requirements.

Provide specific protection for significant, representative, and sensitive ecosystems.

Actively pursue the reclamation of degraded ecosystems.

Goal 4. Increase awareness of the values and importance of Alberta's native prairie ecosystem.

Objectives:

Promote an understanding and appreciation of our native prairie ecosystems among the public.

Promote an understanding and appreciation of our native prairie ecosystems among users of prairie landscapes. Provide information and resources to assist landowners and lessees to conserve native prairie habitats on their land.

Encourage the incorporation of an understanding of prairie ecosystems and prairie conservation within the education curriculum at all levels.

cation of land-use and fragmentation of habitat continue to chip away at remaining native prairie habitat.

The revised Alberta PCAP provides a contemporary statement of what needs to be done to sustain native prairie. There is no other strategic prairie conservation strategy against which to measure our successes and failures and to place in context the efforts that Rowe (1990, 33) calls, "a multitude of uncoordinated finger-in-the-dike programs." There is also no other strategic initiative quite like the PCAP with its applied, specific, grassroots conservation focus.

To be successful, the PCAP will require the support and cooperation of politicians, government agencies, nongovernmental organizations, industry, landholders, and individuals. In the final analysis, the enabling institutions, committed individuals, and teamwork will be the keys to any future successes, as they have been in the past.

CHAPTER 14

A Multiple-Scale Approach to Conservation Planning in the Great Plains

Stephen J. Chaplin
Wayne R. Ostlie
Rick E. Schneider
John S. Kenney

In conservation planning, the issue of scale is important. The protection of all biodiversity, especially the variability within species, eventually demands attention to fine-scale patterns and processes, that is, what stresses are affecting the dynamics of individual populations of threatened species. Conserving the biodiversity of the grasslands of midcontinental North America, however, forces conservation planners to focus on larger geographical areas and assemblages of species or natural communities. Planning at large scales is particularly critical in the Great Plains where few geographic barriers of consequence exist. As a result, species have tended toward widespread distributions frequently determined by climate and large-scale habitat types. The extensive conversion to agricultural uses, especially in the eastern Great Plains (Klopatek et al. 1979; Noss et al. 1995), has resulted in a highly fragmented grassland system. Fragmentation has effectively turned what were nearly continuous populations into metapopulations of semi-independent demes. Excellent examples of this include prairie insects such as the regal fritillary, prairie mole cricket, and several prairie skippers—the Dakota, ottoe, Assiniboin, and poweshiek. Protecting individual demes may be futile unless done within the context of understanding the large-scale functioning of the entire metapopulation (Gilpin 1987).

Another reason to focus at large scales is that some disturbances once operated at that level in the prairies. Two major disturbances, fire (Axelrod 1985; Higgins 1986) and drought (Weaver and Albertson 1956; Weaver 1954), occurred at scales of many thousands of square miles. Herbivore pressure was also a frequent large-scale phenomena. Individual herds of the estimated 60-million-strong presettlement buffalo population intensively impacted large areas as they moved through the landscape (Hanson 1984). At least one black-tailed prairie dog town was estimated to have covered 6,475 km^2 (Bailey 1905).

Conservation Strategy

The Nature Conservancy's Great Plains Program was initiated in 1993. The objectives of the program have been threefold: (1) to compile the information necessary for biodiversity analyses and conservation planning, (2) to consider cooperative actions that would promote the protection of biodiversity while maintaining the economic viability of local communities, and (3) to implement the most promising conservation actions in cooperation with local residents. Each of these objectives needs to be addressed at appropriate scales to succeed in the ultimate goal of the program, which is to preserve the plants, animals, and natural communities that represent the diversity of life within the Great Plains.

For more than two decades, the Nature Conservancy (TNC) and the Natural Heritage Data Centers have employed a "coarse filter–fine filter" paradigm to identify conservation sites (Brown 1991). This approach involves the identification and protection of ecological communities as well as rare species. Identifying and protecting intact representative examples of each ecological community in an area (the coarse filter) assures conservation of a large proportion of the species, biotic interactions, and ecological processes found in the area, including members of poorly studied taxa such as lower plants, microbes, and soil invertebrates. There are species (especially those that are rare) that are likely to be missed if only a few examples of each community type are protected. Protection of these species needs to be addressed individually (the fine filter).

Traditionally, TNC has used the coarse filter–fine filter approach to identify areas for conservation action at the scale of local sites. More recently, with work on bioreserves, its focus has expanded to include conservation at the landscape scale. In addition to these, the Great Plains Program has chosen to focus on two higher levels of scale: the bioregion and the ecoregion.

Bioregion

For the purposes of conservation planning, TNC defines the Great Plains as all of the central North American grassland biome, excluding the prairie peninsula from central Iowa eastward. This boundary includes all or parts of thirteen states and three Canadian provinces covering over 1 million square miles (2.6 x 10^6 km^2). Working at the biome scale will allow an evaluation of conservation needs for wide-ranging species, particularly migratory birds. Also, a number of ecological processes that affect communities and species are best understood from a large-scale perspective (e.g., climate change, some types of disturbance, long-range migration, and material transport via rivers and wind). Understanding these processes should influence our conservation action at smaller scales to ensure long-term protection of the elements of biodiversity (species and natural communities).

Ecoregion

Ecoregions are areas with similar climate, geomorphology, and potential natural vegetation composed of clusters of interacting landscapes. We have designated eight ecoregions in the Great Plains using boundaries delineated by Bailey et al. (1994). These units vary in size from 146×10^3 km^2 to 501×10^3 km^2. A biodiversity analysis of Great Plains ecoregions will serve to target areas for conservation work at the landscape (very coarse filter), community (coarse filter), and species (fine filter) scales.

Landscape

Landscapes are defined as kilometers-wide areas where clusters of interacting vegetative stands are repeated in similar form (Forman and Godron 1986). They are the functional conservation planning unit of a size large enough to encompass ecological processes and species within a mosaic of natural communities. Some landscapes have more of their native biota remaining than others. Our focus is on landscapes of biological significance, which we define as large areas that usually have significant amounts of natural vegetation and concentrations of rare species, high-quality examples of natural communities, or both. Because of their size, none of these landscapes are undisturbed natural areas. All have people living within them and consist of areas with varying levels of human impact and biological significance. Landscapes are the appropriate scale to engage local residents in conservation planning and action. How those residents view and treat the natural system will largely determine the viability of the system.

To identify the landscapes of biological significance in the Plains, TNC developed a set of selection criteria (see table 14.1). Of these nine criteria, greatest emphasis was placed on three: (1) predominant natural vegetation, (2) concentrations of rare element occurrences (species and natural communities), and (3) ecoregion representation. These three were emphasized as a means of identifying the significant landscapes when other biological data may not have been detailed enough to utilize all nine.

Using these criteria, we asked Natural Heritage Data Centers, TNC field offices, local biological experts, or all three to collaborate on the identification and mapping of landscapes within their respective states. A similar process has been initiated in Canada with the Natural Heritage Data Centers' equivalent, Provin-

Table 14.1

Selection and boundary criteria for landscapes of biological significance

For preliminary nomination these criteria are meant to be flexible; landscapes were not required to meet each of the criteria. Due to the nature of some ecoregions in the Plains (e.g., having few rare species or being subject to human disturbances that nearly eliminated natural vegetation) and the general low level of biological information available for many areas of the Plains, very few landscapes could meet all criteria.

1. Continuous Natural Cover: Core areas in a potential landscape should be covered to a great extent by continuous natural vegetation, even if disturbed.

2. Concentrations of Element Occurrences: Landscapes should possess significant concentrations of rare species and high-quality community occurrences.

3. Ecoregion Representation: Landscapes should contain a significant portion of the variability in an ecoregion including geology, topography, soils, and vegetation types.

4. Significant Core Area: A landscape should contain one or more significant core areas—either a very large, high-quality example of a community or a large cluster of several high-quality community occurrences.

5. Area-linked Element Protection: A landscape should be of sufficient size to protect species and communities whose viability is linked to large acreage. Examples include fish and aquatic mollusks dependent on watershed and water quality, riverine community types, top predators, large migratory bird concentrations, and large ungulates.

6. Core Viability Enhancement: Establishment of a landscape should dramatically improve the chances for maintaining the biological integrity of existing managed areas that contain important biodiversity. Landscape protection should help prevent exotic species invasions, increase stewardship options for management and perimeter defense, permit buffer restoration for animals that require interior habitat conditions, enable survival after catastrophic natural disturbances (e.g., tornados, wildfires, landslides), and allow natural disturbance regimes such as existed prior to settlement.

7. Environmental Diversity: A landscape should include the widest range of habitat variety that exists in that part of an ecoregion. For example, it should span the entire moisture gradient that is possible, from river bottoms and marshes to rocky ridges, mountains, and mesas. It should also encompass as many altitudinal changes as possible.

8. Good Design: Design of a landscape should reflect current biological conservation theory. Consideration should be given to the size and proximity of core areas, potential land uses of the matrix surrounding core areas, and the degree of connectivity of core areas within the landscape. In addition, a landscape should be a natural defensible entity (e.g., an entire watershed, an entire mountain, a system of natural corridors and core areas). Landscape borders should follow physiographic boundaries if possible (e.g., edge of the Flint Hills landform; edge of the Turtle Mountains landform; Spearfish Creek–Little Spearfish Creek watersheds).

9. Appropriate Size: The size of a landscape should be appropriate to the ecoregion it may represent. Encompassing an entire ecoregion as a landscape is not appropriate. Depending on the ecoregion, a landscape may range from thousands to hundreds of thousands of acres.

cial Conservation Data Centres (CDCs). Landscape boundaries were plotted on maps (1:500,000) and digitized into a GIS for use in analyses and planning efforts.

To date, sixty-three landscapes have been identified within the U.S. portion of the Great Plains (see table 14.2). Ranging in size from 35 km^2 (San Marcos Springs in Texas) to 62,700 km^2 (Sandhills of Nebraska and South Dakota), they are distributed throughout the Plains (see fig. 14.1) and encompass a wide variety of natural community types. The landscapes average 3,250 km^2 in size and together make up 220,000 km^2 or about 10 percent of the total area of the U.S. portion of the Great Plains.

Within the coarse filter–fine filter paradigm for conservation planning, landscapes are the very coarse filter. To serve as very coarse biodiversity filters, landscapes should capture a significant portion of the variability in an ecoregion. However, there are several reasons why the landscapes selected as biologically significant may not represent the full array of landscapes present in presettlement times. The conversion of large portions of the native prairie to agricultural uses has had dramatic effects on the species and natural communities of the Plains. This is particularly true in the tallgrass prairie of the eastern Plains. The vast majority of the fertile mesic prairie has been converted to row crop agriculture over the past 150 years. What remains intact are a few areas where the soil was too rocky or shallow to plow. No landscape-scale mesic prairie complexes remain in the eastern Plains to be identified, and any effort designed to re-create them would require an immense restoration effort.

Inventory bias has also compounded problems with ecoregion representativeness. This is most pronounced in the western two-thirds of the Plains where the perception is that natural vegetation is still relatively intact and is consequently a lower priority for biological inventory. The result is a general lack of data on the most common vegetation types of each ecoregion. Those areas studied to any great degree within the western Plains have been the atypical areas (e.g., sandhills, mesas, riparian areas), which attract inventory due to their concentrations of rare species or unique flora. Enhanced inventory of the western Great Plains and in Canada would further facilitate the identification of landscapes of biological significance.

Site

The finest-scale conservation planning unit is the site, which is defined in terms of the species or natural community occurrences it is designed to protect. It is the area needed to maintain a viable occurrence at least for the short term and is thus related to the concept of minimum area requirement (Shafer 1990). Most individual sites are less than 1,300 ha (Nature Conservancy 1987) but can be larger for wide-ranging species or large continuous populations or communities. Sites

Table 14.2
Landscapes of the Great Plains

Alexandria Moraine (MN)
Arbuckle Uplift (OK)
Arikaree River Sandsage Prairie (CO, KS, NE)
Arkansas River Sandsage (KS)
Big Stone Potholes (MN)
Black Hills and Grasslands
    Black Hills (SD, WY)
    Badlands (SD)
Central Plains Wetlands
    Great Salt Plains (OK)
    Central Kansas Wetlands (KS)
        Cheyenne Bottoms
        Jamestown Marsh
        Lincoln Salt Marsh
        McPherson Wetlands
        Ninescah Marsh
        Quivira Wetlands
        Slate Creek Marsh
        Talmo Salt Marsh
        Tuthill Salt Marsh
    Rain Water Basin (NE)
Central Platte River (NE)
Cimarron River
    Lower Cimarron (KS, OK)
    Upper Cimarron Mesas (CO, KS, NM, OK)
Chalk Breaks (KS)
Clymer Prairie (TX)
Coastal Sand Plain (TX)
Cross Timbers* (OK)
Des Moines River (IA, MN)
Devil's Lake Basin (ND)
Devil's River/Dolan Creek (TX)
Eastern Nebraska Saline Wetlands (NE)
Flint Hills (KS, OK)
Fort Hood (TX)
Fort Worth Prairie (TX)
Glacial Lake Agassiz
    Agassiz Beach Ridges (MN)
    Aspen Parkland (Manitoba, MN)
    Sheyenne Delta (ND)
Great Plains Pine Escarpments
    Cheyenne Table Escarpment (NE)
    Pine Ridge (NE, SD)
    Wildcat Hills (NE)
High Plains Border* (OK, TX, KS)
Killdeer Mountains/Lower Little Missouri River (ND)
Little Missouri Badlands (ND)
Little Sioux River (IA)

Loess Hills (IA, MO)
Minnesota River
    Middle Minnesota River (MN)
    Upper Minnesota River (MN, SD)
Missouri Coteau
    Bijou Hills* (SD)
    Comertown Prairie (MT)
    Lostwood (ND)
    Medicine Lake Sandhills (MT)
    Orient Hills* (SD)
    Ree Heights* (SD)
    Southern Missouri Coteau (ND, SD)
Missouri River
    Unchannelized Missouri River (NE, SD)
    Upper Missouri–Yellowstone (MT, ND)
Neosho River (KS, OK)
Northeast Blackland Prairie (TX)
Osage Cuestas Tallgrass
    Anderson County Tallgrass (KS)
    Eldorado Springs Tallgrass (MO)
    Liberal Tallgrass (MO)
    Lockwood Tallgrass (MO)
    Marais des Cygnes River (KS)
    Marmaton River (KS, MO)
    Sedalia Tallgrass (MO)
Pembina Gorge (Manitoba, ND)
Prairie Coteau
    Prairie Coteau (MN, SD)
    Sisseton Escarpment (MN, SD)
Red Hills (KS, OK)
Rolling Red Prairies (TX)
Sandhills (NE, SD)
Smokey Hills (KS)
Souris River (Manitoba, ND)
Southeastern Sandhills (OK)
Sweetwater Sandhills (OK, TX)
Texas Hill Country
    Balcones Canyonlands, Northeast (TX)
    Balcones Canyonlands, Southwest (TX)
    Comal Springs (TX)
    San Marcos Springs (TX)
Tule Canyon/Palo Duro Canyon (TX)
Turtle Mountains (Manitoba, ND)
Verdigris River (KS, OK)
West Bijou Creek (CO)
Western High Plains Grasslands* (CO, NE, WY)
White Cloud Blufflands (KS, NE)
Wichita–Quartz Mountains Archipelago (OK)

*Boundaries for these landscapes have yet to be determined and consequently the landscapes have not been mapped.

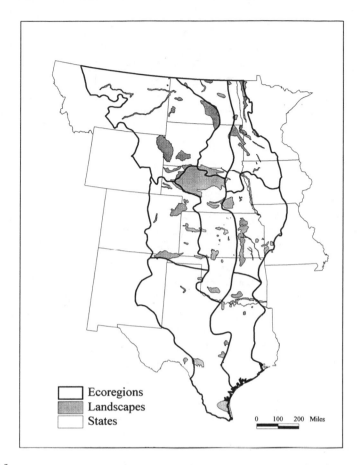

Fig. 14.1 Ecoregions and currently identified landscapes of biological significance in the U.S. portion of the Great Plains. The landscapes depicted do not
necessarily reflect the boundaries of any proposed conservation activity.

can be complex to define if an assemblage of species of conservation interest occurs together. Site boundaries are then a composite of multiple species or community needs. Usually, there will be many sites within a landscape of biological
significance.

Elements of Conservation in the Great Plains

Although conservation analysis and planning can take place at many scales, the
subjects of the planning continue to be ecological communities and species of
concern.

Ecological Communities

Over the last twenty years, communities have been used to help develop conservation priorities only on a state-by-state basis. Community information was systematically collected by biologists from the state Natural Heritage Data Centers. This information was used to develop and refine state-level community classifications and associated conservation ranks. A major obstacle to using communities as conservation units at the regional and national levels was the lack of a consistent, national classification system. To overcome this problem, TNC and the Natural Heritage Data Centers have developed a standardized hierarchical system to facilitate the identification and classification of vegetated terrestrial communities across the United States (Nature Conservancy 1994).

This national classification system was developed primarily for the purposes of conservation planning and biodiversity protection. The intent of the classification system is to provide a complete, standardized listing of all communities that represent the variation in biological diversity and to identify communities that require protection. The classification will be consistent throughout the United States at appropriate scales for the management and long-term monitoring of ecological communities and ecosystems. It is also intended to have applications as a vegetation data layer in landscape and ecosystem characterization.

This terrestrial community classification is hierarchical and combines physiognomy at the highest levels of the hierarchy and floristics at the lowest levels. The physiognomic portion is a modification of United Nations Educational, Scientific, and Cultural Organization (1973) and Driscoll et al. (1984) and utilizes the physical form of the dominant vegetation to organize the units. An important aspect of the classification is that the community elements are related to a set of environmental factors rather than to a particular site. This ensures that the classification has ecological meaning over a broad geographic range. The classification is broadly defined and includes vegetation of uplands, as well as emergent and rooted submerged vegetation of marshes, lakes, ponds, rivers, and marine shores.

The Great Plains section of this national classification has recently been completed. Community data from the respective Natural Heritage Data Centers are being merged into a regional spatial database, which will make possible for the first time a rangewide analysis of Great Plains communities. The distribution of community types will be evaluated in relation to the currently identified landscapes of biological significance to determine what portion of the diversity those landscapes contain. Additional landscapes may be proposed by locating clusters of rare communities, high-quality common communities, or both.

Rangewide analyses will also facilitate more traditional, smaller-scale conservation efforts. Combined with data on occurrence quality, we can target for protection those sites that can best preserve a community type. The rangewide database will also help determine gaps in existing data and target future inventory work.

Within the Great Plains, 663 community types have been identified (see table 14.3). A number of the forest types have their primary range of distribution outside the Great Plains boundary, to the east or west. While nonforest types have their range primarily or entirely within the Great Plains, the vast majority of the rare types are endemic.

The rare communities are imperiled for a variety of reasons. Some, such as the saline marshes of Nebraska, were rare even in presettlement times because they were dependent on an uncommon set of environmental conditions. Human alterations have further reduced these naturally rare communities. Many of the others have become imperiled or degraded due to human actions such as conversion to agricultural or other uses, alteration of disturbance regimes, or the introduction of livestock grazing. The high proportion of rare sparse woodlands (savannas) is due to a combination of conversion and fire suppression. Those savannas that were not cleared and plowed have succeeded to woodlands or forests in the absence of fire. The large number of rare tallgrass communities reflects their nearly complete conversion to farmland. The forb-dominated communities are primarily wetland types. The high percentage of rare types in this category reflects both their natural rarity in the Great Plains and the result of wetland drainage and filling.

Species of Concern

Individual species conservation is the fine filter in biodiversity protection planning. This is the approach often taken under the Endangered Species Act and is

Table 14.3

Terrestrial communities of the Great Plains

	Number of Types	Number of Rare Types	Percent Rare
Forest	144	16	11
Woodland	109	20	18
Sparse woodland	14	8	57
Shrubland	143	19	13
Herbaceous			
Tallgrass	76	17	22
Midgrass	96	13	14
Shortgrass	33	7	21
Forb dominated	18	7	39
TOTAL	633	107	17

Note: Rare communities are those that have been ranked G1 (critically imperiled globally) or G2 (imperiled globally). The graminoid-dominated herbaceous types include a range of hydrologic conditions from marshes and fens to upland prairies.

intended to focus on species that have not been adequately protected at the landscape (very coarse) or community (coarse) levels.

A centralized element occurrence database was compiled of 18,400 records from the Natural Heritage Data Centers in the Great Plains (see fig. 14.2). These programs maintain detailed information on the location and condition of rare species and natural communities occurring within their respective states or provinces (Morse 1993) and rank them by their level of global imperilment. These global ranks range from G1, which indicates a particular element is critically imperiled (less than six viable occurrences or 1,000 individuals), through

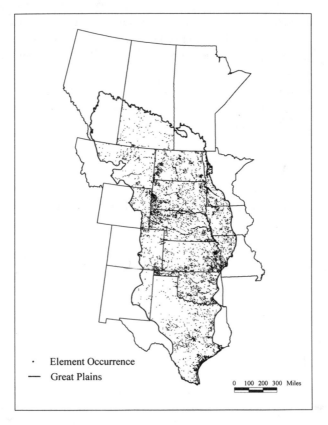

Fig. 14.2 Rare species and high-quality community occurrences in the Great Plains documented by state natural heritage programs and provincial Conservation Data Centres. Data are not yet available in this format from Alberta and Manitoba. The data depicted are not based on a comprehensive inventory of each state or province but rather were compiled from state field surveys, systematics collections, county inventories, publications, expert observation, university research, government agency inventories, and other sources.]

G3 (21 to 100 occurrences or 3,000 to 10,000 individuals) to G5, which indicates that it is demonstrably secure (Master 1991).

Compared to other geographic regions in the United States, the Great Plains is relatively depauperate in globally rare species. Despite the fact that the Plains encompasses approximately 28 percent of the continental U.S. land mass, only 8 percent of G1-G3 species and 6 percent of G1-G3 species occurrences found within the continental United States occur within its borders (see table 14.4). A high proportion of these rare species are endemic to the Great Plains. Of the 285 G1-G3 species tracked by Natural Heritage Data Centers in the Great Plains, 134 are considered to be endemic to the region, and another 57 have the bulk of their range within the Plains (see table 14.5).

Despite the relatively small numbers of rare species, persistent trends in habitat loss throughout the region continue to lead to decreases in once-common species. This includes the many Great Plains endemics that are not currently rare (i.e., ranked G4-G5) but are headed in that direction. For example, recent analyses of trends have shown that grassland nesting birds have exhibited more consistent, widespread, and steeper declines in the past twenty-five years than any other North American bird group (Knopf 1994). Assessments of heritage data show that 59 federally listed species occur in the Plains. Reversing the trends of decline in species not yet listed is a significant conservation objective. It is this group of species that will have the greatest impact on landowners in the future if trends of decline continue and the species become federally endangered or threatened.

Long-term protection of rare and declining species is most likely to be successful if the species are maintained within functioning landscapes. These species have evolved with and are adapted to large-scale processes. While the selected landscapes encompass only 10 percent of the Great Plains, they contain populations of 41 percent of the G1-G3 species. In addition, 32 percent of all the known occurrences of G1-G3 species in the Plains are found within these landscapes (see table 14.4). However, the majority of rare species and their known occurrences are found outside of currently identified landscape areas.

To better understand whether the identified landscapes adequately encompass rare species, spatial analyses of individual taxa need to be undertaken. For example, a number of high-quality occurrences of Mead's milkweed fall within identified landscapes (see fig. 14.3), but other important occurrences are found outside landscape boundaries and will need to be addressed at the site level. Small reserves that maintain individual or assemblages of rare species at least for the short term will continue to play an important role in the mix of conservation efforts to protect biodiversity in the Great Plains and elsewhere (Shafer 1995).

Many of these species-oriented conservation efforts will have to take place on private lands in the Great Plains. A significantly smaller percentage of federally listed species occur on federally owned land in the Great Plains than in the United States as a whole. In the United States, 36 percent of all known occur-

Table 14.4

G1-G3 species and species occurrences tracked by Natural Heritage Data Centers within the Great Plains and currently identified landscapes of biological significance

Taxonomic Group	Number and Percent of U.S. Species Found in the Great Plains				Number and Percent of Great Plains Species Found in Landscapes			
	Species		Species Occurrences		Species		Species Occurrences	
	Number	Percent	Number	Percent	Number	Percent	Number	Percent
Vascular plants	167	7.1	1,954	4.5	55	32.9	450	23.0
Invertebrate animals	42	5.6	696	7.9	25	59.5	370	53.2
Vertebrate animals	76	12.2	2,331	8.8	37	48.7	778	33.4
Amphibians	11	14.7	61	2.9	5	45.5	16	26.2
Birds	17	25.0	1,445	11.9	9	52.9	545	37.7
Fish	27	9.4	592	13.2	12	44.4	159	26.9
Mammals	8	6.9	159	4.7	6	75.0	35	22.0
Reptiles	13	16.3	74	1.7	5	38.5	23	31.1
TOTAL	285	7.5	4,981	6.3	117	41.1	1,598	32.1

Note: Percentages are based on the proportion of G1-G3 species and species occurrences of (1) the continental United States that are found in the Great Plains, and (2) the Great Plains that are found within currently identified landscapes of biological significance.

Table 14.5

Endemism of G1-G3 species tracked by Natural Heritage Data
Centers in the Great Plains

Taxonomic Group	Endemic	Mostly Within Great Plains	Mostly Outside Great Plains	Peripheral	Total
Vascular plants	87	31	29	20	167
Invertebrate animals	13	7	15	7	42
Vertebrate animals	34	19	14	9	76
Amphibians	10	0	0	1	11
Birds	2	8	6	1	17
Fish	11	7	4	5	27
Mammals	4	2	2	0	8
Reptiles	7	2	2	2	13
TOTAL	134	57	58	36	285

Note: A species is considered endemic if all its global distribution is within the Great
Plains, mostly within Great Plains if 50 to 99 percent of its known occurrences are in
the Great Plains, mostly outside Great Plains if 10 to 49 percent of its known occurrences are in the Great Plains, and peripheral if less than 10 percent of its known occurrences are in the Great Plains.

rences of listed species are found on federal land (Natural Heritage Data Centers
Network 1993), whereas in the Great Plains, 19 percent occur on federal lands.
Because most nonfederal lands in the Great Plains are privately owned, the success of any conservation initiative will depend largely upon strong support from
private landowners.

Conservation Goals

Resources for conservation action are always limited. Under such constraints,
one objective is to use the available resources as efficiently as possible to protect
regional Great Plains biodiversity. The Great Plains Program has chosen to adopt
an operational goal that is both concrete and measurable: In each ecoregion, protect multiple viable examples of each natural community and imperiled species
within naturally functioning landscapes.

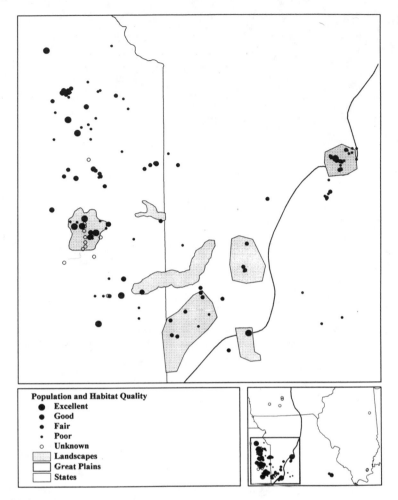

Fig. 14.3 Known distribution of Mead's milkweed with respect to currently identified landscapes of biological significance.

Accomplishing this goal necessitates planning at the ecoregion, landscape, and site scales. We view this operational goal, in a sense, as a hypothesis and will continually test how well it ultimately protects Great Plains biodiversity. This goal does not imply that government entities need necessarily buy nor regulate private land. The nature of land-ownership patterns and public attitudes in the Great Plains is such that public-private partnerships at the local level will probably achieve more rapid and permanent results. It will be important to include local residents in all aspects of the efforts to protect the biodiversity of their local surroundings.

Acknowledgments

We greatly benefited from discussions with Kerry Herndon, Brenda Groskinsky, Rob McKim, Jo Clark, Peter Buesseler, Hal Watson, Shawn Canady, and other participants in the Great Plains Partnership. Data and many helpful suggestions were provided by the Colorado Natural Heritage Program, Iowa Natural Areas Inventory, Kansas Natural Heritage Inventory, Minnesota Natural Heritage and Nongame Research Program, Missouri Natural Heritage Database, Montana Natural Heritage Program, Nebraska Natural Heritage Program, New Mexico Natural Heritage Program, North Dakota Natural Heritage Inventory, Oklahoma Natural Heritage Inventory, South Dakota Natural Heritage Data Base, Texas Parks and Wildlife Department, Wyoming Natural Diversity Database, Manitoba Conservation Data Centre, and Saskatchewan Conservation Data Centre.

Although the research described in this paper has been funded in part by the U.S. Environmental Protection Agency under assistance agreement number X007803-01-1 to The Nature Conservancy, it has not been subjected to the agency's peer and administrative review and therefore may not necessarily reflect the views of the agency, and no official endorsement should be inferred. Neither should the interpretations, conclusions, and recommendations of this paper be necessarily attributed to the Great Plains Partnership, Natural Heritage Data Centers, or provincial Conservation Data Centres.

Working Partnerships for Conserving the Nation's Prairie Pothole Ecosystem: The U.S. Prairie Pothole Joint Venture

Steven J. Kresl
James T. Leach
Carol A. Lively
Ronald E. Reynolds

A Commitment to Waterfowl Conservation

Over the last century, committed sportsmen and women, wildlife managers, and others have worked to reverse the continuing decline of duck numbers and habitat in North America. Despite these efforts, in 1985 continental duck numbers reached their lowest point in forty years (Caithamer et al. 1994). As a result, biologists, managers, and conservationists in the United States and Canada sounded a rallying cry to increase efforts to conserve North America's waterfowl resources. To be successful, they recognized that large areas of critical habitat need to be conserved in the primary breeding and wintering areas of waterfowl. Because most land in the primary breeding range of ducks is in private ownership, efforts to increase productivity would require cooperation of landowners.

Planning and coordination came together in 1986 when the North American Waterfowl Management Plan (NAWMP) was signed by the United States and Canada (U.S. Fish and Wildlife Service 1986). This historic agreement revealed a vision for continental waterfowl and wetland conservation and built a course of action for both countries to secure by the year 2000. In 1989, Mexico, with its many critical wintering areas, became a signatory to the NAWMP and in 1994 became a full partner. Thus, a continental approach to waterfowl management was complete, facilitating an integrated approach to planning, coordination, and implementation of wetland conservation activities.

NAWMP recognized loss and degradation of habitat as the major waterfowl management problem in North America; identified the need for long-term pro-

Prairie Conservation
Island Press (Washington, DC • Covelo, CA)

tection, restoration, and enhancement of habitat on an ecosystem basis, including long-term land-use changes; and set goals and objectives for the protection and improvement of habitat. One goal of NAWMP was to reach a breeding duck population of about 62 million birds that would produce an average fall flight of about 100 million ducks, a level last seen in the 1970s. To accomplish this, NAWMP called for the protection, restoration, and enhancement of 2.4 million ha at a cost of $1.5 billion, while recognizing such objectives require an unprecedented partnership of public and private organizations from a wide spectrum of society.

NAWMP identified thirty-four important habitats, including breeding, migrating, and wintering areas in six key geographical areas (the potholes and parklands on the northern Great Plains and south-central Canada, lower Mississippi valley, Gulf Coast, central valley of California, Great Lakes and St. Lawrence Basin, and Atlantic Coast); acknowledged the job of protecting and enhancing such vast areas requires more than one agency or organization; and recommended the development of coalitions, known as joint ventures, to carry out this enormous task.

Joint ventures are composed of federal, state, and local government agencies, conservation organizations, sportsmen's groups, and private landowners. The concept is to blend resources from a geographic area to maximize financial, organizational, and other support toward common objectives. A management board is established to define priorities and direction for each joint venture, with steering committees established to carry out habitat improvement projects at local levels. By 1988 joint ventures were in operation in each of the six key geographic areas, including the U.S. Prairie Pothole Joint Venture (PPJV).

A Working Partnership

Three states, North Dakota, South Dakota, and eastern Montana, contain most of the PPJV, an area of approximately 24.3 million ha. The PPJV is located in what was, historically, a large grassland ecosystem and is dotted with millions of small wetlands formed when glaciers advanced and retreated across the prairies. Now the area is intensively farmed, and wetland drainage, largely for agricultural purposes, has reduced the historic area of wetlands (2.9 million ha) by over 40 percent (1.75 million ha). Native prairie, reduced in area by 75 percent (6 million ha) (U.S. Fish and Wildlife Service 1994a), is heavily grazed by domestic livestock. Changes in land-use and wetland drainage have accelerated downstream flooding, negatively impacted water quality, and degraded fish and wildlife habitat.

Despite changes to the landscape, the PPJV area remains rich in wildlife. For example, mallards produced there are harvested in all four waterfowl flyways and several provinces in Canada (Munro and Kimball 1982). PPJV area further provides breeding and migratory habitats for over two hundred species of nongame migratory birds and several species listed as endangered, threatened, or candidates under the Endangered Species Act (1973): the piping plover, black tern, bald eagle, peregrine falcon, whooping crane, and interior least tern.

The 1993 PPJV Accomplishment Report (U.S. Prairie Pothole Joint Venture 1994a) describes (1) protection, restoration, or enhancement of 768,930 ha of wetland and grassland habitats, (2) origination of $140 million in support of PPJV habitat activities, (3) matching nonfederal support by the North American Wetlands Conservation Act Funds ($30 million), and (4) specific habitat improvement projects carried out by agencies, landowners, conservation organizations, and other concerned citizen-conservationists who live and work in the PPJV area. In 1994 the PPJV reflected on its accomplishments, updated the implementation plan (U.S. Prairie Pothole Joint Venture 1994b), and added three objectives:

1. To increase waterfowl populations (6.8 million breeding ducks that produce a fall flight of 9.5 million ducks under average environmental conditions), through habitat conservation projects that improve natural diversity

2. To stabilize or increase populations of declining wetland and grassland wildlife species, with special emphasis on nonwaterfowl migratory birds

3. To conserve, through integrated conservation planning and management, other wildlife, in particular wetland-grassland migratory birds and threatened and endangered species, through new partnerships, specifically with Wetlands for the Americas and Partners in Flight

Implementation

The fundamental problems impeding attainment of the objectives for the PPJV are habitat loss and degradation. Primarily as a result of agricultural activities, loss of wetlands through drainage and agricultural conversion of native grasslands to cropland has negatively impacted migratory bird populations that depend on the prairie grassland and wetland ecosystem. Habitat fragmentation has caused the amount of secure nesting cover to be drastically reduced. This has resulted in increased negative impacts of nest predators on ground-nesting birds to the point where populations often cannot be maintained.

The primary means by which PPJV partners will improve waterfowl and other grassland wildlife populations is through the protection and restoration of grasslands in association with wetland complexes. Implementation strategies utilized by the partners address PPJV objectives, which in turn support the goal of the PPJV. All implementation strategies continue to emphasize waterfowl production and recruitment, while providing additional benefits to other grassland-wetland-associated wildlife. For example, where large grassland areas are secure, wetlands are being restored, enhanced, or created to increase wildlife production. In areas where intensive cultivation is the dominant land-use, more intensive conservation strategies are implemented. These may include a combination of practices such as grassland or wetland easement-leases, nesting structures, and predator management.

The PPJV partners recognize that the majority of wetlands, grasslands, waterfowl, and other wildlife occur on private lands. In fact, over 95 percent of the prairie pothole region (PPR) is privately owned (U.S. Fish and Wildlife Service 1994b). Provision of habitat and production of waterfowl and other wildlife must include adequate compensation or benefits for private landowners, who in most cases, must maintain profitable agricultural enterprises. This approach remains the key to the success of the PPJV.

Because agriculture is the predominant land-use within the PPJV, USDA conservation programs are vital to achieving PPJV objectives. Traditional natural resource management agencies and organizations have limited impact on agricultural land practices. Conservation provisions in U.S. farm bills and individual landowner practices provide the greatest opportunities for habitat improvements on private lands in the PPJV. Though not targeted specifically toward wildlife, the USDA's Conservation Reserve Program (CRP) lands are providing important habitat for many species of grassland birds, including some whose populations have declined over the last several decades (Johnson and Schwartz 1993). Nest success for ducks nesting in the CRP cover is well above that needed for population growth, and populations of several nonwaterfowl bird species nesting in North Dakota are increasing (Reynolds et. al 1994). The Wetland Reserve Program and CRP have also played a key role in the attainment of PPJV habitat objectives.

Working Partnerships for Conservation

Because success in meeting the goals of the NAWMP requires local partnerships, several flagship projects were developed in each PPJV state. The intent was to achieve certain habitat objectives with maximized partner participation, support, and awareness. Following is a synopsis of flagship projects in each state in the PPJV.

North Dakota—The Chase Lake Prairie Project (CLPP)

The CLPP is centered in the Missouri Coteau physiographic region (Bluemle 1977), an area of prime importance to waterfowl. The CLPP area covers 2.2 million ha of Coteau in south-central North Dakota. Running northwest to southeast from Saskatchewan through Montana and North and South Dakota (Kantrud et al. 1989a), the Missouri Coteau is a 16- to 80-km-wide band of sharply rolling hills dotted with thousands of various size wetlands whose densities sometimes exceed 38.6 km² (H. A. Kantrud, personal communication).

Considered by many to be the most productive waterfowl habitat in the PPR and an important migration habitat for many species of birds, the Coteau has retained a substantial amount of its original grassland and wetland habitat base and continues to support an impressive array of wildlife, especially birds, several of which are listed as endangered, threatened, or are under consideration for listing. Still, wetland drainage and grassland conversion to cropland have taken their toll on wildlife productivity in this area.

Prepared in 1989, the CLPP plan of action (U.S. Fish and Wildlife Service 1989b) contains thirty-eight action items aimed at enhancing wildlife on public and private lands by providing landowner incentives, increased public awareness, and recreation opportunities. The CLPP is also designed to deliver systematic application of the latest waterfowl management techniques over the area.

Since 1989 over seven hundred agreements beneficial to wildlife were secured from landowners. Nearly twelve hundred wetland basins totaling 890 ha were restored within privately owned CRP acres. In addition, over one hundred wetland basins were restored or created on other private lands, totaling about 142 ha. Approximately 12,141 ha of previously overgrazed, privately owned native prairie were placed under rotational grazing systems to the benefit of both livestock and ground-nesting birds. Over forty nesting islands and six peninsula cutoffs have been constructed to provide secure nesting habitat for migratory birds. Over 2,428 ha of land have been purchased by the U.S. Fish and Wildlife Service in fee title, and another 1,093 ha by other partners. The more than $5 million expended in the CLPP has brought welcome economic gain to the rural, sparsely populated area (U.S. Fish and Wildlife Service 1994a).

Montana—Northeast Montana PPJV Project

This project encompasses Sheridan, Daniels, and Roosevelt Counties in the northeast corner of the state. The core area of the project lies within the northeast corner of Sheridan County, a part of the Coteau with high wetland densities. Projects within the more intensively farmed core area focus on nesting cover enhancement, predator management, and wetland restoration. Outside the core area, where there is more grassland cover, wetland development is the focus. Partnerships with Montana's farmers and ranchers are essential.

South Dakota—Lake Thompson Project

The Lake Thompson Project is located in the heart of a 1,310-km² watershed that encompasses parts of Kingsbury, Lake, Miner, Clark, and Hamlin Counties in eastern South Dakota. Before 1985 Lake Thompson was a shallow 3,642-ha marsh. Heavy rains in the heavily drained watershed turned the marsh into South Dakota's largest natural lake. A task force commissioned by the late governor, George Mickelson, called for an extensive wetland restoration effort in the upstream watershed to prevent future flooding. The PPJV activities in the Lake Thompson area have generated new partnerships. In 1993 over one thousand people attended the Lake Thompson Waterfowl and Wetlands Festival, an integral part of the project's education and outreach effort.

The partnerships forged at Lake Thompson have spawned additional projects, including the South Dakota Ponds Program. This program is a unique partnership of fifty farmer- and rancher-directed conservation districts; four Native American tribes; three federal agencies; the South Dakota Department of Agriculture; the South Dakota Department of Game, Fish, and Parks; Ducks Unlimited; and other private conservation organizations. The goal of the pond coalition is to create six hundred new wetlands with approximately 81 ha of adjacent, protected upland cover. Through improved grazing management, an additional 48,564 ha of enhanced wildlife habitat is expected.

Minnesota—Heron Lake Project

Located in southwestern Minnesota, Heron Lake is internationally recognized as a canvasback breeding and staging area. The lake is also important for many colonial nesting waterbirds, including Franklin's gull, black tern, and black-crowned night heron. The long-term goal of the project is to restore Heron Lake to its historical status as an internationally significant migratory water bird production and staging area. Implementation strategies include wetland and grassland acquisition, restoration, and enhancement. Since 1989 over 3.6 million ha have been acquired, restored, and enhanced by project partners.

Iowa—Iowa Great Lakes Project

For decades, the Iowa Great Lakes, located in northwestern Iowa, have fascinated outdoor enthusiasts. Runoff from surrounding residential and agricultural lands significantly degraded lake water quality. The economic, environmental, and recreational viability of this region revolves around the water quality of these lakes. Partners involved with this project have focused on the protection of existing wetlands, restoration of drained wetlands and converted grasslands, and the seeding of grassed filter strips along streams and watercourses. These practices, combined with other soil and water conservation techniques, have reduced the quantity of pollutants entering the lake system, while also providing essential wildlife habitat.

U.S. Prairie Pothole Joint Venture–Looking to the Future

The strength of the PPJV is in its grassroots organization and in the dedication of the organizations, agencies, and individuals that are willing to work at bringing about long-term protection for wetland and upland habitat in the prairies. The PPJV is also a forward-looking coalition, incorporating the latest technology and stimulating new coordinated efforts in order to better integrate waterfowl and nonwaterfowl migratory bird needs.

The Canadian Prairie Habitat Joint Venture (PHJV) and the PPJV have shared information and loosely cooperated in the organizational stages. However, there has never been a formal mechanism for encouraging specific areas of cooperation. With approval of the 1994 PPJV Update, new windows of opportunity began to open. Broad outlines of an integrated PHJV-PPJV approach were laid out in September 1994 at a joint management board meeting held to discuss cooperation at an international level. There was unanimous consent to develop a small working group to facilitate increased communication and coordination. Further, it was agreed that one joint meeting would be held each year to explore possible joint projects. Informal steps have also been taken to combine certain evaluation and assessment projects. The future will bring increased cooperation and coordination with the PHJV provinces of Manitoba, Saskatchewan, and Alberta and an international outlook for shared ecosystem concerns. Cooperative approaches toward conservation planning and the development of information on certain nonwaterfowl birds is being considered for the future. The proposed outcome will be an ecosystem-level plan that identifies habitat priorities and strategies to benefit priority wetland- and grassland-associated species.

The use of new technologies, such as geographic information systems and GAP analysis, along with tools such as migratory bird population models and multiagency planning and evaluation, will help guide the PPJV into the future. A priority for the PPJV will be increasing the focus on and support for conservation legislation. Agriculture programs such as CRP, if supported and funded, will provide major benefits to wildlife. Working with private landowners will continue to be the backbone of the PPJV. Continuation of essential private lands programs like the Fish and Wildlife Service's Partners for Wildlife program will ensure that the needs of private landowners are considered, along with the needs of waterfowl and other migratory birds.

One of the greatest challenges for the PPJV will be maintaining and increasing funding for projects on the prairies. Strategies in the 1994 PPJV Update will be expensive to implement. The cost of these large investments must continue to be shared by a multitude of agencies, organizations, and individuals dedicated to the goals of the PPJV and the NAWMP. New sources of funding must be secured, and new partners must be recruited if the PPJV is to meet its goals.

The PPJV has reached middle age in an organizational sense. Its early years saw the implementation of many projects and a tremendous amount of enthusiasm and support generated by those projects. Enthusiasm and support remain high in 1995. The vision for the PPJV includes integration, innovation, and continuation of its status as the highest priority joint venture of the NAWMP in the United States. This includes maintaining and expanding leadership in conserving one of North America's most valuable ecosystems.

Summary

The strength of the PPJV is found in its diverse partnerships and common goals. What keeps this broad-based partnership alive is success. Success in the form of acres of grassland and wetland habitat protected, restored, or enhanced and the generation of added awareness and appreciation for the PPR. All partners, including landowners, conservation organizations, corporations, and public agencies, must continue their commitment to the PPJV and NAWMP. Without the continued success of on-the-ground accomplishments and the commitment and appreciation for the resource, interest and participation will fade. All PPJV partners can be proud of their accomplishments to date but need to realize that there is much more to be done.

The PPJV will continue to provide the opportunity for all partners to participate in planning, implementation, and evaluation. Most importantly, PPJV partners will continue working closely with private landowners to integrate wildlife conservation practices that support a profitable agricultural operation. This cornerstone philosophy has been practiced throughout the PPJV. It is the foundation of our past accomplishments and the key to our future successes.

C H A P T E R 16

Northern Grassland Conservation and the Prairie Joint Ventures

Michael G. Anderson
Rod B. Fowler
Jeffrey W. Nelson

The grasslands of central North America have been altered by humans as extensively as any ecosystem on the continent (Bird 1961; Kiel et al. 1972; World Wildlife Fund 1989; Trottier 1992; Knopf 1994). In North Dakota, for example, more than 72 percent of the original prairie now in private ownership has been converted to cropland or tame pasture (from data in Knopf 1994). In Canada, nearly 76 percent of the mixed-grass prairie and more than 95 percent of the tallgrass and fescue prairies have been cultivated (Trottier 1992), and most all of the remaining grasslands are subject to haying or to grazing by domestic animals. At the same time, a large proportion of the original prairie wetlands have been drained, filled, or cultivated (Tiner 1984; Kiel et al. 1972; Ducks Unlimited 1994).

Because of these impacts, a broad consensus has emerged regarding the need for grassland conservation; however, there are multiple views about what specific actions are needed. Protection and enhancement of remnant native prairie are vital, of course, because restoration of complete grassland ecosystems is not possible, and remnant prairie supports many threatened or endangered species (World Wildlife Fund 1989). Moreover, because of the scale of grassland loss and the wide-ranging nature of most animal species adapted to the dynamic prairie environment, we believe that the recovery of many plant and most animal species associated with native grasslands will require management of large landscapes in ways other than full restoration to pre-agricultural conditions, specifically, in ways compatible with agriculture.

Prairie Conservation
Island Press (Washington, DC • Covelo, CA)

Prairie farmers and ranchers produce crops that help sustain North America's human population and provide products for export around the world. They own most of the land we are concerned about and depend on that land for their livelihood. Therefore, they will have to be centrally involved in any large-scale conservation successes. Conservation interests will be able to reclaim certain of these private lands, especially more marginal lands that from the viewpoint of agronomic sustainability should not have been developed in the first place. It is not reasonable, however, to assume that many large tracts of productive, privately owned agricultural lands can be retired from agricultural use. It is necessary, therefore, to focus on combinations of activities, including protection and enhancement of native parcels, managed grazing systems, sound soil and water conservation farming practices, forage production on marginal land, and the like, which should, collectively, provide for a more diverse, wildlife-friendly prairie landscape than exists over much of central North America today. This philosophy, both hopeful and pragmatic, was an important component of the consensus that began to emerge among wildlife interests across the northern prairies during the early 1980s and helped give rise to the prairie joint ventures of the North American Waterfowl Management Plan (NAWMP) (U.S. Fish and Wildlife Service 1986).

NAWMP and the Prairie Joint Ventures

Little progress was made in arresting the trend of habitat loss on the northern prairies until the late 1980s. An important catalyst for change took shape with the signing of the NAWMP agreement between the governments of Canada and the United States in 1986. (Mexico joined the partnership in 1994.) The plan established population goals and provided an overall framework for the management of waterfowl in North America. Strong emphasis was placed on massive new habitat conservation efforts in degraded ecosystems important to waterfowl. The plan recognized that conservation and restoration of wetlands and associated grasslands in the prairie pothole region of central North America was critical to achieve the continental population goals set forth in the plan. The plan considered the prairie pothole region to be a 778,000-km^2 arc of mainly mixed-grass prairie and aspen parkland extending from northern Iowa to the Alberta foothills (Batt et al. 1989). Within this region, NAWMP partners established two parallel joint ventures, the Prairie Pothole Joint Venture (PPJV) in the United States and the Prairie Habitat Joint Venture (PHJV) in Canada.

Development of significant new sources of funds for conservation presented a formidable challenge. Now, nearly nine years since the signing of the NAWMP and six to seven years since the inception of the prairie joint ventures, the actions of federal, state, provincial, and nonprofit partners have resulted in the protection of 109,355 ha of idle native parkland and other grasslands; conversion of 86,586 ha of cropland to grassland; conversion of 168,466 ha of continuously grazed native pastures to deferred or rest-rotation grazing systems; and the protection, restoration, or enhancement of 164,938 ha of prairie wetlands (see table 16.1).

Conservation Strategies

In Canada, where losses of grassland and aspen parkland have been extreme, most recent waterfowl conservation work has focused on reestablishing permanent grass in parkland landscapes with numerous and relatively permanent existing wetlands. Less work has been done in the more drought-prone mixed-

Table 16.1

Grassland conservation accomplishments (hectares) of the NAWMP prairie joint ventures, 1986 through 15 June 1995 (PPJV) and 1 July 1995 (PHJV)

NAWMP Program	PPJV[a]	PHJV[b]	Total Hectares
Grassland Protection			
Native Grasses		23,490 (33% Perpetual)	
Tame Grasses		9,463 (17% Perpetual)	
TOTAL	76,406	32,953 (29% Perpetual)	109,359
Grassland Restoration			
Native Grasses	17,483	15,728 (90% Perpetual)	33,211
Tame Grasses	25,050	28,325 (19% Perpetual)	53,375
TOTAL	42,533	44,053 (45% Perpetual)	86,586
Enhanced Agricultural Management			
Hayland	9,389	14,815	24,204
Pasture	59,814	108,652	168,466
Other Uplands	10,117	1,374	11,491
TOTAL	79,320	124,841	204,161
Wetland Protection/Restoration	83,529	81,409	164,938

[a]PPJV (Prairie Pothole Joint Venture) includes portions of Montana, North Dakota, South Dakota, Minnesota, and Iowa.

[b]PHJV (Prairie Habitat Joint Venture) includes most of the agricultural zones of Alberta, Saskatchewan, and Manitoba.

grass prairies where upland projects have been designed to moderate the use of existing native grasslands by cattle. In cooperation with irrigation districts in southern Alberta and ranchers in the western United States, these initiatives involve provision of reliable water to natural wetland basins and establishment of grazing systems that feature large paddocks where grazing is deferred for most or all of a growing season. Such deferral benefits certain cool-season native grass species by reducing grazing pressure in spring and early summer while providing undisturbed habitat for nesting birds. In addition to NAWMP initiatives, Agriculture Canada's Permanent Cover Program has recently converted some 489,000 ha of erodible marginal cropland to permanent cover in the prairie provinces, partly in priority NAWMP areas.

In the United States, where public lands dedicated to wildlife are more extensive and the USDA Conservation Reserve Program (CRP) has converted 16.2 million ha of cropland to mostly idle grass, some 1.6 million ha in the prairie pothole region, the PPJV has invested relatively more in wetland restoration (see table 16.1). In addition, establishment of waterfowl production areas and other wetland easements by the U.S. Fish and Wildlife Service has protected more than 809,000 ha of wetlands and grasslands in the PPJV region (Nelson and Connolly, in press). Restoration of prairie wetlands is usually accomplished by installing small earthen ditch plugs or removing existing drainage tiles (Galatowitsch and van der Valk 1994). This wetland work is increasingly targeted at areas where CRP or publicly managed grasslands provide extensive upland nesting cover. Significant tracts of grasslands also have been restored or protected by the PPJV (see table 16.1), and nearly 60,000 ha of pasture have been converted to managed grazing systems designed to reduce the impact of cattle on grassland-nesting birds.

In both joint ventures, program planners have attempted to maximize waterfowl benefits while secondarily assisting the conservation of many other native species. Programs are targeted at landscapes, typically 104 to 518 km^2 in size, with high densities of remaining wetlands that should attract large numbers of breeding pairs and adequately support waterfowl broods. Designs for each landscape are developed with the aid of computer models that optimize cover mixes for breeding ducks, tempered by local experience with the acceptability of specific programs to landowners. Such plans involve some combination of land purchase or lease, with establishment of dense nesting cover on formerly cropped land, protection of existing native pasture or idle land, deferred grazing systems, and delayed hay cutting. This variety of programs has the added benefit of creating a diversity of habitat types.

Increasingly, native grass varieties are being seeded in such cover plantings in mixtures designed to establish appropriate species on wet soils, dry soils, eroded knolls, and other specific sites. This is done to maximize cover establishment across whole fields, improve plant species diversity, enhance wildlife diversity, and minimize long-term management costs. Where grass is reestablished, wetland restoration is usually conducted on small temporary and seasonal basins that have been lost from the prairie landscape at disproportionately high rates (Stewart and Kantrud 1971).

Wildlife Responses

Responses by many species of migratory birds and resident wildlife to these landscape interventions have been encouraging. For example, use of newly established grass stands by bobolinks, sedge wrens, LeConte's sparrows, grasshopper sparrows, clay-colored sparrows, and other grassland birds has been extensive (Higgins et al. 1984; Dale 1994; C. de Sobrino and T. Arnold, personal communication 1994). Several studies are now underway to assess the productivity of waterfowl and songbirds that have responded to this cover in order to learn how managers might maximize wildlife benefits.

While accomplishments have been substantial, important questions remain about the biological effectiveness of joint venture programs. These include concerns about the effects of grassland patch size and juxtaposition on breeding success of migratory birds, the attractiveness and safety for breeding birds of various types of planted cover, and the effects of joint venture projects on less-common species. Impacts of joint venture programs on the biology of prairie carnivores are also poorly understood and potentially of great significance to breeding birds. Another important question for managers is the degree to which some agricultural use, for example, periodic haying or grazing, might be permitted on lands owned or leased for wildlife purposes without jeopardizing wildlife habitat values. There is pressure to accommodate such use, both to increase the acceptance of programs within rural communities and to generate revenues to help offset long-term management costs for lands dedicated to wildlife.

Reflections on Joint Venture Partnerships

The significant accomplishments of the PHJV and PPJV have been achieved through partnerships of federal agencies, provincial and state governments, and multiple nongovernmental wildlife organizations. For there to be value in a partnership, it must be facilitating rather than restrictive so that more might be accomplished collectively, not less. The fact that both joint ventures remain active today attests to the positive nature of these partnerships. We, individually, and Ducks Unlimited have been active participants in the joint ventures since their inception. We offer some personal reflections on what has made these partnerships work well, while noting a few of their limitations. Our intent is not to be critical of any joint venture or agency, but to provide helpful thoughts for improving future partnerships.

Undoubtedly the greatest benefit of the partnerships has been the merging of numerous agencies with diverse capabilities to pursue common goals. Full knowledge of land-use issues, problems with waterfowl recruitment, reasons for declines of grassland nesting birds, potential solutions, and program ideas were not resident in any single agency. Therefore, development of the original joint venture plans benefited greatly from the various perspectives that were available from the staff of many agencies. It also quickly became clear that there was no single solution to restoring northern grasslands and wetlands. Due to funding limitations, mission focus, or limitations of operational policy, no single agency was in a position to deliver the full range of proposed solutions. However, because agencies forming these partnerships possessed different skills and interests, most program needs were filled.

Partnerships also have been essential in maintaining financial and political support for the NAWMP. In these times of financial restraint, when government agencies are hard-pressed to maintain funding for wildlife programs, the NAWMP has fared better than most because of the financial leverage achieved by each agency's contribution.

In hindsight, it seems that such partnerships are typically infused with energy during the planning phase, when the commitment and excitement of forging a common vision are driving forces. Once joint ventures face the hard realities of program funding and delivery, however, new difficulties predictably emerge, and partnerships can undergo fundamental change. An early challenge was for partners to understand and accommodate the different cultures and requirements of others on the team, particularly the fundamental differences between the public and private sectors. Nonprofit organizations are strongly mission focused. Their missions are sometimes rather narrow, but usually long-term in outlook. Most government partners have much broader missions and the imperative of responding to short-term political pressures. These differences can lead to conflicts over program emphasis, geographic location, and the time frame in which results must be achieved and demonstrated. For some partners the joint venture is only one of many, perhaps transient, programs; for others, the joint venture's goals are congruent with their organization's mission. Understandably this can result in varying agency commitments to the joint ventures and varying satisfaction with program results.

Another complication is the nature of management and decision making in joint ventures. Most agencies, public or private, are hierarchial, with a single point of authority. Joint ventures, on the other hand, are horizontal, and decisions are made by consensus without a central point of authority. Such partnerships are inherently inefficient at decision making, with progress tending to be controlled by the pace of the most reluctant partner. Furthermore, because of the informal nature of the joint ventures, decisions of the group are not binding on individual members. This necessary independence can be frustrating for other partners who are anxious to influence joint venture programs and policies.

Everyone, it seems, wants to be involved in the delivery of conservation programs. This can result in either turf wars or duplication of efforts, neither of which is constructive. Unfortunately, important activities of lower profile, such as evaluation and policy work, have been slower to get underway. For evaluation in particular, consensus on what is needed has been slow to develop because of differing partner needs and expectations. This is unfortunate because feedback on program effectiveness is essential for successful implementation. As evaluations develop, good communication between researchers and managers is imperative.

The marketing of joint ventures can also create challenges, especially when some organizations depend on public exposure in order to raise funds essential for their continued involvement in partnership programs. Multiagency marketing campaigns are difficult to construct, in our view, and are challenged to deliver programs to the public in a clear and understandable way. In Canada at least, public communication programs also should be broadened in order to increase public support for grassland conservation. Because of the drought-related decline of ducks in the last decade and the relatively high public profile of waterfowl, the PHJV has benefited from the understanding and support of many people. There have been insufficient efforts, however, to educate the general public about the massive loss of prairie grasslands and the significant declines of many other species that depend on this habitat.

One of the biggest challenges for the joint ventures is sustaining agency interest and funding over the long term. Some agencies with broad mandates can spend little on any single program, and agencies that are not active on a daily basis lose touch and consequently can lose interest in what is taking place. They tend to be involved only when there are major issues, and because these are usually major problems, it is easy to develop a negative image of the program. For the joint ventures to maintain successful, durable, and evolving programs, the enthusiasm of partners must be sustained. All partners must learn what is happening on a regular basis. They must hear about the success stories. They must get out on the land and see projects firsthand. They must hear the results of evaluations and understand what is working and what is not. They must sense the need for progressive change. In short, all partners must feel a strong sense of ownership in what is being accomplished.

In an innovative effort to enhance its partnerships, the PHJV recently retained an independent consultant to undertake a broad program evaluation. This exercise is providing an independent assessment of various components of the joint venture and may become an effective catalyst for rekindling the commitment of all partners.

The Need for Expanded Partnerships

The current joint venture partnerships need to be nurtured, but they also need to be expanded. With the major exception of funds from the North American Wetlands Conservation Act and, to a lesser degree, Canadian federal funds, the majority of recent support for habitat conservation on the Canadian prairies has come from traditional waterfowl interests such as Ducks Unlimited and state wildlife agencies. In the United States, several federal, state, and private conservation organizations have targeted resources to restore and protect components of the prairie ecosystem. It seems unlikely, however, that the current partnerships and programs alone, which are primarily wildlife driven, will result in the restoration of a significant proportion of the original northern grasslands. Initiatives with broader-based political and financial support than the joint ventures of today will be needed to ensure success in conserving these grasslands and their associated wildlife.

Most importantly, in our view, conservation organizations must find new ways of working cooperatively with agriculture. Land-use, and thus grassland conservation, is affected far more by agricultural markets, policies, and programs than all other forces combined. Prairie wildlife needs a conservation-friendly U.S. farm bill and revamped agricultural policies in Canada as much as the programs of wildlife organizations. The prairie joint venture leaders understand this, and policy and legislative initiatives are receiving more attention than in the past. However, we believe that the emphasis on this work is still not commensurate with its probable importance in the eventual success or failure of the joint ventures.

The Canadian Wheat Board estimates that some 39.1 million ha of land were cultivated in prairie Canada in 1994. Of these lands, the federal Prairie Farm Rehabilitation Administration estimates that 5.4 million ha are classified as marginal or fragile (L. Moats, personal communication 1995) and arguably never should have been broken. Many of the decisions to break this land were not made in response to market forces, but were fostered by government policies that based delivery quotas for specific commodities on cultivated acres. Prairie grain and oilseed producers recently received annual subsidies of some $110 per ha when cropping marginal lands (Patterson 1993). The only solution that is likely to cause major landscape change is for governments to eliminate or at least realign agricultural subsidies to discourage inappropriate land-use or provide incentives for adjustments to more sustainable agriculture, such as the conversion of marginal cropland to grassland. In the United States, demand to enroll marginal farmland in idled grass exceeds available funds, while the CRP provides an estimated $13 billion worth of environmental benefits at no net cost over what government would have paid to subsidize cropping on the same lands.

Fortunately, changes are coming. In March 1995 the Canadian federal government announced the elimination of the Western Grain Transportation Act, which subsidized grain-shipping costs for prairie farmers and thus encouraged grain production on marginally productive land. A combination of deficit reduction pressures on governments and required compliance with the GATT and NAFTA treaties seems likely to gradually reduce most subsidies for annual cropping. If so, farmers will then make land-use decisions based more on market forces and land capability. Although we believe that this will lead to an increase in grassland, there remains great uncertainty about how much conversion will take place and how those new grasslands might be managed. Therefore, some conservation incentive programs still may be required.

Fifteen years ago, when planning was initiated for the NAWMP, interaction between the agricultural and conservation communities was difficult. Planners of the prairie joint ventures hoped that if conservation programs were implemented in cooperation with farmers and ranchers, trust and cooperation would develop between the two sectors and lead to mutually beneficial changes in agricultural policies and practices. Today, that relationship is greatly improved. In fact, acceptance of conservation incentives by landowners has exceeded the capabilities of joint venture partners. In some programs, farmers have realized increased net profits above and beyond incentive payments, and this has proven to be an important factor in program acceptance. In addition, good working relationships have been established between wildlife interests and several state, provincial, regional, and national agricultural organizations. At a time when global trade and public support for agriculture are in a state of flux, the fact that our two sectors are working together toward more sustainable land-use gives us some reason for cautious optimism about the future conservation of northern grassland wildlife.

Recommendations

We offer the following specific recommendations for enhancing the effectiveness of prairie conservation through joint ventures:

1. Enlist the participation and support of additional agricultural interests in the prairie joint ventures.

2. Promote changes in public policy that will discourage further conversion of marginal agricultural lands for commodity production and encourage grassland conservation and restoration.

3. Encourage the financial participation of other grassland conservation interest groups in the shaping, delivery, and evaluation of NAWMP prairie programs.

4. Enlist broad support from the general public for prairie conservation by communicating the loss of North American grasslands, the impact on many species of wildlife, and potential long-term solutions including the adoption of more sustainable agricultural production systems.

5. Continue to improve operational aspects of the prairie joint ventures in order to enhance efficiency, minimize conflicts, and enable individual partners to work as productively as possible under the joint-venture umbrella.

6. Strengthen communications between the PPJV and the PHJV, especially in the area of biological evaluations.

Preservation and restoration of large ecosystems such as the northern grasslands are long-term challenges. Even the most pragmatic goals may require decades to attain. It is essential for the prairie joint venture partners to recognize that they have begun a very long journey, to commit to making the trip, and to ensure that there are systems in place to provide constructive feedback along with way. Long-term vision and strong leadership will be essential for sustaining the journey.

Acknowledgments

We thank M. Koneff (U.S. Fish and Wildlife Service), R. Clay (Ducks Unlimited Canada), and Danielle Bridgett (Canadian Wildlife Service), for providing summary data on joint venture habitat accomplishments. C. Lively, A. J. Macaulay, J. Patterson, and R. Wettlaufer offered helpful comments on an earlier draft of the manuscript, but we are responsible for the opinions expressed here. Of course, without our joint venture partners we would have far fewer accomplishments to highlight or constructive thoughts to offer on the conservation process.

CHAPTER 17

Sustainable Grassland Utilization and Conservation in Prairie Agricultural Landscapes

Peter J. Buesseler

What can be done with expiring Conservation Reserve Program (CRP) lands in the Glacial Lake Agassiz Interbeach Area? How do we protect and manage the region's native prairie lands and species before they're gone forever? How do we balance an expanding gravel-mining industry with environmental protection?

In the past, agencies and organizations took on these challenges as lists of problems to be solved. Conventional wisdom guided us to isolate the problem, assign responsibility to an appropriate institution, figure out the right answer, and build public support for the solution. Many of these efforts have had preliminary success addressing first-order problems. However, there is an emerging awareness that we need additional tools and strategies to address a whole new set of second-order problems that have emerged.

These new types of approaches include sustainable development, holistic resource management, adaptive management, and ecosystem-based management. Rather than isolating one concern from another, these approaches make connections between them. Situations are not fragmented into independent problems to solve but addressed as webs of multiple views, dilemmas, and interdependencies. While considerable work is being done in agriculture to adapt specific land management practices to these ideas, there are few cases where these practices are being integrated into regionwide, collaborative change.

Organizations and agencies participating in the Great Plains Partnership (GPP) have recognized the Glacial Lake Agassiz Interbeach Area as one of the most important areas in the Great Plains for strengthening coordinated, ecosystem-based management. A number of activities are underway or planned here to improve

Prairie Conservation
Island Press (Washington, DC • Covelo, CA)

ecosystem management. This reflects a long history of collaboration in the Red River Basin around land and water stewardship issues. The Glacial Lake Agassiz Interbeach Area offers an opportunity to explore how landowners and communities can put ecosystem-based management into operation and how our institutions can best assist them in that effort. Current project cooperators include the Minnesota Forage and Grassland Council, the International Coalition, University of Minnesota Extension, Minnesota Department of Agriculture, Minnesota Department of Natural Resources, Great Plains Partnership, The Nature Conservancy, USDA Natural Resources Conservation Service of North Dakota and Minnesota, local resource conservation and development councils, communities, and landowners.

Glacial Lake Agassiz Interbeach Area

The Glacial Lake Agassiz Interbeach Area is located in the northern portion of one of North America's most productive and intensively utilized ecosystems, the tallgrass prairie (see fig. 17.1). The Interbeach Area itself is characterized by relatively less fertile soils that formed on the beach ridges and deltas of the former glacial lake. There are three major grassland landscape areas in the Interbeach area: the Lake Agassiz Beach Ridges in northwestern Minnesota, Aspen Parkland in southeast Manitoba and northwest Minnesota, and the Sheyenne River Delta in southeastern North Dakota. Together these harbor the largest acreage of grassland and wetland habitat left in the northern tallgrass prairie ecosystem.

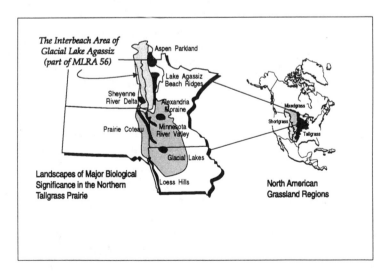

Fig. 17.1 Glacial Lake Agassiz Interbeach Area.

In Minnesota alone these grassland landscape areas include 77,000 ha of pastureland, almost 40 percent of the state's CRP lands (about 303,500 ha) and an estimated 30,350 ha of native prairie. Statewide, forage and grasslands contribute 15 to 20 percent of cash farm income, provide the primary habitat for many wildlife species, are important in reducing soil erosion, and are home to over 40 percent of Minnesota's rare and endangered species. The continuing decline of grass- and forage-based agriculture in the region, upcoming expiration of CRP contracts, and other major changes have brought us to a crossroads in the future of this ecosystem.

Trends

The majority of land in Glacial Lake Agassiz Interbeach Area is privately owned. Approximately three-quarters of the area is in row crop agriculture, with wheat and other small grains being the predominant crops. Except for CRP, the long-term trend for grassland acres continues to go down. In USDA major land resource area (MLRA) 56, over 40 percent of noncrop pasture acres have been converted to other uses in the past ten years. Most of the grassland acres in MLRA 56 occur in the Interbeach area. Livestock production and the total number of beef-dairy farms have declined significantly over the past few decades as well.

Increased use of center-pivot irrigation and continued expansion of aggregate mineral (gravel) mining have put additional pressure on the Lake Agassiz Interbeach grassland landscape areas. In addition to directly reducing grassland acres, both of these activities are having an impact on the hydrology of the many seeps, springs, and fens, as well as groundwater quality issues. Encroachment of exotic plant species (leafy spurge, smooth brome, Canada thistle, etc.) is another significant problem. Herbicide spraying to control spurge and other noxious weeds adds to farm operation costs and further reduces grassland diversity. The U.S. Forest Service estimates that 12,140 ha of the Sheyenne National Grasslands is currently affected by leafy spurge.

One result of these trends has been an accelerating loss of the region's biological diversity. Grasslands in the Interbeach area provide habitat for two federally protected species and two federal candidate species. In Minnesota, these grassland landscapes harbor thirty-four state-listed rare plant species and twenty-three rare animal species. Proactive, integrated action now could prevent future "environmental train wrecks," but such efforts are hampered by two state boundaries, several federal agency boundaries, and an international boundary, as well as a large number of county and local jurisdictions.

Public Concern

A public opinion survey of residents in the Glacial Lake Agassiz Beach Ridges shows strong concern for water quality and quantity issues in the area, and there has been considerable effort over the past ten to fifteen years to address these. However, there does not appear to be a shared understanding and concern about grassland and other interrelated ecosystem issues.

The Great Plains Partnership is conducting a series of focus groups with citizens from communities in potential "hot spots" like Glacial Lake Agassiz Interbeach Area to better understand how people think about ecosystems and ecosystem management. Based on their first four discussions, a few overarching themes seem to be emerging (Harwood Group 1995).

1. People feel they are losing control of their lives and are anxious about what the future holds for them and their communities. These concerns drive many of their views on ecosystems and ecosystem management.

2. People see ecosystems as directly related to their quality of life, specifically to their personnel health and economic security. People approach ecosystem management from a perspective of maintaining their way of life and seek a balance between ecosystems and maintaining economic security.

3. People want the ethic of rights and responsibility, not laws and mandates, to drive ecosystem management. They believe everyone has a responsibility to protect ecosystems but within the context of individual freedom.

4. People want an active role in addressing ecosystem issues. They do not want outsiders to dictate what they should do—something they often feel happens now.

A Conceptual Framework for Ecosystem Stewardship

The terms "sustainability," "ecosystem-based management," and "ecological integrity" are closely related and often used interchangeably. Ecosystem stewardship integrates all of them. To understand their relationship, one can view them as follows:

Sustainability—A Desired Outcome. Sustainability is the achievement of economic and social well-being without damaging the planet's resource base. It is as much a political concept as it is scientific in that it represents what people value in ecosystems.

Ecosystem-based Management—A Methodology. Ecosystem-based management is a geographically targeted, whole systems approach to achieve sustainability for both natural and human communities. Interrelated problems are considered on multiple scales in a partnership approach. This is different from the model of "multiple use," where stakeholders and agencies work in isolation or competition with each other to improve individual resources.

Ecological Integrity—A Measurement. Ecological integrity is an ecosystem state that maintains a capacity to produce desired goods and services on a sustainable basis. Environmental indicators are analogous to the vital signs used in human health. (Minnesota Department of Natural Resources 1995)

Framing Problems from a Public Point of View

To have a productive discussion of problems confronting the Interbeach area requires that we incorporate the full diversity of perspectives into the framing. Each perspective offers a distinctive view of what causes the problem, what should be done about it, and who should do what. Taken together, these perspectives constitute a public framing of the problem. This does not make it easier for us to make the choice that confronts us—just the opposite. It does, however, make the discussion more productive and hence progress more likely.

For the Glacial Lake Agassiz Interbeach Project, the Harwood Group's policy dialogue approach to engaging citizens and decision makers on ecosystem issues will be used (Harwood Group 1994). This approach is built on eight key principles.

1. Understand people's starting points. Engage citizens by exploring how various people initially approach the issue.

2. Engage citizens, not just stakeholders. Public decision makers and agencies should develop an understanding for the political will that is shared by citizens, rather than just the disparate views of various stakeholder groups.

3. Create opportunities for citizens to make judgments. Finding sustainable ways to communicate requires an understanding of what people believe and why and how they came to hold their views.

4. Work through choices and trade-offs. Sustainable change requires setting priorities and working through choices and trade-offs.

5. Seek common ground. Help people discover the values and goals they share.

6. Forge new compacts. Focus on new compacts to create new possibilities for change.

7. Link citizens with decision makers. Decision making is an iterative process and citizen input is gathered throughout the process.

8. Produce results. Create a realistic agenda for change.

Network Structure

The Glacial Lake Agassiz Interbeach Project is being designed and managed as a network. This approach seeks to strengthen the connections between people and programs in existing organizations. It includes and improves these institutions rather than simply adding a new one to the situation. Networks benefit from five organizing principles: a unifying purpose, independent members, voluntary links, multiple leaders, and integrated levels (Lipnack and Stamps 1994).

Accommodating the Diversity of Needs

A strategic framework for action in the Interbeach area needs to accommodate an array of action levels, with triggering mechanisms that reward local initiative and local capacity for effective implementation. Various groups and regions have very different capacities for change and adaptation. Policy makers can be ambitious about ecosystem management initiatives, but communities will capitalize on assistance only if their local leadership and infrastructure capacity have reached a critical mass. The key principles that will guide action are (1) actions should help people help themselves; (2) actions should encourage, support and reward local initiative; and (3) actions should accommodate an array of readiness levels.

Basic Readiness Actions

Where there is a need for substantial assistance to understand interrelated problems, prospects, and options, and to organize a response, the network's focus will be to help build a basic level of readiness. Examples include synthesis and integration of existing data, identification of critical landscapes and ecosystems, convening of cross-disciplinary and jurisdictional meetings, promotion of best management practices (BMPs), species recovery projects, and outreach. These actions help people understand their current situation and meet the most basic needs before moving ahead.

Adaptation Actions

Where there is a good understanding of problems, prospects, and options, and people are sufficiently organized to take strong actions, the network's focus will be to facilitate adaptation. Examples include: cross-jurisdictional–institutional cooperation agreements, expansion or redesign of existing programs, integrated collaborative landscape projects, multidisciplinary planning. These actions enable stakeholders to expand their current activities, organize more effectively, and try new approaches.

Redesign Actions

Where there are special comparative advantages, a proven capacity for innovation and leadership, and a consensus for action, the network's focus will be to create opportunities for expanding beyond traditional approaches. Examples include policy or market incentives for "environmental services" of land, planning and budgeting systems for service sharing, and common project funding based on ecological and economic boundaries of problems (the problemshed). These actions create new directions for integrating economic and environmental issues.

Improving Ecosystem Stewardship in the Interbeach Area

Current Activities

The Glacial Lake Agassiz Interbeach Area has been identified by the Great Plains Partnership as one of the best opportunities and most important areas in the Great Plains for improving ecosystem stewardship. This reflects increasing concern about the Interbeach area as well as the level of activity already underway in the region. These activities provide a strong foundation for strengthening coordinated, ecosystem-based approaches to the issues facing the Glacial Lake Agassiz Interbeach Area. Examples include the following:

1. A relatively extensive, biological survey of the Interbeach counties in Minnesota has been completed by the Minnesota Department of Natural Resources, and a Conservation Data Centre has recently been established in Manitoba. Analysis and presentation of much of these biodiversity data are now available through GIS systems.

2. The U.S. Fish and Wildlife Service has included this part of the tallgrass prairie as one of its top priorities in its Upper Mississippi Tallgrass Prairie Ecosystem Management Plan. It has also received approval to begin the planning and National Environmental Policy Act compliance phase of a northern tallgrass prairie habitat preservation area refuge. This project would seek to protect and manage the remaining tallgrass prairie through a concerted effort by a variety of agencies, organizations, and individuals.

3. The Army Corps of Engineers is involved in a major environmental impact statement describing water retention basins in the Red River Basin.

4. The U.S. Geological Survey has included the Red River of the north basin as a study unit under the National Water Quality Assessment program. In northwestern Minnesota a number of joint landowner-agency work groups are exploring and developing alternatives for CRP lands as part of the Minnesota CRP Investment Initiative led by the Minnesota Department of Agriculture.

5. Through funding from the Critical Wildlife Habitat Program, Manitoba is creating a network of protected tallgrass prairie lands in the Tolstoi-Gardenton area of southeastern Manitoba. This will include variously owned, leased, and voluntarily protected lands. In a message delivered to the GPP Executive Coun-

cil 19 January 1995, Premier Filmon committed his government to developing a partnership with Minnesota and North Dakota to demonstrate sustainable development in the region.

6. In July 1995 the State of Minnesota will initiate two, two-year projects: Sustainable Grassland Conservation and Utilization targeting the Agassiz Beach Ridges, and Glacial Lake Agassiz Beach Ridges: Mining and Protection, a long-term plan to balance protection of native prairies with a sustainable aggregate industry in Clay County, Minnesota.

7. The USDA Natural Resources Conservation Service and President's Interagency Ecosystem Management Task Force has targeted the Glacial Lake Agassiz Interbeach Area as a New Initiative Laboratory project. Funding will be used to expand ecosystem planning and coordination efforts to the whole region. This will bring in many more stakeholders and substantially broaden interagency and cross-regional support, and promises to provide the foundation for a more ecosystem-wide improvement effort.

Future Activities

Build Up Regional and Local Capacity for Addressing Integrated Issues

The goal is to improve ways for engaging people in the area to work together in addressing ecosystem challenges. This will mean seeking appropriate mechanisms for involvement, framing issues in ways that make sense and are meaningful to people, and creating an environment to overcome obstacles. Planned activities include organizing resource information from Minnesota, North Dakota, and Manitoba for the Interbeach area; conducting more extensive public research work to gain greater clarity for how people think about the area and its ecosystems, the values they hold when it comes to these issues, and how people in the area respond to strategies for ecosystem; and accelerating and broadening participation in a regional, Internet-based communication-information network—the Red River Basin Information Network. This will provide participants with sufficient knowledge about the area and interrelated issues to be able to effectively discuss them. Materials will be prepared in such a way as to stimulate discussion. They will help clarify what the issues or points of concern are that lead people to view the problems from different perspectives.

Develop Integrated Strategies for the Glacial Lake Agassiz Interbeach Area

Locally based, interdisciplinary teams (including landowners, government agencies, extension services, farm consultants, agribusiness, credit lenders, and conservation and agricultural organizations) will be offered the assistance of regional and statewide experts and agency representatives to develop a strategy for grass-lands and forages within the Interbeach area. Their task will be to create collab-orative and individual strategies for positioning landowners, businesses, financial and government institutions, and other affected interests to better address grass-land opportunities and needs. The work groups will explore and develop inte-grated strategies for addressing key issues such as the future of CRP lands and protection of biodiversity. Participating agencies and organizations will then de-velop individual or joint action plans to carry out the strategies. Planned activi-ties will build and support a strong, broad-based ecosystem management im-provement team, prepare landscape strategies, and prepare individual and joint agency-organization action plans. Use of landscape work groups gives ownership and responsibility of the project to those closest to the real needs. These are the landowners, as well as the many public and private programs and services these landowners turn to for help. If successful, the project will create a climate of dy-namic collaboration and enthusiasm among these diverse stakeholders and de-velop the local leadership to address needs in the future.

Initiate Integrated Grassland Projects

This part of the project gives landscape teams the opportunity, and responsibil-ity, to tailor projects to address the specific needs and opportunities of the area. This could include trying out new management practices, establishing grazing associations, conducting workshops, or doing whatever else is needed to take full advantage of its grassland and forage resources in the future. Planned activities include on-site applied research-demonstration projects; participatory projects linking the research team, agency-program staff, and landowners; targeted pro-jects to help individuals and local institutions address obstacles; and an intera-gency, cross-regional federal budget proposal for the Interbeach area. The pri-mary purpose is to provide stakeholders within the local area an opportunity to cooperatively design and implement innovative projects to enhance grass and forage lands. The work groups will have the flexibility, and accountability, to look at all aspects of the economic, social, and environmental system and target what they feel are the best opportunities for improving the situation. Projects will help prepare individuals and local institutions for future sustainable grassland utiliza-tion and conservation within the landscape.

Conduct Farm- and Community-Level Economic Analyses

Evaluate individual and community-level implications of post-CRP decisions. Two activities are planned. First, conduct farm and community assessments. Twenty to twenty-five landowners (forty to fifty with New Initiative Laboratory funding) will be assisted with working their whole farms (not just CRP areas) through University of Minnesota farm management and financing decision aid. Several landscape management alternatives will be tried with each farm. The participant will select a desired management option, along with a pre-CRP baseline, for subsequent analysis. Aggregate results of the farm-level studies will be used to assess community-level economic implications of different management schemes through changes in farm input demand and farm outputs.

Second, conduct lender assessments. Cooperating agricultural lenders will use in-place credit scoring and environmental liability checklists to rate each farmer's post-CRP land management option with respect to its creditworthiness. Each farm plan will be examined by credit officers from three different financial institutions. Their credit opinions should provide a better idea about how environmentally sound management options are viewed by the financial community.

Through this analysis, changes at farm level will be scaled up to examine community-level implications and effects on financial institutions. Collectively, these results will give a better understanding of the financial implications of post-CRP management decisions throughout a community and hence their likelihood of adoption. Together with existing state and county land resource information and GIS capabilities, participants will learn how to better integrate environmental and economic goals and strategies for the area.

CHAPTER 18

Ecosystem Management and the National Grasslands

Stanley E. Senner
Brian D. Ladd

The national grasslands of the United States contain about one-third of approximately 4.5 million acres of lands first purchased in the 1930s by the federal government, mostly from destitute farmers and settlers. There are currently twenty national grassland units, ranging in size from about 600 ha to over 400,000 ha. Most of these units are in the Great Plains region; three are west of the Rocky Mountains.

The national grasslands have been under U.S. Forest Service management since 1953. The federal government's original intent in acquiring devastated lands in the Dust Bowl years was to slow soil erosion, increase forage production, and provide relief and economic stability for remaining rural residents (West 1990). Soon after various federal restoration and conservation projects began, it became clear that change in both public and private land management would be required if another Dust Bowl were to be prevented. The federal government thus took on the task of demonstrating proper grassland agriculture to private landowners, with the hope that federal demonstrations would be seen and adopted on private lands (Hurt 1985). Federal advisers who supported the demonstration function believed that private landowners had an obligation to future generations to protect the land.

Grazing has always been the dominant use of the national grasslands. In the 1930s prospective public land ranchers formed associations in order to negotiate access to the newly available lands. The value of the new public grasslands for other uses (such as outdoor recreation, wildlife habitat, and watershed protection) was also recognized early in their history, and specific projects were initi-

Prairie Conservation
Island Press (Washington, DC • Covelo, CA)

ated to enable these uses. But for many years most uses other than grazing—and in some areas oil, gas, and mineral development—did not have strong public support.

This began to change in the early 1970s, when an increased emphasis was placed on managing the national grasslands for multiple uses. The drive for industrial growth in the cause of the public good and increased concern for the environmental consequences of such growth focused attention on the use of public lands. Although the national grasslands have not received the attention from either the federal government or nongovernmental conservation organizations that many high-profile national forests have received, there has been a growing awareness of the importance of these lands to the conservation of the biotic integrity of the Great Plains. The Great Plains Initiative, sponsored by the Western Governors' Association in cooperation with several federal agencies, is a modern example of how concern for prairie conservation has resulted in a project that recognizes the incongruity of political boundaries with biogeographic provinces and is attempting to deal with this incongruity. The purpose of the Great Plains Initiative is to identify and coordinate research on critical conservation problems affecting the natural resources of the Great Plains ecoregion.

In 1994 the National Audubon Society began a review of management on the national grasslands in relation to the distribution and habitat requirements of primarily nongame grassland birds. Attention was focused on species of birds endemic to the Great Plains or having a strong association with grassland habitats. The National Audubon study included a questionnaire circulated to Department of Agriculture Forest Service district rangers responsible for national grasslands, a review of grassland management plans and budgets, and an assessment of the degree and significance of fragmentation in land ownership within grassland administrative boundaries (Senner and Ladd 1995).

The Current Situation

Management Plans, Expertise, and Resources

National forest land and resource management plans (forest plans) provide the objectives and general direction for national grasslands management. To focus management activities, many grasslands are divided into management areas that focus on a priority use or value (e.g., livestock grazing, wildlife, developed recreation sites, watershed integrity) while still providing for other uses and values. Management indicator species are frequently designated in those areas where the primary emphasis is on management for wildlife habitat. These species, which can include birds, mammals, reptiles, and amphibians, are chosen on the basis of how well their response to habitat change represents the response of all the species that use the habitat similarly. Management indicator species also may be designated for their high interest to hunters or other consumptive users. Such species are often defined at a regional level, however, which can obscure the local variability of a species' response to habitat change and make cause-and-effect attributions difficult (Knopf, in press). Wildlife species listed at state or federal levels as threatened or endangered also receive special attention in national grassland wildlife management plans.

The objectives of wildlife management on national grasslands as described by forest plans are typically straightforward: maintain viable populations of all existing vertebrate wildlife species by protecting a minimum percentage of the maximum possible habitat for each species on the grassland (U.S. Department of Agriculture 1985). Habitat is to be provided for threatened and endangered species according to their needs for recovery. In none of the plans reviewed was an operational definition of viable population provided.

Since the time of Peek and Risser's 1979 study, which also was supported by the National Audubon Society, several wildlife biologists have been hired and dedicated to individual national grassland units, and the expertise is now mostly available to conduct, for example, wildlife inventory and monitoring projects. However, funding and sufficient labor are not readily available. Wildlife management budgets have increased on all units since the late 1970s, but the proportion that the wildlife allocation comprises within the total budget for each grassland has in many cases not increased. Actual expenditures on wildlife vary considerably from unit to unit and year to year, and direct comparisons between grasslands are difficult to make. Those units with the smallest percentages of their budgets designated for wildlife management are not necessarily the least active in managing for wildlife; for example, Crooked River National Grassland in Oregon has one of the smallest wildlife allocations for 1995 but is probably the most extensively engaged in cost-share projects and volunteer habitat restoration initiatives. Many units have ongoing research or restoration projects that are funded by state wildlife agencies, universities, and a variety of nongovernmental organizations. The result of the current management, personnel, and funding situation for grassland birds is that the baseline population and distribution data and historical ecological perspectives needed for making habitat management decisions that protect the biotic integrity of the landscape are not available or not readily made available.

The Role of Grazing

Allotment management plans (AMPs) or comparable documents provide the direction for management of grazing by domestic livestock (primarily cattle) on national grasslands. AMPs typically recognize the importance of maintaining wildlife habitat and prescribe manipulation of grazing timing and intensity as the main tool for habitat improvement. Terms specified within AMPs, however, are usually not enforceable. Stocking rates for cattle have remained relatively constant on most units over the past twenty years, and most grassland managers report that goals on grazing allotments are being met. The vegetative response to range management measures is regularly evaluated and stocking rates adjusted accordingly, but the response of the wildlife assemblage of the area to these same management activities is not regularly evaluated.

Information on the level of grazing by wild herbivores, such as deer, elk, antelope, and prairie dogs is available for only a few grasslands. Some units set acreage limits for prairie dog towns and prevent town expansion beyond these limits. Most units report that forage consumption by wild herbivores is small compared to that of cattle.

Grazing associations are active on about three-quarters of the national grassland units. Units without grazing associations are located in Texas, New Mexico, and Oklahoma. Some grasslands issue individual or "direct" permits in addition to the general permits issued to grazing associations. Grazing associations exert a powerful influence on the management of national grasslands; Peek and Risser reported in the late 1970s that in the absence of clear management policies from the Forest Service, grazing associations not only administer the livestock operation but often assume the role of developing policies. The relationship between grazing associations, grassland staff, and others interested in the national grasslands is fairly congenial on some units and adversarial on others. This was reflected in comments reported in the 1995 Audubon study about the receptivity of public land ranchers to educational presentations related to grassland birds. Some respondents indicated that grassland ranchers perceived conservation interests as a threat to the ranching livelihood and would not be interested, while one unit reported that the grazing association with whom they contracted was "one of the most open-minded and progressive grazing associations around"— and would likely welcome such presentations (Senner and Ladd 1995).

Grazing on the national grasslands has numerous implications for grassland birds. Grazing is a special concern on the national grasslands because of the multiple-use objectives of these public lands, the historical precedent of using grazing manipulations as the primary wildlife habitat management tool, and the potential significance of the national grasslands in the conservation of the biotic integrity of the regions in which they are located. In the absence of solid data on the response of wildlife to range management activities, however, knowledge of the impacts of grazing on grassland birds will be limited to site-specific observations.

Oil and Gas Development and Recreational Use

Values indicating the level of oil and gas development and recreational use on the national grasslands were obtained from most grasslands in the 1995 Audubon study. Proximity and access to large urban centers is a good indicator of the level of recreational use a grassland will receive. The Caddo and Lyndon B. Johnson units in Texas illustrate this. Together these units make up only about 4 percent of the area of Little Missouri National Grassland in North Dakota but realize about twice the dispersed recreational use of Little Missouri. Caddo and Lyndon B. Johnson are located close to Dallas–Fort Worth, while Little Missouri is located in sparsely populated southwestern North Dakota. Hunting and fishing account for a large share of the recreational use of the grasslands. Some recreational uses (e.g., off-road vehicle use) on the national grasslands may pose direct threats to wildlife such as nesting birds, but such uses are typically controlled and monitored on grassland units (Senner and Ladd 1995).

Oil and gas development, which includes the physical plant of wells, pipelines, and access roads, is most intense on the Little Missouri, Thunder Basin, Caddo, and Lyndon B. Johnson National Grasslands, but absent entirely from about half of the units. The construction of oil and gas wells, access roads, and pipelines causes temporary disturbance to the landscape and has largely undocumented impacts on local bird populations. Grasslands having oil or gas development usually have a policy prescribing the minimum distance that a construction must be from the nesting or display areas of birds.

Fragmentation and the Demonstration Function

Land-ownership fragmentation is much more severe on some national grasslands than on others. In the 1995 Audubon study, two measures of fragmentation were used to compare grassland units: the ratio of linear boundary miles of federal parcels to the total area of those parcels on each unit, and the number of federal parcels of land less than 2.6 km² contained within the administrative boundaries of each grassland. Two grasslands, McClellan Creek in Texas and Butte Valley in northern California, are each basically one intact block. In contrast, Thunder Basin National Grassland in Wyoming, one of the largest units, contains 338 land fragments of less than 2.6 km², and Cedar River National Grassland in North Dakota, one of the smallest units, has the highest km-of-boundary-to-total-area ratio. In landscapes of highly fragmented land ownership and land-use, conserving native biodiversity is a tremendous undertaking (Senner and Ladd 1995).

Fragmentation of ownership affects the demonstration function (i.e., the promotion of grassland agriculture and sustained yield management of grassland resources) of the national grasslands. Units that are highly fragmented have difficulty demonstrating on all federal holdings, since small holdings may be isolated, lacking road access, or embedded within private lands. Units that are relatively intact can more easily coordinate demonstrations, but the effectiveness of these demonstrations may still be low because private landowners have become used to having public lands available for grazing and thus use their own lands for intensive forage or food crop agriculture.

National grassland demonstrations may be passive or active, and the 1995 Audubon review attempted to evaluate the effectiveness of both. Passive demonstrations have their effect simply by their mere presence: the adjacent landowner sees what is being done on the federal land and imitates it. Active demonstrations involve an observable demonstration on federal land, as well as education and extension activities intended to raise awareness and facilitate adoption of the demonstrated management practices by private landowners. A land-use survey conducted as part of the Audubon study concluded that passive demonstrations were having only limited impact, and responses of district rangers responsible for national grasslands indicated that active demonstrations were also having only limited success in influencing adjacent private landowners. The demonstration function, long held as a distinctive and central element of the national grasslands mission, must be critically examined and criteria developed for the evaluation of its effectiveness if it is to contribute to the conservation of grassland bird habitat of sufficient quality and quantity (Senner and Ladd 1995).

One way to facilitate management for all uses on the national grasslands is by consolidating federal holdings to reduce the number of isolated or inaccessible parcels of land. Such action would be consistent with the conservation of beta diversity both within national grassland units and between national grassland units and neighboring areas. Beta diversity refers to the change in biological diversity that occurs across landscape gradients or between disjunct landscape parcels (Knopf and Samson, in press). A more fragmented landscape may mean greater numbers of species (i.e., greater alpha diversity), but a less fragmented landscape would facilitate management for conserving the native flora and fauna of the area and ensuring their ecological distinctness from the biota of neighboring areas.

Most units remain open to land adjustment opportunities when they arise. Some adjustments have been made since the national grasslands became a public entity in 1960, but the patchwork situation on most units continues. This condition is especially relevant for grassland birds that are area-sensitive and that require a certain minimum area of quality habitat before using that habitat. Species such as the Savannah sparrow and Henslow's sparrow are more likely to occur in habitat patches of at least 40 ha and 55 ha, respectively (Herkert 1994a). The areas required for maintaining genetically diverse populations of bird species that can survive environmental events such as drought, fire, and invasions of exotic species over several decades or centuries may be several times larger.

Opportunities for Volunteers

Information on volunteer opportunities on the national grasslands was sought in the 1995 Audubon review. Needs were indicated in several areas, but most consistently in the area of wildlife inventory and monitoring. Needs for volunteer assistance with habitat restoration projects and public education were mentioned less frequently; they included presentations at nature centers and engagements with local schools, community groups, and grazing associations. A grassland employee at Buffalo Gap National Grassland mentioned the need for volunteers to advocate grassland values that extend beyond the local economy, while the Sheyenne unit indicated a need for volunteer assistance with determining specific prairie management goals in conjunction with the Forest Service and the grazing associations. The most salient point about roles for volunteers on the national grasslands was the Forest Service's recognition that assistance is needed for the development and maintenance of a relevant wildlife database. The implication for grassland bird conservation, and the conservation of native prairie ecosystems more broadly, is simply that contributions from volunteers will be needed if public land management is going to be sensitive and responsive to ecological realities. Maintenance of rangeland productivity could not be done by the Forest Service without the work of the grazing associations, and this need for the assistance of grassland users holds for other grassland values as well (Senner and Ladd 1995).

Ecosystem Management and Recommendations

The national grasslands provide a testing ground for management from an ecosystem perspective. Grassland birds will be an important focus in an ecosystem approach because they are (1) sensitive indicators of environmental quality, (2) able to disperse widely (and thereby able to show how regional losses in biological diversity may originate with localized decisions), and (3) relatively easily monitored and identified (Knopf and Samson, in press). Without recognition of the regional context of each unit—as well as the role of intermingled, nonfederal lands in conserving habitat of sufficient size, quality, and diversity to protect native wildlife populations—single-species management or management directed to single uses of the grasslands will continue. Conservation of grassland bird species will provide effective protection for other birds and wildlife associated with the Great Plains ecoregion.

To increase capability for ecosystem management on the national grasslands, we recommend the following:

1. Develop a list of volunteer opportunities (especially in the areas of wildlife inventory and monitoring, habitat restoration, and prairie conservation education and planning) for each national grassland and distribute this to residents of local communities, to schools, colleges, and universities, to local conservation and recreation interests, and to national organizations with interests in the national grasslands.

2. Develop a national framework for ecosystem conservation and management for each national grassland. Historical perspectives on important ecological processes for each unit will be needed for the development of such a plan.

3. Clarify and raise the profile of the national grasslands within the Forest Service, specifically, and the federal government, generally, to encourage greater def-

inition of purpose and allocation of resources commensurate with the ecological importance of these lands. Federal funding of national grasslands wildlife conservation and public education, and expansion of opportunities for public involvement in the management of the national grasslands will be elements of this process.

4. Develop or activate land consolidation plans that have as an objective the blocking up of federal landholdings in order to simplify management, improve access, reduce costs, and protect blocks of native grassland that are important to area-sensitive wildlife.

5. Identify lands enrolled in the Conservation Reserve Program (CRP) that are currently intermingled with federal lands and explore post-CRP contract land-use options with enrollees as part of a lands consolidation plan. The Conservation Reserve Program, a voluntary, publicly funded cropland set-aside program that has helped restore wildlife habitat, was first authorized under the Food Security Act of 1985.

6. Review the basis for the demonstration function as part of the national grasslands mission. What has been learned on the national grasslands about efforts to influence private landowners through passive or active demonstration?

7. Coordinate research on each national grassland unit to answer the following ecological and policy questions related to the needs of grassland birds: (a) What is the breeding status and estimated population size of grassland bird species on the unit? (b) How do the habitats present on the unit compare with those required and preferred by grassland birds? (c) Is there justification for continuing and strengthening the demonstration function of the national grasslands? (d) On which units is the adversarial nature of the relationship between ranchers and conservation interests (and perhaps the Forest Service) greatest, and how can this be addressed?

8. Standardize and maintain checklists for birds found on the national grasslands.

9. Develop (on those units currently without a plan) comprehensive wildlife management plans at the grassland unit level that aim to protect the native flora and fauna of the region. Plans may involve two or more units in conjunction when these units are in close proximity and are ecologically similar.

10. Promote awareness among ranchers on the national grasslands of the objectives and interests of wildlife conservation advocates—and vice versa. Work to remove stereotypes and false dichotomy of beliefs where these exist, and clarify real differences where these exist.

11. Examine the assumptions underlying current Forest Service wildlife management approaches, especially single-species and ecosystem management. This

would include critical review of the feasibility, desirability, goals, and philosophical bases of these approaches.

12. Determine the current state of knowledge about wildlife habitat and wildlife population trends on national grassland units and work toward completion of a comprehensive floral and faunal inventory that can be kept current through ongoing monitoring.

13. Identify areas of special ecological significance for grassland birds on the national grasslands and manage for the protection of habitat in these areas.

14. Assign a person representing wildlife interests to each grazing association with which the national grasslands are in contact or, preferably, establish a representative advisory board for each unit that would help evaluate programs and determine management direction.

15. Evaluate the response of a representative sample of the entire wildlife assemblage to any range and wildlife habitat improvements and changes in management and grazing practices.

16. To increase accountability, keep national grassland budgets separate from those of national forest units.

CHAPTER 19

The Sandhill Management Plan: A Partnership Initiative

Gene D. Mack

The Sandhill Management Plan is a resource management approach authored by the sandhills' people and the U.S. Fish and Wildlife Service (FWS). It reflects the belief that an ecosystem approach, including the people, is necessary to sustain a healthy environment.

For years, resource management aligned a specific sandhill resource with its respective agency. Wildlife agencies managed for wildlife (primarily game species); Soil Conservation Service controlled erosion; water resource agencies focused on irrigation; and agricultural groups worked to improve production. Each group concentrated on its mission from its perspective, with little outside interaction. As agencies enacted land-use regulations, landowners perceived them as a loss of property rights. Opposition to such regulations reinforced the belief that the regulations were necessary. Ongoing contention between conservation and agricultural groups has caused opportunities and solutions to be missed, or to be judged on a win or lose basis.

Past actions by the FWS in the sandhills illustrate this relationship. In the early 1980s the FWS drafted a wetland acquisition plan to protect the region's wetlands from land development. Local residents strongly opposed any type of government acquisition or involvement with sandhill wetlands. They claimed the program addressed wildlife needs but ignored the needs of the people. The acquisition plan was abandoned and no alternatives were considered.

In the sandhills, landowners control more than 90 percent of the land (Bose 1977). Local involvement is needed to provide guidance on what management can be done with the support of the people. In the 1990s the FWS began an ecosystem approach aimed at reducing contention between environmental and agricultural groups and combining local and professional knowledge of the resources to develop a workable plan.

Landscape Description

The sandhills is 49,970 km² of grass-covered sand dunes stretching across north-central Nebraska and southwestern South Dakota. The region is a continuous expanse of grasslands and wetlands largely on privately owned ranches. Approximately 526,000 ha of wetlands are scattered throughout (Rundquist et al. 1981). Ecological connections between dunes, hydrology, plants, and people have created diverse habitats. Dunes vary from high, steep hills in the western region to small mounds in the east. Wetlands exist in the interdunal valleys and include hyperalkaline lakes, freshwater lakes and marshes, wet meadows, and fens. Plant communities range from extensive short- and tallgrass prairies to isolated deciduous and coniferous forests. Plants associated with arid conditions inhabit the tops of dunes, while lush stands of aquatic plants are found in the valleys a few hundred yards away. The species of plants present are dependent on the ranching style of individual ranchers.

Ranching dominates the economy. Population is sparse, with fewer than 0.6 people per square kilometer (U.S. Bureau of the Census 1993). Large ranches, 1,600 to 2,400 ha, are needed to sustain a ranch family. The grasses on the dunes are used for summer grazing, and the meadows are mowed for winter hay. The amount of winter forage can be the limiting factor in the success of a ranch. Thus, ownership of meadows is critical to the value of a ranch. Overall, ranching has benefited the grasslands. In the semiarid climate, proper grazing has aided decomposition of organic matter, improved compaction in the soft sands, and stimulated plant growth and reproduction.

The hydrology associated with sand dunes affects the landscape and economy of the region. Porous sand quickly captures the 43.2 to 58.4 cm of annual precipitation, allowing little runoff. High infiltration rates, as much as 3 m per day (Bleed 1990), limit plant use of precipitation before it reaches the water table. From 25 to 50 percent of the water becomes part of the Ogallala Aquifer (Lawton 1984). The thickest portion of the Ogallala Aquifer exists under the sandhills and contains nearly one billion acre-feet of water (Dreeszen 1984). During seasonal wet periods, groundwater mounds form under the dunes. The hydraulic head of the mounds slowly releases excess water to the neighboring wetlands and streams (Winter 1986). In turn, lakes and wetlands restrict the release of groundwater. The local water table is maintained at a higher level. Lands with the water table about two feet below the surface produce lush stands of native grasses. As the vertical distance to the water table increases or decreases, vegetation shifts toward more arid or aquatic plants, respectively. Groundwater released to streams provides 90 percent of the water flowing from the sandhills (Bleed 1990). Average annual flow is 2.96 billion m³ (Dreeszen 1984).

Vegetation on the dunes has increased or decreased depending on such factors as climate, overgrazing, and wildfires throughout the sandhills' existence. Since settlement, control of wildfires and managed grazing has increased vegetation on the dunes. About 720 plant species exist (Kaul 1990), including two endangered species. Hayden's penstemon occurs in bare sand on the dunes and prairie fringed orchid inhabits the wet meadows. Resident and migratory wildlife are abundant in the grassland-wetland ecosystem. Migrating along the Central Flyway, thousands of birds use the area for breeding and resting. About 224 species of birds can be found (Labedz 1990).

Human settlement has altered the ecology of the sandhills. In the late 1800s open range grazing abused the federally owned land. Investors overstocked cattle until two harsh winters caused huge livestock losses and forced the investors out of business. By the early 1900s much of the land was in private ownership and grazing was controlled. Eager to develop the land, landowners drained large lakes and wetlands to increase forage production. The alignment of valleys between the linear dunes made it possible to extend miles of ditches from one wetland to the next. The total number of lakes and wetlands lost or altered is unknown. By the 1950s much of the drainage was completed, but the impact on the landscape and hydrology continues today. The drainage lowered the hydraulic head of the basin and increased groundwater discharge (Winter 1988). Increased streamflows eroded the sand bottom of channels, further lowering the water table and drying adjacent lands. Downstream, aggradation and groundwater recharge created new wetlands. These were often dredged to reclaim flooded meadows.

Center-pivot irrigation began booming in the eastern region in the 1970s. Tax laws, irrigation technology, low land values, and high grain prices encouraged investors to convert grassland to cropland. Irrigated cropland increased from 28,550 ha to 87,010 ha in ten years. Nebraska Natural Resource Commission (1993) documented the effect of the additional irrigation on the sandhill ecology. In some areas, pumping caused local water tables to drop. Neighboring wet meadows became drier, and forage production declined. In other areas irrigation water raised local water tables, creating marshes and lakes.

Often the newly formed wetlands were drained, only to create problems downstream. Fertilizer and pesticides leached into the groundwater and contaminated nearby domestic wells. The commission's report concluded that nitrate leaching could not be prevented even with the use of the best fertilizer management practices. Wind erosion was ten times greater on cropland than on grassland. The blowing sand damaged young corn and deposited dunes on neighboring pastures. By 1990 more than 20,200 ha were no longer profitable to farm. Cost of reseeding often exceeded the productive value of the land. The Conservation Reserve Program replanted the areas, but plant communities remain in an early successional stage and provide little forage value.

Management Approach

In 1991 the FWS began an ecosystem approach in the sandhills. It focused on obtaining a better understanding of the ecosystem and developing a program that would benefit people and wildlife. Understanding the people and their need for the resources was very important. FWS personnel made visits to ranchers and organizations to share perspectives and concerns. The interaction helped ranchers and FWS build trust and discover their common ground. Both groups indicated a need to maintain a grassland-wetland ecosystem and improve wildlife numbers; both recognized the role ranching has played in maintaining the diversity and abundance of flora and fauna; and both were concerned about the level and quality of the groundwater.

The Nebraska Cattlemen (NC), a private organization, joined the FWS to design an interacting group of ranchers and government personnel. To ensure broad support, organizations active in the sandhills were asked to recommend leaders from their membership. The group, named the Sandhills Task Force, consisted of thirteen members, eight involved in ranching and five associated with government and private organizations. Members were recommended by NC, Natural Resource Districts (NRD), Nebraska Association of County Officials, North Central Resource Conservation and Development, Nebraska Game and Parks Commission (NGPC), FWS, and Preserve Our Water Resources. Membership was weighted toward sandhill ranchers to ensure their voice was heard.

One year after the task force was formed, the group had drafted the Sandhill Management Plan. Its stated goal is to enhance the sandhill wetland-grassland ecosystem in a way that sustains private ranching, wildlife and vegetative diversity, and associated water supplies. This is to be accomplished by identifying workable management strategies and by building partnerships between landowners, government, and public interests.

The task force recognized that a successful program needs to be sensitive to private property rights and be flexible so solutions can be molded to solve specific problems. Five broad management strategies were identified: education and technical assistance, acquisition, lease agreements, legislation, and financial support. The strategies are not all equal in value or need but provide a full complement of management approaches.

Education and Technical Assistance

Education has been identified as the most positive and effective way to benefit people and land resources. It encourages local pride and active participation in resource management. Educational strategies will be directed toward providing a better understanding of the interrelationship between grasslands, hydrology, livestock, and wildlife in a healthy ecosystem. The task force recognized wetlands as one area where environmental concerns and property rights often cause conflicts. Wetland education would target wetland identification, existing regulations, and wetland benefits to people, ranching, and wildlife. School and public programs will encourage interaction with ranching and wildlife through educational films, literature, seminars, training courses, and field trips.

The task force concluded that the best approaches are one-on-one and small group meetings because information flows in both directions. For example, while a rancher learns about wetland values to groundwater and wildlife, the conservation personnel learns the importance of meadows to a ranching operation.

Technical assistance is an extension of education. Its focus is on solving specific land management problems. This approach would join landowners and resource experts in an effort to design projects that meet the needs of the landowner and the resources. Technical assistance would inform landowners of the necessary steps (legal and planning) and the cost-benefit of projects. Over a broader area, technical assistance will develop a directory of expertise and programs associated with resource agencies. Landowners are often discouraged from doing conservation projects because of not knowing whom to contact or what permits or funding exists. All technical assistance would be voluntary and directed toward building a cooperative attitude between landowners and agencies.

Acquisition

Acquisition was identified as two types: conservation easement and fee title. Its role would be to purchase minimum interest in lands necessary to preserve a specific resource. All acquisition would be voluntary and based on sound biological and ecological criteria. Criteria identified include ecological significance of the site, its value to wildlife, and the threat of change in land-use. Easements would monetarily compensate the landowner for specific rights. One application would be to compensate landowners for abandoning maintained drainage ditches and allowing wetlands to revert to their natural state. Fee acquisition would be a last alternative to ensure that unique ecosystems will remain. Acquisition could be the practical solution to restore a drained fen because the spongy ground and aquatic habitat provide little value to ranching operations.

Lease Agreements

Lease agreements were noted as a short-term compensation to landowners for land-use changes. One example may include leasing meadows to shift plant composition toward warm-season grasses to improve nesting habitat. Such a project would mean a loss of hay production for several years. A lease agreement would provide monetary compensation for the loss. After a period of years, the money and effort would be directed to another site.

Legislation

Legislation was recognized as one strategy that departs from the partnership approach, but it may be needed to protect broad resources. Increasing demand for water in agricultural and populated areas may require legislation to ensure the ecological integrity of the sandhills. Beneficial legislation would consider the impact on local people. It would not undermine property rights and would be flexible enough to benefit unique situations.

Financial Support

Financial support is a strategy to build a coalition of people and agencies to enable the other listed strategies to work effectively. It would help match conservation needs with the available people and financial resources. An example would be combining private conservation funds, agency expertise, and landowner property to restore riparian habitat.

Implementation

The plan recognized a need for a full-time coordinator. The individual would oversee all aspects of an established program, serve as a liaison between landowners and conservation groups, form partnerships in education and technical assistance, and obtain multiple sources of funding. Staff would begin with one individual and increase with the success of the program. A complete staff would include a coordinator, two extension biologists, and one clerk. Funding for staff and projects would depend on partnerships, cost-share programs, and challenge grants.

After the plan was written, draft copies were distributed to hundreds of individuals, mostly ranchers, and organizations for their review. Public meetings were organized and sponsored by local task force members. An FWS representative was invited as their guest. Most of those present were ranchers concerned with government intervention in their lives. During the meetings, the atmosphere shifted from apprehension to vocal support. Questionnaires distributed during the meetings showed nearly 90 percent of the people supported the plan. Other support has included the governor, U.S. senators and representatives, NC, and Natural Resource Districts.

On 21 September 1993, a formal signing ceremony was held on a sandhill ranch. Those present included area ranchers, task force members, regional office personnel of the FWS, congressional and gubernatorial staff, local news media, and reporters for National Public Television and the *Minneapolis Star Tribune*. The group also toured several wildlife projects accomplished through partnerships between ranchers and FWS. The day ended with a steak barbecue. The ceremony reinforced the idea that "win-win" solutions can be found.

Since the signing ceremony, a coordinator has been hired by the FWS and twenty-four partnership projects have been completed. Partners sponsored land management courses for landowners, distribution of resource management information, and enhancement of riparian and wetland habitats. Each project, with its unique problem, combined expertise and funds from various sources. Partners have included landowners, Nebraska Board of Educational Lands and Trusts, Leafy Spurge Task Force, Nebraska Branch of Holistic Resource Management, NC, NRD, NGPC, National Resources Conservation Service, FWS, and county government.

Habitat projects have focused on finding solutions that benefit the landowner and the resources. One project, for example, brought together the landowner, NGPC, and FWS to solve a stream degradation problem. The landowner had lost vehicle access to one side of the stream. He could no longer hay the area and was forced to use the land for pasture. Stream erosion had also drained critical riparian wetlands used by threatened fish species. The partnership arrangement provided the resources to fence out the stream, reestablish willows, construct small control structures, and restore riparian wetlands. The landowner was pleased with his new crossing and the improvement of half a mile of stream.

Some projects have created nontraditional partnerships. One example is a two-day cattlemen-environmental workshop cosponsored by NC and FWS. The workshop focused on viewing traditional management practices in new ways and bringing together cattlemen and conservation personnel to discuss issues and concerns related to property rights and regulations. NC used its communication network to reach the public, and the FWS provided funding and assistance in developing the workshop. The cooperative effort brought together both groups in a nonthreatening, constructive environment.

The Sandhill Management Plan is young but growing. Its progress proves that a grassroots approach can break down barriers and provide solutions. Local involvement has given the people assurance that they have some control in affecting their land and well-being. The two dozen completed projects serve as demonstration areas and proof that win-win solutions can be accomplished between landowners and resource agencies.

CHAPTER 20

Management Challenges for Prairie Grasslands in the Twenty-first Century

Richard K. Baydack
James H. Patterson
Clayton D. Rubec
Allen J. Tyrchniewicz
Ted W. Weins

Landscapes in the prairies have been simplified by agricultural trends toward monoculture, with subsequent negative effects on biodiversity. Although it is tempting to portray agriculture as the villain in this game of grassland conservation, working with farmers and the rural community is vital to the success of conservation efforts.

Traditionally, the problems of grassland conservation and the decline of agriculture have been dealt with separately. Conservation experts have developed strategies for conserving grasslands, while agriculture policies were modified to improve agriculture. In addition, rural development strategies were developed to attempt to arrest the decline of communities. These sectoral approaches fail to recognize that declines of landscapes, economies, and social structures are interrelated, as are policy solutions.

The presence of farming per se has not caused this environmental and economic dislocation. The problems have been brought about by the substantial expansion of cultivated acreage beyond the sustainable land base onto marginal agricultural lands and wetlands. This expansion was not fueled by market forces. In the 1980s and early 1990s, cropping practices responded to agricultural policy and support programs based on the area of land under commodity production. The negative impacts of these policies and programs have resulted from market-distorting and price-depressing international agricultural trade disputes.

This chapter offers recommendations designed to ensure the future sustainability of the prairies in an agriculturally dominated environment. The strategies

Prairie Conservation
Island Press (Washington, DC • Covelo, CA)

link ecology, economics, and social factors in achieving sustainable development principles.

Policy Influences

Canada

Canadians are the custodians of a substantial proportion of the earth's northern latitude ecosystems. Since settlement, the diversity and richness of these ecosystems have collectively been tied to the prosperity and well-being of its peoples. Many of Canada's natural landscapes have been altered, resulting in the emergence of a growing number of modified ecosystems, with natural ecosystems often isolated and uncommon. Debate in Canada is currently focused on the establishment of a national, systematic plan for the protection of representative examples and the biological diversity of these landscapes, including the prairies.

Policy development on the Canadian prairies has occurred for the objective of primary production. The settlement processes in Canada and the United States were very similar, relying on the free homestead system, preemptions, school and railway land grants. In Canada, some policies used to develop the Canadian prairies are still in existence today in some form.

Just recently, the Crow's Nest Pass Agreement of 1897 was rescinded. It ensured rail rates on grain shipments out and agricultural inputs onto the prairies. However, it sent the wrong message to producers. Many farmers on the Canadian prairies were expanding production into marginal cropping land to export raw material out of Canada in response to the revised form of the agreement known as the Western Grain Transportation Act. Developed to meet a short-term need, it represents how policies must change with economic, social, and ecological conditions on the prairies. With the elimination of the Crow's Nest Pass Agreement, projections for Manitoba indicate that about 9 percent of the marginal land will be taken out of production.

Since producers react to government agriculture policies to ensure good returns from production, it is on the prairies that the influences of policy can be best examined. In response to strong commodity prices and increasing export markets, cultivated acreage expanded on the Canadian prairies in the 1970s and 1980s (Statistics Canada 1992). The prolonged drought of the 1980s revealed agriculture production was not sustainable on much of the marginal land. Government support programs did not provide adequate risk protection, so new safety net programs were designed, again based on acreage, yield, and commodity prices. International trade disputes have resulted in depressed market prices and production is being driven more by government programs than by market demands.

Agriculture policies in Canada reward expansion of cropland and make land-use changes difficult (Girt 1990). Canadian Wheat Board policies, for example, base incentive payments on area cultivated. Increased cultivation is often at the expense of grasslands considered marginal for agriculture production. Similarly, present grain quota systems in Canada base the amount of grain that a farmer can sell on the number of acres cultivated.

In 1991–1992, just over $8 billion was spent to support agriculture from federal and provincial sources (Statistics Canada 1992). About $3 billion to $5 billion of these expenditures were direct export and production subsidies including direct commodity payments, the Western Grain Transportation Act "Crow Benefit," crop insurance, and financial assistance programs. Canadian grain and oilseed products receive annual subsidies on the order of forty-five dollars per acre. These subsidies have kept agriculture alive, and any producer surplus in the past few years has come from public programs (Lerohl 1990).

In spite of this, farmers have demonstrated a willingness to set aside marginal lands for permanent cover for fifteen to twenty dollars per acre. Projects such as the Permanent Cover Program of Agriculture Canada and Prairie Care by Ducks Unlimited Canada under the North American Waterfowl Management Plan (NAWMP) are success stories in this context. Using the Canadian prairies as an example, a set-aside program targeted at the estimated 8 to 10 million acres of marginal and fragile land in grain production, could reduce costs to the public treasury on the order of $240 million per year. Rural development income options for alternative use of set-aside lands could add to the performance of the rural economy. Payments to landowners that are consistent with international trade arrangements could help sustain farm families during the period of adjustment to alternative land-use options that build new economic activity into the rural community.

The adoption of conservation farming practices through the Prairie Care Program resulted in an average net increase in profits of over $5.25 per ha per year (Josephson 1992). This does not include any incentive or government support payments. Effective prairie conservation programs in Canada's agricultural working landscape, as exemplified by the NAWMP, can only come about through revitalized rural communities. The creation of a sustainable development market force would ensure environmental, economic, and social sustainability.

United States

Agricultural policies in the United States are similarly focused on maintaining agriculture, often at the expense of the long-term productivity of farmland

(Barnes 1993). In particular, Article 9 of the Uniform Commercial Code (UCC) establishes the law for secured financing of personal property in all fifty states. Article 9 provisions emphasize productivity that harms farmland and, ultimately, the entire environment.

The origin of the UCC's provisions in the mid-twentieth century coincided with the return to a normal, postwar economy. Article 9 was meant to reconcile notions of national success, economic opportunity, and economic expansion, using a model from manufacturing industries. Agriculture does not fit well into the UCC paradigm because the basic ingredients of the agricultural process, nurturing and land stewardship, do not conform to the mold of personal property (Barnes 1993). What has resulted is a short-term approach to farming, since the long-term costs such as overproduction or poor management are not factored into UCC provisions. What is needed is a system whereby land management practices can be considered in the determination of whether to extend credit so that "good farming" could be rewarded in every phase of a farmer's financial plan (Nickles 1987).

The United States is in the process of drafting the 1995 Farm Bill, which will have a major effect on agriculture in the United States. As in Canada, many external factors, such as budget, trade agreements, and the environmental movement, are now having a pronounced effect on the development of agriculture policy. There is potential for the 1995 Farm Bill to address issues such as conservation of wetlands, protection of environmentally sensitive areas, watershed management, and animal waste management. Current programs such as the Conservation Reserve Program (CRP) and the Export Enhancement Program (EEP), could also feel the modification of the 1995 Farm Bill.

The CRP, a program designed to take crop land out of production, is due for review in 1995. The program has a cost of $1.8 billion set up as a special line in the budget and not included as part of the agriculture budget. "There is strong environmental interest group support for including the budget line for the conservation reserve in the appropriation for agriculture, if the special line is not continued. This will result in a cost of about $1.8 billion that will have to be absorbed within at best a flat agricultural budget" (Johnson 1995). As well as the budget issue, a significant number of CRP contracts expired in 1995. This program will have significant effects on agriculture in the United States, as well as Canada.

The CRP has been identified as one reason that Canada will increase its exports of grain to the United States. With the United States reducing production by removing land from crop production, domestic prices increase, making a market for Canadian grain. The development of this nearby market increases the demand for Canadian grain.

Designed to facilitate the marketing of grain to U.S. export markets, the EEP has also been identified as an important influence on demand from the United

States for Canadian grain. The program provides a subsidy to importing countries, thereby providing grain farmers with higher domestic prices than that of the world markets. Program funds are used to ensure competition with major grain exporters, and recent research results on the EEP have indicated a sharp reduction and reorientation toward more targeted market development (Johnson 1995). Inefficient transfer of funds to farmers and little market development will have an important impact on U.S. agriculture.

The effect of the EEP was an increase in the domestic price in the U.S. market because most farmers were exporting their grain using EEP. Like the CRP, EEP contributed to increased demand for Canadian grain because of the short supply available domestically.

North American Free Trade Agreement (NAFTA)

NAFTA charts a new course for economic cooperation throughout North America. It also establishes rules and procedures for resolving trade disputes between countries. The North American Commission on Environmental Cooperation (NACEC) offers the opportunity for continental environmental cooperation on an equal scale.

By taking a cooperative and facilitative approach, NACEC could play a significant role in the conservation and protection of North American ecosystems, habitats, and species. The NAWMP is a model of how continental ecosystem conservation can succeed. In the same sense that NAFTA has become a template of continental economic cooperation, the NAWMP has become a template for continental environmental cooperation. In turn, the creation of the NACEC under the North American Agreement on Environmental Cooperation is an opportunity to link these two parallel areas for cooperation. This commission could help promote continental environment and the economy linkages under the aegis of sustainable development and address the imperative for biodiversity conservation on a continental basis (Rubec et al. 1995).

Within an individual country, regulation and protection are important tools for environmental security. However, between countries, regulation and protection are much more restricted in terms of application and more difficult to codify. Transboundary pollution issues in North America, such as acid rain impacts or water quality degradation, have usually required protracted negotiations often resulting in litigation. On the other hand, transboundary conservation issues have traditionally involved cooperative agreements and nonlitigative solutions.

A reality in North America is that governments are hard-pressed to provide significant funding for major new programs. It is widely recognized that NAFTA is not designed as a mechanism for governments to spend money but rather to facilitate wealth generation in the private sector for the benefit of all North Americans. Similarly, an objective of the NACEC should be for governments to facilitate conservation on a continental scale, particularly through environment-economy linkages.

World Policy Influences

Agriculture in Canada and the United States is affected by events in other countries. Both countries are working toward the development of a policy framework for the transformation to environmentally sustainable agriculture. The agriculture and conservation sectors are in general agreement as to what needs to be done to restore the environmental sustainability of the agricultural landscape. However, the negative impact of international trade disputes has created an economic and political environment where changing agricultural policies and programs are perceived as a threat to the industry.

There is a ray of hope on the horizon. Ratification of the Uruguay Round of the General Agreement on Tariffs and Trade (GATT) and of NAFTA should lead to a phased liberalization in agricultural trade. In addition, the UN Conference on Environment and Development 1992 Global Convention on Biodiversity has raised awareness of biodiversity conservation efforts around the world.

Using trade liberalization to fuel sustainable agriculture and prairie conservation makes good environmental and economic sense. Organization for Economic Cooperation and Development countries are currently expending over $350 billion annually on agricultural subsidies. The negative environmental and economic impacts of these subsidies are not just felt domestically but are devastating to developing countries. The World Bank has recently estimated that a 50 percent reduction in trade barriers by Europe and the United States would raise the value of exports from developing countries by $50 billion a year, providing critical resources to address environmental and other problems.

The European Union, the United States, and everyone else in the game of exporting agricultural produce have been major forces through large export and domestic subsidies. The level of subsidies grossly distorts production of commodities, making it profitable to cultivate lands marginal for agriculture. Thus, farmers are making decisions to farm programs not the land. It is the land, water, and wildlife that suffer.

A number of details regarding effects of international trade policies on conservation of native prairies now have an empirical basis in fact that can be demonstrated to farmers, and many farmers have become evangelists for change. While farmers are not revolutionizing the world, they are trying to run it differently. Efficient and effective international trade regulations benefit everyone who lives on this landscape. Also, farmers recognize that if they don't become part of the solution, they are going to be identified as the problem, and the solutions are going to be imposed them. So there is a tremendous vested interest in the farm community, public and private, to get involved right at the beginning, to start rationalizing these various factors and taking into account the environment and sustainability, as well as their ability to make a living.

The GATT negotiations have probably set a world record for length. The greatest stumbling block was agriculture and how to wean the world from this gross distortion of commodity production markets, prices, and land-use decision making. The more money committed to the problem, the greater the problem became. There has been a successful conclusion to the Uruguay Round of GATT and the World Trade Organization is to come into being in 1995.

GATT calls for reductions of 36 percent in export subsidies from all participating members, with about 20 percent reduction in export volumes. In the long run, this will establish criteria and timetables for phasing down export subsidies, a major cause of unsustainable land-use. Similarly, there is a 21 percent reduction called for in domestic subsidies. When negotiations began, Canada was at a level of about $4 billion per year in agricultural subsidies. According to these criteria, $1 billion per year would not be eligible and could either go back to the treasury or be used for green, conservation-friendly programs.

Sustainable Agriculture

Some government policies and programs do not promote agriculture that is sustainable. These policies are inconsistent with the goal of protecting the productive capacity of the land for future generations. Farmers are facing many challenges when modifying current production practices. This modification becomes more feasible, however, when there are tangible economic, social, and ecological benefits. An understanding of the factors that affect economic viability, agricultural production practices, resource use, and ecological resilience is an essential prerequisite to the design of policies, budgets, and agreements for sustainable agriculture and rural development. This range of issues is not accounted for in any decision-making process, at any level.

Sustainable agriculture is a subset of sustainable development. Therefore it is appropriate to revisit the definition of sustainable development to help define the scope of concern. Sustainable development is "a process of change in which the exploitation of resources, the direction of investments, the orientation of technological development, and institutional change are all in harmony and endorse both the current and future potential to meet human needs and operations" (Brundtland 1987).

This definition provides an idea of the scope of sustainable development but does little to explain how sustainable development can be operationalized. The various contributors to development, such as agriculture, manufacturing, service industries, and even lifestyle, have different viewpoints on sustainable development. To narrow the broad definition of sustainable development in order to confine this project to a manageable task, a definition of sustainable agriculture is required. One best suited for the needs of prairie conservation is "that which, over the long term, enhances environmental quality and the resource base on which agriculture depends, provides for basic human food and fibre needs, is economically viable, and enhances the quality of life for farmers and society as a whole" (Vasadava 1991).

While this definition focuses on agriculture, it should be recognized that environmental quality and the resource base are being shared by other activities. A definition of sustainable agriculture does not provide an effective method to evaluate a program or a policy. What is required is some form of measurement that would provide an accurate estimator of the sustainability of policies or programs. A technique is required for policy makers that can outline where policies could be improved to reach the goal of sustainable development. Work at the International Institute for Sustainable Development led to the creation of Principles for Sustainable Agriculture (Tyrchniewicz and Wilson 1994). The principles include the following:

Stewardship. There exists both an individual and collective responsibility to sustain the environment for both our own and future generations. Economic and social activities should be undertaken in such a fashion as to maintain and preferably enhance the capacity of the resources used for the benefit of future generations as well as our own.

Conservation. There is need to maintain biological diversity while strengthening essential ecological processes. Nonrenewable resources must be used wisely, thus involving their recovery after use to the extent economically feasible. The major renewable resources in agriculture, soil and water, must be protected so that productivity is maintained.

Rehabilitation. Where renewable resources such as the soil have been damaged, effort must be expended in their rehabilitation so that to the extent feasible their original productivity is restored or preferably increased. It is recognized that lack of adequate care has contributed to soil and water degradation on the prairies.

Internalization of Costs. In our society certain production inputs and outputs are not priced in terms of their real value. Examples include the air we breathe and the carbon dioxide absorbed by plants. Furthermore, the by-products of production in terms of their environmental damage are not necessarily subject to a monetary penalty. What is required is that the real costs of both presently considered "free goods" or "undervalued goods" be incorporated into the total costs when determining the net returns from production. Such costing, for example, will include the value of any net loss or gain in soil nutrients as a result of crop production.

Scientific and Technological Innovation. Research to enhance the development of technologies that contribute to the maintenance of environmental quality and economic growth must be supported. Such support should extend to provision of educational services that will further the research program while at the same time maintaining social and cultural values.

Economic Viability. Production cannot be sustained over the long run unless it is economically viable. Such viability requires that the net returns from marketing are positive. Unless such returns are adequate within a region, the prairies for example, producers cannot be expected to continue to utilize their available resources for this purpose. The net returns from production should enable adequate standards of living to be maintained while at the same time being sufficient to continue to attract replacement operators.

Trade Policy. Barriers to trade can create impediments to the achievement of sustainability. Consequently, trade liberalization is an important component of progress toward sustainable development. In addition such liberalization leads to greater international efficiency in production. As a result, true comparative advantage should be an objective of trade policy. This implies recognition of the real costs of production and therefore the maintenance of environmental integrity.

Societal Consideration. Economic activity should minimize social costs while maximizing social benefits. At the same time it should not detract from human health and cultural resources or the quality of land and water. Cultural and social diversity should be respected. In agriculture a balance must be struck between the size of production units consistent with technology and a social structure acceptable to all stakeholders including those providing the infrastructure.

Global Responsibility. Ecological interdependence exists among nations as there is no boundary to our environment. Stakeholders in the maintenance of the environment are therefore not necessarily local. How the local environment is treated ultimately impacts other parts of the world and can be expected to haunt those guilty of its mistreatment. There is a responsibility on the part of all nations to "think globally when acting locally."

Future Policy Directions

Effective grassland conservation in the prairie landscape can only come about through revitalized rural communities supported by more diverse and stable income opportunities that are economically and environmentally sustainable. Implementation of both GATT and NAFTA may provide a unique opportunity to work toward rural renewal.

Three potential elements of trade-driven adjustment to sustainable agriculture follow:

Market Forces. As international agricultural production subsidies are reduced, it is anticipated that commodity prices will increase. This should encourage land-use decisions that are more responsive to market forces and to the sustainable capability of the land base.

Nondistorting Commercial Income Support. Agricultural safety net policies and programs can be modified to remove trade and land-use distortions and comply with international trade agreements. Decoupling support from commodity production to broader farm income should encourage a shift from gross production to sustainable productivity.

Conservation Incentives. A portion of the $1 billion trade war peace dividend could be used as financial incentives, specifically for conservation measures, to help rural areas and the agricultural industry adjust to environmentally sustainable and economically sound practices.

Recommendations

Special efforts must be taken in coming years to ensure that perceptions of landowners on the prairies are appropriate with respect to grassland conservation. In many cases, more grasslands will mean more wildlife. In many cases again, this will be viewed as a problem that can only be solved by eliminating the cause—that is, the grassland itself. We urge government agencies to respect these views by ensuring that adequate compensation, and possibly education, are provided. Additional consideration should be given to novel approaches such as ecotourism, rural bed-and-breakfast operations, or nature viewing.

The potential for future conservation of grasslands has been enhanced by the creation of the NACEC. The following recommendations as developed by Rubec et al. (1995) should be adhered to by this agency in the future.

1. Sponsor continental policy to identify and promote adoption of aspects of trade liberalization that have positive impacts on continental conservation.

2. Recognize the NAWMP as a priority continental program contributing to prairie conservation. The NACEC should also facilitate support for other programs that reinforce and augment the achievements of the NAWMP.

3. Support the establishment of innovative funding initiatives that harness market forces and new partnerships in the business sector resulting from NAFTA to foster support for continental biodiversity conservation programs.

4. Serve as, or foster the implementation of, a Continental Round Table on Biodiversity Conservation and Sustainable Development, which would lead to spin-off benefits for grassland conservation.

5. Provide leadership for the establishment and enhancement of a continental approach to prairie conservation and models for biodiversity risk assessment, protected areas and ecosystem science, and biodiversity information integration.

Summary

A New Framework for Prairie Conservation

Paul G. Risser

Prairies, the largest vegetation type in North America, originally covered 300 million ha in the United States, 50 million in Canada, and 20 million in Mexico (Rzedowski 1978; Samson and Knopf 1994). Since European settlement, these grasslands have been reduced dramatically but still occupy more than 130 million ha in North America. This vast grassland is classified into six different types: the tallgrass, mixed-grass, and shortgrass prairies of the central plains; the desert grasslands of the southwestern United States and central Mexico; the California grasslands; and the Palouse prairie in the intermountain region of the northwestern United States and British Columbia, Canada (Risser et al. 1981). Soils of the prairies are variable, usually Mollisols or Aridisols (Boul et al. 1980), with glacial influences predominating north of the Missouri River. The discussion in this summary chapter concerns the tallgrass, mixed-grass, and shortgrass prairies where the continental climate of each has an average annual rainfall of about 100 cm, 50 cm, and 30 cm, respectively. Potential evaporation exceeds precipitation throughout these grasslands except in the eastern tallgrass prairie (Sims et al. 1978).

Prairies have occurred only in the last several million years and, thus, in geological time are relatively recent. As a partial consequence, the prairies contain relatively few unique taxa of plants and animals (Wells 1970; Axelrod 1985). For example, only 11.6 percent of the mammals and 5.3 percent of the birds are endemic (Wells 1970; Axelrod 1985). Climate changes have mixed plants and animals, and as recently as ten thousand years ago, mammals of the central and southern plains were found with species from the tundra. Hunting and climate change during and subsequent to the Ice Age caused the extinction of many large grazing animals, for example, horses, camels, antelope, rhinos, bison, elephants, and tapirs. Today, the prairie community is dominated by species that have colonized the grasslands from surrounding ecosystems (chap. 11 this volume).

Forces Affecting Biological Diversity

North American grasslands contain about seventy-five hundred plant species from about six hundred genera (Hartley 1950). In general, higher numbers of plant species occur in mesic and topographically complex environments, with, for example, about fifty vascular species in a semiarid, nearly level grassland and two hundred species in a subhumid prairie (Coupland 1979; Steiger 1930). These patterns of plant composition and diversity are caused by forces operating at different spatial and temporal scales (chap. 3 this volume). Over large spaces and long time periods, evolution, postglacial migration patterns, and the broad-scale distribution of plants across various topographic gradients determine the distributional patterns. At intermediate scales, local topographic patterns and processes such as drought, fire, and large mammal herbivory are primary forces (chap. 14 this volume). Local patterns of immigration and extinction, interspecific and intraspecific competition, and small-scale disturbances (e.g., ants and small mammals) operate at the smallest spatial and shortest time scales.

Disturbance has always been a critical component in maintaining the plant and animal communities and ecosystem processes of grasslands (chap. 3 this volume). Prehistoric prairies included large herds of migratory antelope, bison, elk, and deer as well as extensive prairie dog towns (Sims 1988). Although the relationship between rainfall and grassland production is positive and generally linear (Lauenroth 1979), primary production and species diversity may be adversely affected by the absence of grazing of the tallgrass and mixed-grass prairies (Risser et al. 1981) (chap. 4 this volume). If grazing is too intense, or occurs at inappropriate times of the year or for too long a time, however, grassland quality may be decreased. For optimum growth, many of the dominant plants require regular grazed and ungrazed periods during the growing season.

Under natural conditions, the prairies probably burned at intervals of five to thirty years (Wright and Bailey 1982) and perhaps as frequently as every two to five years prior to European settlement (Anderson et al. 1989; chap. 4 this volume). With the control of fires during the middle to late 1800s, woody plants encroached on the prairie, and savannas frequently converted to closed forest. Research has provided much information about how fires affect plant species composition, soil and plant nutrients, litter depth, and animal populations (Anderson et al. 1989; Collins and Wallace 1989). In general, periodic burning tends to increase primary production under conditions of average or above-average rainfall but decreases production if fires are too frequent, for example, more often than every five to eight years in the mixed-grass prairie (chap. 4 this volume) and with drought or below-average precipitation (chap. 3 this volume).

The effects of fires on invertebrates is important since insects represent about 90 percent of the terrestrial species, and their biomass regularly exceeds that of vertebrates if domestic livestock are excluded. More than 90 percent of the total arthropod community resides in the soil and litter, including the large component of the aboveground taxa that either nest (e.g., ants) or have significant life stages (e.g., scarab beetles) belowground. Although invertebrates may be harmed with burning at the time of fires, the resulting increased plant growth and habitat heterogeneity from patchy fires are frequently beneficial. Many arthropods are associated with early and late stages of vegetation succession. As long as grazing is not so heavy that key host species are lost, a heterogeneous mixture of species and habitat structure generated by burning and grazing appears to maximize invertebrate diversity. However, more research is needed to fully understand these relationships (chap. 7 this volume).

Specialized Habitats

Rivers, streams, and wetlands are all specialized habitats in the central North American prairies. Rivers in the region frequently depend on periodic inundation of the floodplain for energy (detrital organic matter and primary production), on which much of the fish community depends, and for producing the appropriate physical habitat for spawning and feeding as well as a nursery area for fish and aquatic invertebrates (chap. 8 this volume). Because they provide crucial habitat for waterfowl and breeding habitat for more than two hundred species of nongame migratory birds (chap. 15 this volume), wetlands are particularly important. These wetlands typically cycle through an early stage of annual species that grow during the dry season from the substrate. Then, with the wet season and flooding, perennial plants emerge and eventually senesce or are consumed by herbivores, and the marsh is relatively open until the next dry season. Not all wetlands go through these stages; for example, the playas in the southern part of the plains dry annually and are dominated by species that tolerate those conditions (chap. 4 this volume).

The sandhills in Nebraska are significant as wildlife habitats and livestock rangeland, and they also play an important role in the dynamics of groundwater. Infiltration of rainwater into these sandy soils is rapid, and 25 to 50 percent of it becomes part of the Ogallala Aquifer, where the thickest part of the aquifer is under the sandhills. During wet seasons, groundwater accumulates under the dunes and is slowly released to the neighboring wetlands and streams. In these sandy soils, the water table is held near the surface, supporting productive native grasses on the dunes and in the wetland meadows between the dunes. Grasslands on the dunes are used for summer grazing, while the meadows between the dunes are mowed for winter hay. The amount of winter forage can be the limiting factor in the success of a ranching operation, and, thus, ownership of meadows is critical to the value of a ranch (chap. 19 this volume).

Values of Prairies

Prairies provide many services, such as grazing for livestock, soil conservation and water quality enhancement benefits, wildlife habitat, and recreational experiences, including hunting and access to scenic vistas. When considering conservation of prairies, the question of value always arises—and there are many possible measures of the value of the prairie. Some values can be measured by uses that are paid for in the market, that is, using measures similar in concept to the gross national product, which measures the total value of goods and services provided by the economy in a particular period of time. Consumptive uses, such as grazing livestock, selling prairie hay, or leasing hunting rights, can be quantified by market price. Other nonconsumptive use values (such as research and education) are not as easily quantifiable. There are also more intangible, nonuse values, such as aesthetic benefits and human satisfaction, the perceived existence values that arise from the mere fact that the prairie resource exists, and option values that are those assumed because society may want to preserve the option to use the resource in the future (chap. 2 this volume).

There are several ways of quantifying the value of prairies. For example, both consumptive and nonconsumptive use values can frequently be measured by the market or by surveys of people who quantify their estimate of the benefits of the prairie. Nonuse values can be estimated by similar surveys or by processes such as contingent valuation (i.e., what individuals would be willing to pay to protect the prairie). Experience in establishing prairie preserves clearly demonstrates that price does not always equal value, and values derived outside of a market transaction can be greater than those derived in the market (chap. 2 this volume). As the conservation of prairies and their benefits becomes more sophisticated and involves more partners, additional work will be needed to develop better measures of the multiple values of prairies.

Human Impact on the Prairie

Fragmentation

The central North American grasslands have been severely impacted by human activities. Very little (about 2 percent) of the tallgrass prairie remains (Knopf, in press), considerable portions of the mixed-grass prairie have been modified or lost, and although much of the shortgrass prairie remains, a significant amount of it, too, has been converted to cropland or changed due to domestic grazing or other uses. Where the prairies remain, they are frequently fragmented into smaller and more isolated areas. This fragmentation results in smaller and isolated plant and animal populations, with greater likelihood of species loss by local extinctions, decreased probability of colonization from outside the remaining patch, and genetically isolated and reduced populations that are more likely to suffer the negative impacts of inbreeding and genetic drift (chap. 11 this volume). Stated in terms of population genetics, this fragmentation has effectively turned what were nearly continuous populations into metapopulations of semi-independent demes. Examples of this are prairie insects such as real fritillary, prairie mole cricket, and several prairie skippers—the Dakota, ottoe, Assiniboin, and poweshiek. The severity of prairie fragmentation may be so great that it may be impossible to stem the gradual loss of species and genetic diversity on the smaller prairie tracts, especially with climate change and regional air pollution (chap. 3 this volume).

Changes in Community Structure

In addition to the local loss of species, there has been a significant restructuring of many of the plant and animal communities. For example, with grazing, many of the more palatable native plant species are reduced or eliminated and replaced by weedy exotic species. When the grasslands were widespread and intact, plant populations continually fluctuated in size, and local extinctions were commonly reestablished from nearby populations. However, in the current fragmented landscape, local sources of seeds for reestablishment are limited or absent, leading to the long-term loss of species (chap. 3 this volume). Habitat loss and degradation, as well as competition from edge species and nonnative species and chemical pollution, have been the greatest threats to prairie invertebrates (chap. 7 this volume) and to prairie bird populations (Knopf, in press). For example, fewer than 18 percent of grassland birds show positive continental population trends from 1966 to 1991 (Peterjohn and Sauer 1993), and recent analyses of trends have shown that grassland nesting birds have exhibited more consistent, widespread, and deeper declines in the past twenty-five years than any other North American bird group (Knopf 1994).

The wide-scale destruction of habitat and the alteration of remaining habitat have also led to the extermination or decimation of keystone mammals (e.g., bison and prairie dogs) and locally of several top predators (chap. 16 this volume). Frequently, there is an increase in generalist mammal species such as opossum, raccoon, red fox, coyote, white-tailed deer, white-footed mouse, house mouse, fox squirrel, eastern cottontail, woodchuck, and deer mouse (chap. 11 this volume). The same trend toward generalist species is also true of the fish species in the prairie region (chap. 8 this volume).

Although most human activity has resulted in the loss or degradation of native prairie, there has been some planting of grasslands during the last decade. Through the U.S. Department of Agriculture Conservation Reserve Program (instituted in 1985 and due for reconsideration in 1995), about 5 percent of the cropland in shortgrass prairie states and 3.4 percent in mixed-grass regions has been replanted with native and nonnative grass species. These replanted grasslands have a much simpler structure than native prairies and contain many fewer species. More research is required to understand how these relatively simple, reconstituted prairies will support and affect plants and animals in these prairies themselves and in the surrounding areas (chap. 11 this volume).

Specialized Habitats

Increased agriculture and other development activities have consumed groundwater and surface water, straightened streams and rivers, constructed levees that separate the channel from its floodplain, and built dams that have reduced the ranges of many fishes and cut off historic spawning regions. The average particle size of the substrate in streams and rivers is now smaller; aquatic habitats are more homogenized; both riparian and instream vegetation are much reduced; and there are many more barriers to fish migration. Exotic fish species have been introduced, which has improved sportfishing but also decreased or eliminated native fish species. Water quality has been reduced by pollution, both from domestic and industrial wastes and from agricultural fertilizers and pesticides (chap. 8 this volume).

In the sandhills during the late 1800s, federally owned land was heavily grazed by outside investors until two harsh winters forced them out of this abusive ranching business. With conversion to private ownership in the early 1900s, cattle stocking rates were more reasonable, but much of the wetland area was drained for increased agriculture. By the 1950s most of the drainage was completed, but it had reduced the area of wetlands and lowered the hydraulic head in the region, which further decreased wetlands. In the 1970s investors and landowners converted significant amounts of grassland to cropland (which increased wind erosion about ten times) because of favorable tax laws, the availability of center-point irrigation technology, low land values, and high grain prices. Pumping caused local water tables to drop in some areas; neighboring meadows became drier; and forage production declined. In other areas, irrigation water raised local water tables, creating marshes and lakes, which were then frequently drained. More fertilizers and pesticides were used, which then leached into groundwater and contaminated nearby domestic wells (chap. 19 this volume).

Because of the changed land uses, there was less native grassland in the sandhills and in the prairie pothole region. This reduced nesting cover and concentrated nesting birds in remnant patches, which were easily searched by predators with resulting low nest success of waterfowl (chap. 6 this volume). Reduced habitat and changes in hydrology have continued to plague waterfowl and other species, making them vulnerable not only to predation but also to drought and other changes in weather and climate. During 1985 continental duck numbers reached their lowest point in forty years (Caithamer et al. 1994).

Current Conservation Practices and Programs

At a superficial level, it might be tempting to conclude that the plants and animals of the prairies have been relatively free of impacts from humans. This impression is fed by the relative lack of endemic species, the broad resource requirements of many of its species, the adaptations of several prairie species to tolerate disturbance (chap. 11 this volume), and the fact that no North American prairie species have become extinct. However, large portions of the prairie are simply gone, a number of species are threatened and endangered, many species have been drastically reduced in numbers, and just as importantly, the composition and structure of the remaining prairies have been dramatically changed (chaps. 8 and 14 this volume). Much of this change has significantly diminished the variety found in all components of the prairie, thus reducing both its current value as well as its future capability to produce goods and services, and further jeopardizing the prairie's potential to persist in the future.

Conservation Programs in the Prairies

With the exception of a few areas such as the Flint Hills in Kansas and Oklahoma, most of the remaining tallgrass prairie remnants are relatively small in size and typically isolated. Many of these are managed by local- and state-level organizations, and in general, management techniques of these remnants have been more intensive than in the more widespread remaining mixed-grass and short-grass prairies (chaps. 3 and 5 this volume) Many conservation techniques in all types of prairies are designed to replicate the natural disturbance regime by burning, mowing, or grazing by native or domestic herbivores. Herbivory and burning affect most components of the prairie, including species composition, amounts of litter, soil surface conditions, seed distribution and seedling establishment, the rapidity of nutrient cycling, and the likelihood of fires. Both fires and mowing reduce the encroachment of woody plants into the prairie. Since most tallgrass prairies are small and isolated, they are subject to invasion by exotic species and also to loss of species by local extinction, with little possibility of recolonization from adjacent areas. Thus, many conservation programs in these prairies include the physical removal of exotic species and the artificial planting of native species.

In the mixed-grass and shortgrass region, the Conservation Reserve Program (CRP), with over 12 million acres (4.86 million ha) of grass habitat in the north-central plains, has had a major impact on prairie conservation. Under this program, 5 percent of the cropland in shortgrass prairie states and 3.4 percent in mixed-grass regions have been replanted to grasslands with native or nonnative grasses (chap. 11 this volume). These are far simpler grasslands than native rangelands. Since the CRP has been underway for only a decade, there is not enough research to provide sufficient information about the nature and persistence of the resulting prairie communities nor just how these replanted grasslands will interact with the surrounding native prairies.

Over the past decade or so, the need to preserve grasslands has been recognized not only by several federal agencies and nongovernmental organizations, such as The Nature Conservancy, but also by states and tribes. As a result, many programs are now underway, and a major challenge is to coordinate among these programs and organize them to be compatible with and supported by local residents and private landowners (chap. 12 this volume).

The Great Plains Partnership (GPP), formerly the Great Plains Initiative, began in the United States and Canada in 1986 (Mexico joined in 1994) with several partners: the Western Governors' Association and its international counterparts, the Nature Conservancy, the International Association of Fish and Wildlife Agencies, the U.S. Fish and Wildlife Service, the U.S. Environmental Protection Agency, the province of Manitoba (with the Prairie Habitat Joint Venture), and the State of Minnesota. By demonstrating that both economic and environmental interests can be served by preventing the decline of species and their ecosystems, the goal is to support the well-being of the Great Plains. The GPP is an experimental program working with and through private landowners. Since its origin, it has quickly expanded to involve other government and nongovernmental entities, including several agricultural organizations that have designated representatives to participate (chap. 12 this volume).

In Canada, the Prairie Conservation Action Plan (PCAP) 1989–1994, was established to conserve biological diversity on the Canadian prairies. This plan used a large, multiparty Prairie Conservation Coordinating Committee, mostly to encourage other organizations to inventory remaining prairie and parkland, identify areas that should be protected, develop recovery plans for selected prairie species, and develop public awareness programs. Although some successes were achieved, the agricultural community was not actively involved. In 1994 the plan was revised, with a larger group of stakeholders who are empowered at the local level to develop initiatives using local knowledge and expertise. This reorganization is based on the belief that the most effective conservation action occurs where there is an applied focus at the community level (chap. 13 this volume).

Wetlands

Wetlands were drained in large numbers with the spread of agriculture into the prairies in the late 1800s and early 1900s. Subsequently, there was great concern about decreased waterfowl populations, and the Migratory Bird Waterfowl Hunting Stamp Act was passed in 1934 to make funds available for waterfowl habitat restoration. The private organization Ducks Unlimited was established in 1937 for similar purposes. Although there has been other significant habitat restoration and conservation legislation, some with provisions for international cooperation and transfer of funds among the United States, Canada, and Mexico, the North American Waterfowl Management Plan (NAWMP) is the most comprehensive (chap. 6 this volume). The NAWMP includes several joint ventures: the Prairie Pothole Joint Venture (the Prairie Habitat Joint Venture in Canada), the Rainwater Basin Joint Venture, and the Playa Lakes Joint Venture. These are "habitat" joint ventures in which waterfowl population goals are translated into habitat objectives, and habitat is recognized as having both wetland and upland components. Thus, with the implementation of the NAWMP for waterfowl, major contributions are being made to both the wetland and upland components of the prairie ecosystem (chap. 15 this volume).

The dramatic increase in waterfowl from 1985 to 1994 was due to plentiful rainfall in the summer of 1993, record snows during the winter, and high rainfall in the spring of 1994, but this success resulted also from activities under the NAWMP. In addition, the CRP and many other efforts of protection, restoration, and enhancement of wetlands and grasslands on public and private lands have contributed. Restoration of prairie wetlands is increasingly targeted to coincide with areas where CRP or publicly managed grasslands provide upland nesting cover (chap. 16 this volume). All these activities point to the success of resource conservation by numerous collaborating agencies, organizations, and individuals focusing on the habitat and involving landowners (chaps. 6 and 15 this volume).

Conservation of streams and rivers in the North American central grasslands has largely been conducted as part of conserving prairies. However, some states have instream flow standards that specify the minimum flow required to maintain the biotic integrity of the stream. Other states are just beginning to develop such standards, have only weak ones, or have none at all (chap. 8 this volume).

Sandhills

More than 90 percent of the Nebraska Sandhills is in private ownership. For many years, the management of the region's natural resources was on an agency-resource basis; that is, the U.S. Soil Conservation Service attempted to control erosion, while the U.S. Fish and Wildlife Service focused on wildlife programs. As these agencies enacted regulations and programs, individual landowners perceived them as a loss of property rights. Opposition to agency regulations by the landowners reinforced the agencies' belief in the need for these regulations. This downward spiral between conservation and agricultural interests was arrested in 1991 when the Fish and Wildlife Service began to bring the two groups together. There was agreement on (1) a need to maintain grassland-wetland ecosystems and improve wildlife numbers, (2) recognition of the role ranching has played in maintaining the diversity and abundance of plants and animals, and (3) a continuing concern about the level and quality of the groundwater. After preliminary discussions, a joint group drafted the Sandhill Management Plan, with five broad management strategies: education and technical assistance, acquisition (conservation easements and fee title), lease agreements, legislation, and financial support. This plan was signed in 1993, a coordinator was hired, and twenty-four partnership projects have been completed. Projects were selected to focus on finding solutions that benefit both the landowner and the natural resources (chap. 19 this volume).

Framework for Prairie Conservation

Much is known about the prairies of North America, including their distribution, composition, and behavior, as well as their goods, services, and values. Significant portions of these prairies have been lost or severely changed since the time of European settlement. Although there are certainly gaps in our understanding of prairies, a major challenge today is to use what is known about the prairies to conserve and maintain the remaining ones. Additional research must be conducted on key questions about prairies, especially to find answers that will guide us in conserving prairies under changing natural and human-dominated influences.

Conservation programs in North American prairies have involved many participants. For example, especially in the tallgrass prairie, numerous remnants have been preserved by the actions of small groups of people and organizations that have undertaken the responsibility of maintaining the prairie, frequently by laboriously planting native species, eradicating exotics, and burning or mowing to simulate the natural disturbance regimes. On the other hand, other conservation efforts have involved large-scale activities such as statewide or regional inventories, comprehensive research programs on the structure and function of grasslands ecosystems, and major policy directives affecting lands under the responsibility of federal or state agencies from one or more countries.

Conservation of prairies in the future will continue to involve many participants combined in different organizational ways and operating at different scales. The chapters in this book, however, suggest that we can now provide a much stronger framework for conservation of North American prairies. This framework is based on the following four principles:

1. Prairie conservation can now be based on relatively sophisticated research results, thereby supporting stronger conservation programs and providing a scientific rationale for conservation management decisions.

2. Prairies provide many goods, services, and values to many constituents, and conservation programs must seek to serve as many constituents as possible while always recognizing that perpetuation of the prairie resource is the first priority.

3. Prairie conservation efforts will be successful only if they consciously incorporate considerations of short-term and long-term economic values of the prairies.

4. Landowners must be actively involved in all dimensions of prairie conservation, and all conservation partnerships must recognize that some goods, services, and values of prairies extend beyond property boundaries.

Several existing prairie conservation programs have already begun to formalize this new framework. For example, while recognizing that working with private landowners will continue to be the backbone of the project, the Prairie Pothole Joint Venture (PPJV) will use new technologies, such as geographic information systems and GAP analysis, along with other tools such as migratory bird population models and multiagency planning and evaluation processes. The strengths of the PPJV are its diverse partnerships and common goals of prairie habitat. Partners will continue working closely with private landowners to integrate wildlife conservation practices that support a profitable agricultural operation. There will be an immediate focus on conservation legislation (chap. 15 this volume), since many of the current laws, regulations, and government programs obstruct prairie conservation (chap. 12 this volume).

The experience of the Great Plains Partnership demonstrates that meeting the four principles of the framework is sometimes difficult. Although the project began with good intentions, the state and federal agencies sometimes had difficulty working with landowners and local residents. Landowners care about the future of the land and, as a consequence, have great fear about private property rights. This apprehension is so great that they will frequently defend property rights even if abuses to the prairies are occurring. From discussions with landowners, there was obvious support for bottom-up partnerships to work on projects on the ground. Part of the difficulty, however, was that agencies did not know how to relate bottom-up initiatives to existing organizational top-down planning and regulations. Moreover, terminology and bureaucratic procedures are frequently a hindrance. For example, landowners identify closely with the ideas of land stewardship but not with such terms as "ecosystem management," which are now a routine part of agency terminology (chap. 17 this volume).

The experience of the Great Plains Partnership (GPP) also demonstrates the need for closer coordination between prairie research and conservation activities.

When the GPP attempted to bring the best scientific knowledge to bear on prairie conservation, it discovered that most of the biologically important areas were known, but when it tried to pull together science and data regarding the plains, it ran into a variety of problems. Federal databases, with the exception of GAP analysis, which is still incomplete, were almost useless for doing broad-scale assessment. They were too large and complex, too specific, not well cataloged, and were using different scales, definitions, classifications, hardware, and software. The Nature Conservancy's State Heritage Program systems had common scales, definitions, and classifications, but these databases did not include all the relevant information. To develop a practical mechanism for data exchange involving federal and state agencies, the GPP created the Great Plains International Data Network (chap. 12 this volume). In general, the challenges are to enlist the use of existing data or to collect needed data so that they are meaningful for specific locations on the landscape and so that data can be shared among programs. Moreover, it is clear even from the discussion in this book that the most up-to-date research data are not always accessible to and usable by those making decisions on the land.

Science will not provide all the answers. As Clark states: "There is no question that conserving the biological and social resources of the Plains is a large challenge. Habitat, species, communities, ways of life, quality of life, science, resources, institutions, communications, economies, governance—those are a lot of ingredients to mix together over a thousand-mile range and boundaries that are not just geographic but jurisdictional and discipline-based as well as social (chap. 12 this volume)." Nevertheless, research is needed to identify key characteristics of prairie ecosystems that are necessary for sustaining the grasslands, to guide management decisions, to set priorities and to clarify the implications of different choices, to serve as a basis for changing or modifying management techniques, and to define what is possible for prairie conservation. Effective prairie conservation will be guided by integrated and land-based databases, constructed in ways that allow individuals and institutions to understand not only the behavior of the prairie but of each other (chap. 12 this volume).

Efforts to conserve the prairie will necessarily focus on many spatial scales. The protection of all species, including variability within species, eventually demands attention at the smallest scales to describe the dynamics of the species, threats to them, and their requirements for survival (chaps. 7 and 14 this volume). On the other hand, disturbances such as fire, drought, and herbivore pressure have occurred at generally larger scales, as do other important processes such as climate change, long-range migration, and material transport via wind and rivers. Thus, conservation efforts must also focus on larger geographical areas and assemblages of species or natural communities (chaps. 4 and 14 this volume).

There are several reasons why redoubled efforts in prairie conservation are particularly important at this time in our history. First, much of the prairie has been lost or substantially altered; thus, there is an imperative to take the necessary action to conserve the North American prairies. Second, concerted and collaborative actions have resulted in recognizable successes (e.g., the high numbers of waterfowl found in the northern regions during 1994). Third, specific programs have meaningfully involved landowners (e.g., NAWMP and the GPP) and have been successful. Stakeholders have been empowered at the local community level to work toward prairie conservation initiatives, using local knowledge and expertise. This principle reflects the belief that most effective conservation action occurs where there is an applied focus at the local community level (chaps. 13 and 17 this volume). Fourth, although many prairie species have declined in number and genetic diversity, none is extinct, leading to the possibility of recovery through cooperative, proactive conservation strategies designed to meet the needs of both people and wildlife (chaps. 6 and 12 this volume). Fifth, considerable research has been accomplished, and much is known about these continental grasslands. Moreover, successful prairie conservation programs have identified specific issues that need to be addressed in future research. Example research questions include identifying the effects of grassland fragmentation on the success of plants and animals as well as the interactions among them; understanding how the grasslands will respond to climate change and other broad-scale disturbances; specifying conditions that will lead to the recovery of rare or declining species; ascertaining the effectiveness of planted and reconstituted prairies in providing habitat for prairie species; identifying the key characteristics of prairies, which ensure their ability to provide specified goods, services, and other values over sustained periods of time; and quantifying the degree to which agricultural uses (e.g., periodic haying or grazing) can be permitted without jeopardizing the multiple uses and values of the prairies (chaps. 15 and 16 this volume).

The proposed framework for prairie conservation emphasizes that not all of the most significant research questions have been answered, and a significant portion of what is known about the prairies is currently not accessible and therefore not being used by decision makers. However, much information is known and is being used to effectively manage the North American prairie. Our challenge today is to use the framework for a broader approach, namely, one that recognizes that declines of landscapes, economies, and social structures are interrelated, as are policy solutions (chap. 20 this volume).

It is unreasonable to assume that the prairie can be conserved by setting aside and managing many large tracts of productive, privately owned agricultural lands. Moreover, conservation programs within rural communities will be acceptable only if there are built-in mechanisms for providing or generating revenues to help offset long-term management costs for lands dedicated to conservation. Thus, future successful conservation actions will require combinations of activities, including protection and enhancement of native prairie landscapes, managed grazing systems that are economically sound but also retain multiple grassland values, local and regional soil and water conservation practices that support the perpetuation of grassland plants and animals, selected forage production on marginal land to provide wildlife habitat, economically sound rotating use of key habitats, and other activities that collectively conserve the North American prairie while also supporting landowners and the regional culture (chap. 16 this volume). Thus, research in the future will not be directed toward biological processes in isolation from the economic and cultural setting. Rather, future research will combine biological information with such topics as computer models that optimize productive habitat types and uses for both economic and ecological purposes and will consider policy options such as leases and economic incentive programs that are acceptable to and encouraged by landowners.

Species List

Bacterium

fowl cholera   *Pasteurella multocida*

Plants

annual sunflower   *Helianthus annuus*
Arizona cottontop   *Digitaria californica*
aromatic aster   *Aster oblongifolius*
Ashe juniper   *Juniperus ashei*
big bluestem   *Andropogon gerardii*
bitterweed   *Hymenoxys odorata*
blanket flower   *Gaillardia* spp.
blowout penstemon   *Penstemon haydenii*
blue grama   *Bouteloua gracilis*
broomweed   *Gutierrezia dracunculoides*
buffalo grass   *Buchloë dactyloides*
Canada thistle   *Cirsium arvense*
Canada wildrye   *Elymus canadensis*
Carolina canary grass   *Phalaris caroliniana*
cheatgrass   *Bromus secalinus*
common curlymesquite   *Hilaria belangeri*
crested wheatgrass   *Agropyron cristatum*
curlycup gumweed   *Grindelia squarrosa*
dotted gayfeather   *Liatris punctata*
downy brome   *Bromus tectorum*
eastern redcedar   *Juniperus virginiana*
evening primrose   *Oenothera* spp.
false flax   *Camelina microcarpa*
foxtail barley   *Hordeum jubatum*

fringed sagebrush	*Artemisia frigida*
globe mallow	*Sphaeralcea coccinea*
goat's beard	*Tragopogon dubius*
green needlegrass	*Stipa viridula*
hairy grama	*Bouteloua hirsuta*
heath aster	*Aster ericoides*
honey mesquite	*Prosopis glandulosa*
indian grass	*Sorghastrum nutans*
ironweed	*Vernonia baldwinii*
Japanese brome	*Bromus japonicus*
junegrass	*Koeleria pyramidata*
Kentucky bluegrass	*Poa pratensis*
lamb's quarter	*Chenopodium* spp.
lazy daisy	*Aphanostephus* spp.
leadplant	*Amorpha canescens*
leafy spurge	*Euphorbia esula*
little barley	*Hordeum pusillum*
little bluestem	*Schizachyrium scoparium*
locoweed	*Astragalus* and *Oxytropis* spp.
lotebush	*Ziziphus obtusifolia*
Mead's milkweed	*Asclepias meadii*
Missouri spurge	*Euphorbia missurica*
needle-and-thread	*Stipa comata*
nightshade	*Solanum*
pale alyssum	*Alyssum alyssoides*
patagonian plantain	*Plantago patagonica*
plains bristlegrass	*Setaria leucopila*
plains sunflower	*Helianthus petiolaris*
porcupine grass	*Stipa spartea*
prairie fringed orchid	*Platanthera praeclara*
prairie sandreed	*Calamovilfa longifolia*
prickly lettuce	*Lactuca seriola*
redberry juniper	*Juniperus pinchoti*
rough dropseed	*Sporobolus asper*
Russian-olive	*Elaeagnus angustifolia*
sagebrush	*Artemisia* spp.
sand bluestem	*Andropogon hallii*
sand dropseed	*Sporobolus cryptandrus*

sand lovegrass	*Eragrostis trichodes*
sand skeletonweed	*Lygodesmia* spp.
sandhill muhly	*Muhlenbergia pungens*
scarlet gaura	*Gaura coccinea*
scarlet globe-mallow	*Sphaeralcea coccinea*
Scribners panicum	*Dicanthelium oligosanthes* var. *scribnerianum*
Siberian elm	*Ulmus pumila*
sideoats grama	*Bouteloua curtipendula*
small soapweed	*Yucca glauca*
smooth brome	*Bromus inermis*
spotted knapweed	*Centaurea maculosa*
squirreltail	*Sitanion hystrix*
switchgrass	*Panicum virgatum*
tall dropseed	*Sporobolus heterolepis*
Texas cupgrass	*Eriochloa sericea*
Texas wintergrass	*Stipa leucotricha*
thread leafed sedge	*Carex filifolia*
tobosagrass	*Hilaria mutica*
vine-mesquite	*Panicum obtusum*
western ragweed	*Ambrosia psilostachya*
western snowberry	*Symphoricarpos occidentalis*
western wheatgrass	*Agropyron smithii*
white spruce	*Picea glauca*
yellow sweetclover	*Melilotus officinalis*

Animals

Insects

Assiniboin skipper	*Hesperia assiniboia*
Dakota skipper	*Hesperia dacotae*
orange sulphur	*Colias eurytheme*
ottoe skipper	*Hesperia ottoe*
poweshiek skipper	*Oarisma poweshiek*
prairie mole cricket	*Gryllotalpa major*
regal fritillary	*Speyeria idalia*
sachem	*Atalopedes campestris*

Fish

Arkansas darter	*Etheostoma cragini*
Arkansas River shiner	*Notropis girardi*
banded killifish	*Fundulus diaphanus*
bighead carp	*Hypophthalmichthys nobilis*
bigmouth buffalo	*Ictiobus cyprinellus*
black bullhead	*Ictalurus melas*
blacknose shiner	*Notropis heterolepis*
bluegill	*Lepomis macrochirus*
blue sucker	*Cyceptus elongatus*
carpsucker	*Carpiodes carpio*
chub shiner	*Notropis potteri*
common carp	*Cyprinus carpio*
common shiner	*Luxilis cornutus*
creek chub	*Semotilus atromaculatus*
emerald shiner	*Notropis atherinoides*
flathead catfish	*Pylodictis olivaris*
flathead chub	*Hybopsis gracilis*
flathead minnow	*Pimephales promelas*
fountain darter	*Etheostoma fonticola*
gizzard shad	*Dorosoma cepedianum*
goldeye	*Hiodon alosoides*
grass carp	*Ctenopharyngodon idella*
green sunfish	*Lepomis cyanellus*
guadalupe bass	*Micropterus treculi*
hornyhead chub	*Nocomis biguttatus*
lake sturgeon	*Acipenser fulvescens*
longear sunfish	*Lepomis cyanellus*
longnose gar	*Lepisosteus osseus*
Neosho madtom	*Notrus placidus*
orangethroat darter	*Etheostoma spectabile*
Ozark minnow	*Notropis nubilus*
paddlefish	*Polyodon spathula*
pallid sturgeon	*Scaphirhynchus albus*
plains minnow	*Hybognathus placitus*
plains topminnow	*Fundulus sciadicus*
red shiner	*Cyprinella lutrensis*
sauger	*Stizostedion canadense*

sharpnose shiner	*Notropis oxyrhynchus*
shortnose gar	*Lepisosteus platostomus*
sicklefin chub	*Macrhybopsis meeki*
silverband shiner·	*Notropis shumardi*
silver carp	*Hypophthalmichthys molitrix*
silver chub	*Macrhybopsis storeriana*
skipjack herring	*Alosa chrysochloris*
smalleye shiner	*Notropis buccula*
southern redbelly dace	*Phoxinus erythrogaster*
speckled chub	*Hybopsis aestivalis*
spotted bass	*Micrpterus punctulatus*
striped bass	*Morone saxatilis*
sturgeon chub	*Macrhybopsis gelida*
Topeka shiner	*Notropis topeka*
western silvery minnow	*Hybognathus argyritis*
white bass	*Morone chrysops*
white crappie	*Pomoxis annularis*

Herpetofauna

brazos water snake	*Nerodia harteri*
bullsnake	*Pituophis catenifer*
common kingsnake	*Lampropeltis getula*
Concho water snake	*Nerodia paucimaculata*
dunes sagebrush lizard	*Sceloporus arenicolus*
gray treefrog	*Hyla versicolor*
ground skink	*Scincella lateralis*
lined snake	*Tropidoclonion lineatum*
northern leopard frog	*Rana pipiens*
ornate box turtle	*Terrapene ornata*
plains garter snake	*Thamnophis radix*
plains spadefoot	*Spea bombifrons*
prairie rattlesnake	*Crotalus viridis*
prairie skink	*Eumeces septentrionalis*
racer	*Coluber constrictor*
spiny softshell	*Apalone spinifera*
Texas horned lizard	*Phrynosoma cornutum*
Texas map turtle	*Graptemys versa*

tiger salamander	*Ambystoma tigrinum*
western diamondback rattlesnake	*Crotalus atrox*
western hognose snake	*Heterodon nasicus*
Woodhouse's toad	*Bufo woodhousii*
Wyoming toad	*Bufo hemiophrys baxteri*
yellow mud turtle	*Kinosternon flavescens*

Birds

American white pelican	*Pelicanus erythrorhynchos*
American wigeon	*Anas americana*
Baird's sandpiper	*Calidris bairdii*
Baird's sparrow	*Ammodramus bairdii*
bald eagle	*Haliaeetus leucocephalus*
black tern	*Chlidonias niger*
black-billed magpie	*Pica pica*
black-crowned night heron	*Nycticorax nycticorax*
blue-winged teal	*Anas discors*
bobolink	*Dolichonyx oryzivorus*
burrowing owl	*Athene cunicularia*
Canada goose	*Branta canadensis*
canvasback	*Aythya valisineria*
Cassin's sparrow	*Aimophila cassinii*
chestnut-collared longspur	*Calcarius ornatus*
cinnamon teal	*Anas cyanoptera*
clay-colored sparrow	*Spizella pallida*
common grackle	*Quiscalus quiscula*
common yellowthroat	*Geothlypis trichas*
dickcissel	*Spiza americana*
eastern meadowlark	*Sturnella magna*
ferruginous hawk	*Buteo regalis*
Franklin's gull	*Larus pipixcan*
gadwall	*Anas strepera*
giant Canada goose	*Branta canadensis maxima*
grasshopper sparrow	*Ammodramus savannarum*
greater prairie-chicken	*Tympanuchus cupido*
greater white-fronted goose	*Anser albifrons frontalis*
green-winged teal	*Anas carolinensis*

Henslow's sparrow	*Ammodramus henslowii*
horned lark	*Eremophila alpestris*
house finch	*Carpodacus mexicanus*
interior least tern	*Sterna antillarum athalassos*
lark bunting	*Calamospiza melanocorys*
lark sparrow	*Chondestes grammacus*
least tern	*Sterna antillarum*
LeConte's sparrows	*Ammodramus leconteii*
lesser prairie-chicken	*Tympanuchus pallidicinctus*
lesser scaup	*Aythya affinis*
lesser snow goose	*Chen caerulescens caerulescens*
loggerhead shrike	*Lanius ludovicianus*
long-billed curlew	*Numenius americanus*
long-billed dowitcher	*Limnodromus scolopaceus*
mallard	*Anas platyrhynchos*
McCown's longspur	*Calcarius mccownii*
Mississippi kite	*Ictinia mississippiensis*
mottled duck	*Anas fulvigula*
mountain plover	*Charadrius montanus*
northern harrier	*Circus cyaneus*
northern pintail	*Anas acuta*
northern shoveler	*Anas clypeata*
peregrine falcon	*Falco peregrinus*
piping plover	*Charadrius melodus*
prairie falcon	*Falco mexicanus*
red-winged blackbird	*Agelaius phoeniceus*
redhead	*Aythya americana*
ring-necked duck	*Aythya collaris*
ruddy duck	*Oxyura jamaicensis*
sandhill crane	*Grus canadensis*
Savannah sparrow	*Passerculus sandwichensis*
sedge wren	*Cistothorus platensis*
sharp-tailed grouse	*Tympanuchus phasianellus*
short-eared owl	*Asio flammeus*
snowy plover	*Charadrius alexandrinus*
Sprague's pipit	*Anthus spragueii*
stilt sandpiper	*Calidris himantopus*
Swainson's hawk	*Buteo swainsoni*

trumpeter swan	*Cygnus buccinator*
upland sandpiper	*Bartramia longicauda*
vesper sparrow	*Pooecetes gramineus*
western meadowlark	*Sturnella neglecta*
white-rumped sandpiper	*Calidris fuscicollis*
whooping crane	*Grus americana*
Wilson's phalarope	*Phalaropus tricolor*

Mammals

badger	*Taxidea taxus*
beaver	*Castor canadensis*
big brown bat	*Eptesicus fuscus*
bison	*Bison bison*
black bear	*Ursus americanus*
black rat	*Rattus rattus*
black-footed ferret	*Mustela nigripes*
black-tailed jackrabbit	*Lepus californicus*
black-tailed prairie dog	*Cynomys ludovicianus*
caribou	*Rangifer tarandus*
coyote	*Canis latrans*
deer mouse	*Peromyscus maniculatus*
domestic cat	*Felis silvelstris*
domestic cow	*Bos taurus*
domestic dog	*Canis lupus familiaris*
eastern chipmunk	*Tamias striatus*
eastern cottontail	*Sylvilagus floridanus*
eastern spotted skunk	*Spilogale putorius*
eastern woodrat	*Neotoma floridana*
fox squirrel	*Sciurus niger*
Franklin's ground squirrel	*Spermophilus franklinii*
gray fox	*Urocyon cinereoargenteus*
gray squirrel	*Sciurus carolinensis*
gray wolf	*Canis lupus nubilus*
grizzly bear	*Ursus arctos*
heather vole	*Phenacomys intermedius*
hispid cotton rat	*Sigmodon hispidus*
hispid pocket mouse	*Chaetodipus hispidus*

hoary bat	*Lasiurus cinereus*
house mouse	*Mus musculus*
kangaroo rat	*Dipodomys* spp.
least weasel	*Mustela nivalis*
lynx	*Lynx canadensis*
mammoth	*Mammuthus* spp.
masked shrew	*Sorex cinereus*
mastodon	*Mastodon* spp.
meadow jumping mouse	*Zapus hudsonius*
meadow vole	*Microtus pennslyvanicus*
mountain lion	*Puma concolor*
mountain sheep	*Ovis canadensis audubonii*
mule deer	*Odocoileus hemionus*
nine-banded armadillo	*Dasypus novemcinctus*
northern grasshopper mouse	*Onychomys leucogaster*
Norway rat	*Rattus norvegicus*
olive-backed pocket mouse	*Perognathus faciatus*
opossum	*Didelphis virginiana*
Ord's kangaroo rat	*Dipodomys ordii*
plains harvest mouse	*Reithrodontomys montanus*
plains pocket gopher	*Geomys bursarius*
plains pocket mouse	*Perognathus flavescens*
prairie vole	*Microtus ochrogaster*
pronghorn	*Antilocapra americana*
raccoon	*Procyon lotor*
red bat	*Lasiurus borealis*
red fox	*Vulpes vulpes*
red squirrel	*Tamiasciurus hudsonicus*
red-backed vole	*Clethrionomys gapperi*
Richardson's ground squirrel	*Spermophilus richardsonii*
river otter	*Lutra canadensis*
southern flying squirrel	*Glaucomys volans*
spotted skunk	*Spilogale* spp.
swift fox	*Vulpes velox*
thirteen-lined ground squirrel	*Spermophilus tridecimlineatus*
Tule elk	*Cervus elaphus nannodes*
wapiti	*Cervus elaphus*
western harvest mouse	*Reithrodontomys megalotis*

white-footed mouse	*Peromyscus leucopus*
white-tailed deer	*Odocoileus virginianus*
white-tailed jackrabbit	*Lepus townsendii*
white-tailed prairie dog	*Cynomys leucurus*
wolf	*Canis lupus*
wolverine	*Gulo gulo*
woodchuck	*Marmota monax*
woodland vole	*Microtus pinetorum*

Literature Cited

Abrams, M. D. 1988. "Effects of Burning Regime on Viable Seed Pools and Canopy Coverage in a Northeast Kansas Tallgrass Prairie." *Southwest Naturalist* 33:65–70.

Adams, D. E., and L. L. Wallace. 1985. "Nutrient and Biomass Allocation in Five Grass Species in an Oklahoma Tallgrass Prairie." *American Midland Naturalist* 113:170–81.

Adams, G. D. 1988. "Wetlands of the Prairies of Canada." In *Wetlands of Canada*. Ecological Land Classification Series, no. 24, 155–94. Ottawa: Sustainable Development Branch, Environment Canada.

Agriculture Canada and Alberta Agriculture, Food, and Rural Development. 1994. *Canada-Alberta Soil Conservation Initiative. Canada-Alberta Soil Conservation Service Report, 1989–1993*. Edmonton: Agriculture Canada and Alberta Agriculture, Food, and Rural Development.

Ahlgren, I. F. 1974. "The Effect of Fire on Soil Organisms." In *Fire and Ecosystems*, edited by T. T. Kozlowski and C. E. Ahlgren, 47–72. New York: Academic Press.

Alberta Environmental Protection. 1995. *Alberta's State of the Environment Comprehensive Report*. Edmonton: Alberta Environmental Protection. 119 p.

Albertson, F. W., and G. W. Tomanek. 1965. "Vegetation Changes during a Thirty-year Period in Grassland Communities near Hayes, Kansas." *Ecology* 46:714–20.

Allen, T. F. H., and T. W. Hoekstra. 1992. *Towards a Unified Ecology*. New York: Columbia University Press.

Allen T. F. H., and T. B. Starr. 1982. *Hierarchy: Perspectives for Ecological Complexity*. Chicago: University of Chicago Press.

Anderson, K. L., et al. 1970. "Burning Bluestem Range." *Journal of Range Management* 23:81–92.

Anderson, R. C. 1982. "An Evolutionary Model Summarizing the Roles of Fire, Climate, and Grazing Animals in the Origin and Maintenance of Grasslands: An End Paper." In *Grasses and Grasslands: Systematics and Ecology*, edited by J. R. Estes et al., 297–308. Norman: University of Oklahoma Press.

———. 1989. "The Historic Role of Fire in the North American Grassland." In *Fire in the North American Tallgrass Prairie*, edited by S. L. Collins and L. L. Wallace, 8–18. Norman: University of Oklahoma Press.

Anderson, R. C., et al. 1989. "Numbers and Biomass of Selected Insect Groups on Burned and Unburned Sand Prairie." *American Midland Naturalist* 122:151–62.

Andren, H. 1994. "Effects of Habitat Fragmentation on Birds and Mammals in Landscapes with Different Proportions of Suitable Habitat." *Oikos* 71:355–66.

Archer, S. 1989. "Have Southern Texas Savannas Been Converted to Woodlands in Recent History?" *American Naturalist* 134:545–61.

Arenz, C. L. In press. "Initiation of a Butterfly Monitoring Program at the Tallgrass Prairie Preserve, Osage County, Oklahoma." *Proceedings of the Oklahoma Academy of Science.*

Armstrong, D. M., et al. 1986. "Distributional Patterns of Mammals in the Plains States." *Occasional Papers of the Museum, Texas Tech University* 105:1–27.

Askins, R. A. 1993. "Population Trends in Grassland, Shrubland, and Forest Birds in Eastern North America." *Current Ornithology* 11:1–34.

Axelrod, D. I. 1985. "Rise of the Grassland Biome, Central North America." *Botanical Review* 51:163–201.

Bailey, R. G., et al. 1994. *Ecoregions and Subregions of the United States.* Map. Washington, D.C.: U.S. Department of Agriculture, Forest Service.

Bailey, R. M., and M. O. Allum. 1962. *Fishes of South Dakota.* Miscellaneous Publication 119. Ann Arbor: University of Michigan, Museum of Zoology.

Bailey, V. 1905. *Biological Survey of Texas.* North American Fauna 25. Washington, D.C.: U.S. Biological Survey.

Ballinger, R. E., and S. M. Jones. 1985. "Ecological Disturbance in a Sandhills Prairie: Impact and Importance to the Lizard Community on Arapaho Prairie in Western Nebraska." *Prairie Naturalist* 17:91–100.

Bamforth, D. B. 1987. "Historical Documentation and Bison Ecology on the Great Plains." *Plains Anthropologist* 32:1–16.

Barker, W., and W. Whitman. 1988. "Vegetation of the Northern Great Plains." *Rangelands* 10:266–72.

Barkley, T., ed. 1977. *The Atlas of the Flora of the Great Plains.* Ames: Iowa State University Press.

Barnes, A. M. 1983. "A Review of Plague and Its Relevance to Prairie Dog Populations and the Black-footed Ferret." In *Management of Prairie Dog Complexes for the Reintroduction of the Black-footed Ferret,* edited by J. L. Oldemeyer et al., 28–37. Washington, D.C.: U.S. Department of the Interior, Fish and Wildlife Service.

Barnes, P. W., and A. T. Harrison. 1982. "Species Distribution and Community Organization in a Nebraska Sandhills Mixed Prairie as Influenced by Plant/Soil-Water Relationships." *Oecologia* 52:192–201.

Barnes, P. W., et al. 1983. "Distribution, Production, and Diversity of C3- and C4-dominated Communities in a Mixed Prairie." *Canadian Journal of Botany* 61:741–51.

———. 1984. "Vegetation Patterns in Relation to Topography and Edaphic Variation in Nebraska Sandhills Prairie." *Prairie Naturalist* 16:145–58.

Barnes, R. L. 1993. "The U.C.C.'s Insidious Preference for Agronomy over Ecology in Farm Lending Decisions." *University of Colorado Law Review* 64:457–512.

Batt, B. D. J., et al. 1989. "The Use of Prairie Potholes by North American Ducks." In *Northern Prairie Wetlands,* edited by A. G. van der Valk, 204–27. Ames: Iowa State University Press.

Beiswenger, R. E. 1986. "An Endangered Species, the Wyoming Toad *Bufo hemiophrys baxteri*—The Importance of an Early Warning System." *Biological Conservation* 37:59–71.

Benke, A. C. 1990. "A Perspective on America's Vanishing Streams." *Journal of the North American Benthological Society* 9:77–88.

Bianchet, F., et al. 1994. "Mountain Goat Recruitment: Kid Production and Survival to Breeding Age." *Canadian Journal of Zoology* 72:22–27.

Bicak, T. K, et al. 1982. "Effects of Grazing on Long-billed Curlew (*Numenius Americanus*) Breeding Behavior and Ecology in Southwestern Idaho." In *Wildlife-Livestock Relationships Symposium: Proceedings 10,* edited by J. M. Peek and P. D. Dalke, 74–85. Moscow: University of Idaho, Forest, Wildlife, and Range Experiment Station.

Biddy, C. J., et al., eds. 1992. *Putting Biodiversity on the Map: Priority Areas for Global Conservation.* Cambridge: International Council for Bird Preservation.

Biondini, M. E., et al. 1989. "Seasonal Fire Effects on the Diversity Patterns, Spatial Distribution, and Community Structure of Forbs in the Northern Mixed Prairie." *Vegetatio* 85:21–31.

Bird, R. D. 1961. "Ecology of the Aspen Parkland of Western Canada in Relation to Land Use." Canadian Department of Agriculture Research Branch Publication 1066. Ottawa: Canadian Department of Agricultural Research.

Birney, E. C., et al. 1976. "Importance of Vegetative Cover to Cycles of *Microtus* Populations." *Ecology* 57:1043–51.

Bisby, G. F. et al., eds. *Designs for a Global Plant Species Information System.* Systematics Association, Special Volume, no. 48. Oxford: Clarendon Press.

Blanchard, F. N. 1923. "The Amphibians and Reptiles of Dickinson County, Iowa." *University of Iowa Studies in Natural History, Lakeside Laboratory Studies* 10:19–26.

Blaustein, A. R. 1994. "Chicken Little or Nero's Fiddle? A Perspective on Declining Amphibian Populations." *Herpetologica* 50:85–97.

Bleed, A. 1990. "Groundwater." In *An Atlas of the Sand Hills,* edited by A. Bleed and C. Flowerday, 67–92. Lincoln: University of Nebraska, Conservation and Survey Division.

Bloemink, B. 1995. *Georgia O'Keeffe: Canyon Suite.* New York: George Braziller.

Blouet, B., and M. Lawson. 1975. *Images of the Plains: The Role of Human Nature in Settlement.* Lincoln: University of Nebraska Press.

Bluemle, J. P. 1977. *The Face of North Dakota Geological Story.* Bismarck: North Dakota Geological Survey Education, series 2.

Bock, C. E., and B. Webb. 1984. "Birds as Grazing Indicator Species in Southeastern Arizona." *Journal of Wildlife Management* 48:1045–49.

Bock, C. E., et al. 1990. "The Effect of Livestock Grazing upon Abundance of the Lizard, *Sceloporus scalaris*, in Southeastern Arizona." *Journal of Herpetology* 24:445–46.

Bogan M. A., et al. 1995. "A Portrait of Faunal and Floral Change in the Sandhills of Northern Nebraska." In *A Biological Survey of Fort Niobrara and Valentine National Wildlife Refuges,* edited by M. A. Bogan, 6–24. Fort Collins: U.S. Department of the Interior, National Biological Service.

Bolen, E. G., et al. 1989a. "Playa Lakes: Prairie Wetlands of the Southern High Plains." *BioScience* 39:615–23.

———. 1989b. "Playa Lakes." In *Habitat Management for Migrating and Wintering Waterfowl in North America,* edited by L. M. Smith et al., 341–65. Lubbock: Texas Tech University.

Bonner, J. T. 1988. *The Evolution of Complexity.* Princeton: Princeton University Press.

Bose, D. R. 1977. *Rangeland Resources of Nebraska.* Denver: Society of Range Management.

Boul, S. W., et al. 1980. *Soil Genesis and Classification.* Ames: Iowa State University Press.

Boulding, K. 1966. "The Economics of the Coming Spaceship Earth." In *Environmental Quality in a Growing Economy,* edited by H. Jarrett, 3–14. Baltimore: Johns Hopkins University Press.

Bowles, J. B. 1981. "Iowa's Mammal Fauna: An Era of Decline." *Proceedings of the Iowa Academy of Science* 88:38–42.

Box, T. 1967. "Range Deterioration in West Texas." *Southwestern Historical Quarterly* 71:39–45.

Bradley, C., and C. Wallis, 1996. "Prairie Ecosystem Management: An Alberta Perspective." In *Proceedings of the Fourth Prairie Conservation and Endangered Species Workshop.* Natural History Occasional Paper No. 23. Edmonton: Provincial Museum of Alberta.

Bragg, A. N. 1960. "Population Fluctuation in the Amphibian Fauna of Cleveland County, Oklahoma, during the Past Twenty-five Years." *Southwestern Naturalist* 5:165–69.

Bragg, T. B. 1978. "Effects of Burning, Cattle Grazing, and Topography on Vegetation of the Choppysands Range Site in the Nebraska Sandhills Prairie." In *Proceedings of the First International Rangeland Congress,* edited by D. N. Hyder, 248–53. Denver: Colorado.

———. 1982. "Seasonal Variations in Fuel and Fuel Consumption by Fires in a Bluestem Prairie." *Ecology* 63:7–11.

———. 1986. "Fire History of a North American Sandhills Prairie." Abstract. *Program of the Fourth International Congress of Ecology.* Syracuse: State University of New York.

———. In press. "The Physical Environment of Great Plains Grasslands." In *The*

Changing Prairie: North American Grasslands, edited by A. Joern and K. H. Keeler. New York: Oxford University Press.

Bragg, T. B., and L. C. Hulbert. 1976. "Woody Plant Invasion of Unburned Kansas Bluestem Prairie." *Journal of Range Management* 29:19–24.

Brand, M. D., and H. Goetz. 1986. "Vegetation of Exclosures in Southwestern North Dakota." *Journal of Range Management* 39:434–37.

Branson, F., and J. E. Weaver. 1953. "Quantitative Study of Degeneration of Mixed Prairie." *Botanical Gazette* 114:397–416.

Briske, D. D., and A. M. Wilson. 1977. "Temperature Effects on Adventitous Root Developments in Blue Grama Seedlings. *Journal of Range Management.* 30:276–280.

Broach, E. 1992. "Angels, Architecture, and Erosion: The Dakota Badlands as Cultural Symbol." *North Dakota History* 59:2–15.

Brooks, D., and D. A. McClennan. 1991. *Phylogeny, Ecology, and Behavior.* Chicago: University of Chicago Press.

Brown, B. A. 1991. "Landscape Protection and the Nature Conservancy." In *Landscape Linkages and Biodiversity,* edited by W. E. Hudson, 66–71. Washington, D.C.: Island Press.

Brown, D. E., and R. A. Minnich. 1986. "Fire and Creosote Bush Scrub of the Western Sonoran Desert, California." *American Midland Naturalist* 116:411–22.

Brown, J. R., and S. Archer. 1987. "Woody Plant Seed Dispersal and Gap Formation in a North American Subtropical Savanna Woodland: The Role of Domestic Herbivores." *Vegetatio* 73:73–80.

Brown, K. S., Jr. 1991. "Conservation of Neotropical Environments: Insects as Indicators." In *The Conservation of Insects and Their Habitats,* edited by N. M. Collins and J. A. Thomas, 349–404. Fifteenth Symposium of the Royal Entomological Society of London. San Diego: Academic Press.

Brown, V. K. 1985. "Insect Herbivores and Plant Succession." *Oikos* 44:17–22.

Brundtland, G. H. 1987. *Our Common Future.* World Commission on Environment and Development. New York: Oxford University Press.

Bryant, F. C., et al. 1983. "Controlling Mature Ashe Juniper in Texas with Crown Fires." *Journal of Range Management* 36:165–68.

Bryan, L. 1991. *The Buffalo People: Prehistoric Archeology on the Canadian Plains.* Edmonton: University of Alberta Press.

Bryson, R. A., and F. K. Hare, eds. 1974. *World Survey of Climatology.* Vol. 11, *Climates of North America.* New York: Elsevier Scientific Publishing.

Bryson, R. A., et al. 1970. "The Character of Late-glacial and Post-glacial Climatic Changes." In *Pleistocene and Recent Environments of the Central Great Plains,* edited by W. Dort Jr. and J. K. Jones Jr., 53–74. Lawrence: University Press of Kansas.

Burgess, P. 1992. *A New Vision for the Heartland: The Great Plains in Transition.* Denver: Center for the New West.

Burr, B. M., and L. M. Page. 1986. "Zoogeography of Fishes of the Lower Ohio–Upper Mississippi Basin." In *The Zoogeography of North American Freshwa-*

ter Fishes, edited by C. H. Hocutt and E. O. Wiley, 287–324. New York: John Wiley and Sons.

Bury, R. B., et al. 1991. "Regional Comparisons of Terrestrial Amphibian Communities in Oregon and Washington." In Wildlife and Vegetation of Unmanaged Douglas-Fir Forests, edited by L. F. Ruggiero et al., 341–50. Portland: U.S. Department of Agriculture, Forest Service.

Busack, S. D., and R. B. Bury. 1974. "Some Effects of Off-road Vehicles and Sheep Grazing on Lizard Populations in the Mojave Desert." Biological Conservation 6:179–83.

Butler, J. L., and D. D. Briske. 1988. "Population Structure and Tiller Demography of the Bunchgrass Schizachyrium scoparium in Response to Herbivory." Oikos 51:306–12.

Cabeza de Vaca, A. N. 1984. "Relation that Alvar Nunez Cabeza de Vaca gave . . . from the Year 1527 to the Year 1536." In Spanish Explorers in the Southern United States, 1528–1543, edited by F. Hodge and T. Lewis, 12–126. Austin: Texas State Historical Association.

Caithamer, D. F., et al. 1994. Waterfowl Population Status, 1994. Washington, D.C.: U.S. Department of the Interior, Fish and Wildlife Service.

Campbell, J. A., et al. 1989. "Potential Impact of Rattlesnake Roundups on Natural Populations." Texas Journal of Science 41:301–17.

Canada-Alberta Environmentally Sustainable Agriculture Agreement. November 1992.

Canada-Alberta Soil Conservation Initiative. 1994. CASCI Final Report, 1989–1993. Agriculture Canada and Alberta Agriculture Food and Rural Development, 20 pp.

Canadian Wildlife Service. 1995. Draft Highlights of Accomplishments towards the Goals of the Prairie Conservation Action Plan, 1989–1994. Edmonton: Canadian Wildlife Service, 19 pp.

Capinera, J. L., and T. S. Sechrist. 1982. "Grasshopper (Acrididae)-host Plant Associations: Response of Grasshopper Populations to Cattle Grazing Intensity." Canadian Entomologist 114:1055–62.

Carr, S. M., et al. 1986. "Mitochondrial DNA Analysis of Hybridization between Sympatric White-tailed Deer and Mule Deer in West Texas." Proceedings of the National Academy of Science 83:9576–80.

Castaneda, P. 1984. "Account of the [Coronado] Expedition to Cibola which took place in the year 1540 . . ." In Spanish Explorers in the Southern United States, 1528–1543, edited by F. Hodge and T. Lewis, 281–387. Austin: Texas State Historical Association.

Catlin, G. 1973. Letters and Notes on the Manners, Customs, and Conditions of North American Indians. New York: Dover Press.

Clark, T. W., et al. 1982. "Prairie Dog Colony Attributes and Associated Vertebrate Species." Great Basin Naturalist 42:572–82.

Clements, F. E. 1916. *Plant Succession: An Analysis of the Development of Vegetation.* Carnegie Institute Publication 242. Washington, D.C.: Carnegie Institute.

———. 1936. "Nature and Structure of the Climax." *Journal of Ecology* 24:252–84.

Clinton, W. 1994. "Environmentally and Economically Beneficial Practices on Federal Landscaped Grounds." Memo to Heads of Executive Departments and Agencies. Washington, D.C.: The White House.

Coffin, D. P., and W. K. Lauenroth. 1988. "The Effects of Disturbance Size and Frequency on a Shortgrass Plant Community." *Ecology* 1609–17.

———. 1990. "Vegetation Associated with Nest Sites of Western Harvester Ants (*Pogononyrex occidentalis* Cresson) in a Semiarid Grassland." *American Midland Naturalist* 123:226–35.

Coleman, D. C., et al. 1977. "An Analysis of Rhizosphere-Saprophage Interactions in Terrestrial Ecosystems." In *Soil Organisms as Components of Ecosystems.* Ecological Bulletin (Stockholm), no. 25:299–309.

Collins, A. R., et al. 1984. "An Economic Analysis of the Black-tailed Prairie Dog (*Cynomys ludovicianus*) Control." *Journal of Range Management* 37:358–61.

Collins J. T. 1982. *Amphibians and Reptiles in Kansas.* Public Education Series, no. 8. Lawrence: University of Kansas Museum of Natural History.

Collins, S. L. 1987. "Interaction of Disturbances in a Tallgrass Prairie: A Field Experiment." *Ecology* 68:1243–50.

———. 1992. "Fire Frequency and Community Heterogeneity in Tallgrass Prairie Vegetation." *Ecology* 73:2001–6

Collins, S. L., and S. C. Barber. 1985. "Effects of Disturbance on Diversity in Mixedgrass Prairie." *Vegetatio* 64:87–94.

Collins, S. L., and D. J. Gibson. 1990. "Effects of Fire Structure in Mixed- and Tallgrass Prairies." In *Fire in the North American Tallgrass Prairie,* edited by S. L. Collins and L. L. Wallace, 81–98. Norman: University of Oklahoma Press.

Collins, S. L., and G. E. Uno. 1983. "The Effects of Early Spring Burning on Vegetation in Buffalo Wallows." *Bulletin of the Torrey Botanical Club* 110:474–81.

———. 1985. "Seed Predation, Seed Dispersal, and Disturbance in Grasslands: A Comment." *American Naturalist* 125:866–72.

Collins, S. L., and L. L. Wallace, eds. 1990. *Fire in North American Tallgrass Prairie.* Norman: University of Oklahoma Press.

Committee on the Status of Endangered Wildlife in Canada. 1994. *Endangered Species in Canada List.* Ottawa: Committee on the Status of Endangered Wildlife in Canada.

Conant, R., and J. T. Collins. 1991. *A Field Guide to Reptiles and Amphibians of Eastern and Central North America.* Boston: Houghton Mifflin.

Condra, G. E. 1939. *An Outline of the Principal Natural Resources of Nebraska and Their Conservation.* Bulletin 20. Lincoln: University of Nebraska Conservation and Survey Division.

Conner, J. V., and R. D. Suttkus. 1986. "Zoogeography of Freshwater Fishes of the Western Gulf Slope of North America." In *The Zoogeography of North American Freshwater Fishes,* edited by C. H. Hocutt and E. O. Wiley, 413–56. New York: John Wiley and Sons.

Coppock, D. L., and J. K. Detling. 1986. "Alteration of Bison and Black-tailed Prairie Dog Grazing Interaction by Prescribed Burning." *Journal of Wildlife Management* 50:452–55.

Coppock, D. L., et al. 1983a. "Plant-Herbivore Interactions in a North American Mixed-grass Prairie: 1. Effects of Black-tailed Prairie Dogs on Intraseasonal Aboveground Plant Biomass and Nutrient Dynamics and Plant Species Diversity." *Oecologia* 56:1–9.

———. 1983b. "Plant-Herbivore Interactions in a North American Mixed-grass Prairie: 2. Responses of Bison to Modification of Vegetation by Prairie Dogs." *Oecologia* 56:10–15.

Corn, P. S. 1994. "What We Know and Don't Know about Amphibian Declines in the West." In *Sustainable Ecological Systems: Implementing an Ecological Approach to Land Management,* edited by W. W. Covington and L. F. DeBano, 59–67. Fort Collins: U.S. Department of Agriculture, Forest Service.

Corn P. S., et al. 1995. "Amphibians and Reptiles." In *A Biological Survey of Fort Niobrara and Valentine National Wildlife Refuges,* edited by M. A. Bogan, 32–59. Fort Collins: U.S. Department of the Interior, National Biological Service.

COSEWIC. 1994. *Endangered Species in Canada List.* Committee on the Status of Endangered Wildlife in Canada.

Coupland, R. T. 1961. "A Reconsideration of Grassland Classification in the North American Great Plains of North America." *Journal of Ecology* 49:135–67.

———. 1973. "Producers: 1. Dynamics of Above-ground Standing Crop." Matador Project, Canadian IBP Program. Technical Report 27. Saskatoon, Saskatchewan: Canadian IBP Program.

———. 1979. *Grassland Ecosystems of the World: Analysis of Grasslands and Their Uses.* New York: Cambridge University Press.

———. 1992. "Mixed Prairie." In *Ecosystems of the World.* Vol. 8A, *Natural Grasslands, Introduction and Western Hemisphere,* edited by R. T. Coupland, 151–82. Amsterdam: Elsevier Scientific Publishing.

Cowardin, L. M., et al. 1979. "Classification of Wetlands and Deep Water Habitats of the United States." Washington, D.C.: U.S. Government Printing Office.

———. 1985. *Mallard Recruitment in the Agricultural Environment of North Dakota.* Wildlife Society Monograph 92.

Cross, F. B., and J. T. Collins. 1975. *Fishes in Kansas.* Lawrence: University of Kansas Museum of Natural History and State Biological Survey.

Cross, F. B., and R. E. Moss. 1987. "Historic Changes in Communities and Aquatic Habitats of Plains Streams in Kansas." In *Community and Evolutionary Ecology of*

North American Stream Fishes, edited by W. J. Matthews and D. C. Heins, 155–65. Norman: University of Oklahoma Press.

Cross, F. B., et al. 1986. "Fishes in the Western Mississippi Drainage." In *The Zoogeography of North American Freshwater Fishes,* edited by C. H. Hocutt and E. O. Wiley, 363–412. New York: John Wiley and Sons.

Crosson, P. R. 1992. "Sustainable Agriculture." *Resources* 106:14–17.

Curry, J. P. 1994. *Grassland Invertebrates: Ecology, Influence on Soil Fertility, and Effects on Plant Growth.* London: Chapman and Hall.

Dahl, T. E. 1990. *Wetlands Losses in the United States 1780s to 1980s.* Washington, D.C.: U.S. Department of the Interior, Fish and Wildlife Service.

Dale, B. 1994. *Summary of the North American Waterfowl Management Plan Assessment for Nongame Species in the Three Prairie Provinces in 1993.* Ottawa: Environment Canada, Canadian Wildlife Service.

Daly, H. E. 1991. "Toward an Environmental Macroeconomics." *Land Economics* 67:255–59.

Danks, H. V. 1994. "Regional Diversity of Insects in North America." *American Entomologist* (spring): 50–55.

D'Antonio, C. M., and P. M. Vitousek. 1992. "Biological Invasions by Exotic Grasses, the Grass/Fire Cycle, and Global Change." *Annual Review of Ecology and Systematics* 23:63–87.

Daubenmire, R. 1968. "Ecology of Fire in Grasslands." *Advances in Ecological Research* 5:209–66.

———. 1978. *Plant Geography.* New York: Academic Press.

Day, T. A., and J. K. Detling. 1990. "Grassland Patch Dynamics and Herbivore Grazing Preference Following Urine Deposition." *Ecology* 71:180–88.

Degenhardt, W. G., et al. In press. *Amphibians and Reptiles of New Mexico.* Albuquerque: University of New Mexico Press.

DeJong, E., and K. B. MacDonald. 1975. "The Soil Moisture Regime under Native Grassland." *Geoderma* 14:207–21.

Dempster, J. P. 1991. "Fragmentation, Isolation, and Mobility of Insect Populations." In *The Conservation of Insects and Their Habitats,* edited by N. M. Collins and J. A. Thomas, 143–53. Fifteenth Symposium of the Royal Entomological Society of London. San Diego: Academic Press.

Denevan, W. 1992. "The Pristine Myth: The Landscapes of the Americas in 1492." *Transactions of the Association of American Geographers* 82:369–85.

Derksen, D. U., and W. D. Eldridge. 1980. "Drought-displacement of Pintails to the Arctic Coastal Plain, Alaska." *Journal of Wildlife Management* 44:224–29.

Dix, R. L. 1960. "The Effects of Burning on the Mulch Structure and Species Composition of Grasslands in Western North Dakota." *Ecology* 41:49–56.

Dixon, J. A., and P. B. Sherman. 1990. *Economics of Protected Areas: A New Look at Benefits and Costs.* Washington, D.C.: Island Press.

Dormaar, J. F., and W. D. Willms. 1990. "Effect of Grazing and Cultivation on Some Chemical Properties in the Mixed Prairie." *Journal of Range Management* 43:456–60.

Dreeszen, V. 1984. "Overview of Nebraska and the Sandhills." In *The Sandhills of Nebraska: Yesterday, Today, and Tomorrow,* 1–15. Proceedings of the 1984 Water Resources Seminar Series. Lincoln: University of Nebraska.

Driscoll, R. S., et al. 1984. *An Ecological Land Classification Framework for the United States.* Miscellaneous Publication 1439. Washington, D.C.: U.S. Department of Agriculture, Forest Service.

Ducks Unlimited. 1994. *Ducks Unlimited Continental Conservation Plan.* Parts 1–3. Memphis: Ducks Unlimited, Ducks Unlimited Canada, and Ducks Unlimited de Mexico.

Dugan, J. T., et al. 1994. *Water Level Changes in the High Plains Aquifer—Predevelopment to 1992.* Investigations Report 94–4027. Lincoln: U.S. Department of the Interior, Geological Survey Water Resources.

Dunbar, W. 1806. "Journal of a Voyage Up the Washita River, in Louisiana, in the Years 1804–1805." In Thomas Jefferson, *Message from the President of the United States Communicating Discoveries Made in Exploring the Missouri, Red River, and Washita.* New York: Hopkins and Seymour.

Duvall, D., et al. 1985. "Behavioral Ecology and Ethology of the Prairie Rattlesnake." *National Geographic Research* 1:80–111.

Ehrenreich, J. H., and J. M. Aikman. 1963. "An Ecological Study of the Effect of Certain Management Practices on Native Prairie in Iowa." *Ecological Monographs* 33:113–30.

Elliott, E. T., and D. C. Coleman. 1977. "Soil Protozoan Dynamics of a Colorado Shortgrass Prairie." *Soil Biology and Biochemistry* 9:113–18.

England, R. E., and A. DeVos. 1969. "Influence of Animals on Pristine Conditions on the Canadian Grasslands." *Journal of Range Management* 22:87–94.

Engle, D. M., and P. M. Bultsma. 1984. "Burning of Northern Mixed Prairie during Drought." *Journal of Range Management* 37:398–401.

Erhardt, A., and J. A. Thomas. 1991. "Lepidoptera as Indicators of Change in the Semi-natural Grasslands of Lowland and Upland Europe." In *The Conservation of Insects and Their Habitats,* edited by N. M. Collins and J. A. Thomas, 213–36. Fifteenth Symposium of the Royal Entomological Society of London. San Diego: Academic Press.

Erickson, N. E., and D. M. Leslie Jr. 1987. *Soil-Vegetation Correlations in the Sandhills and Rainwater Basin Wetlands in Nebraska.* Biological Report 87. Washington, D.C.: U.S. Department of the Interior, Fish and Wildlife Service.

Erwin, W. J., and R. H. Stasiak. 1979. "Vertebrate Mortality during the Burning of a Reestablished Prairie in Nebraska." *American Midland Naturalist* 101:247–49.

Evans, E. W. 1984. "Fire as a Natural Disturbance to Grasshopper Assemblages of Tallgrass Prairie." *Oikos* 43:9–16.

———. 1988. "Grasshopper (Insecta: Orthoptera: Acrididae) Assemblages in Tallgrass Prairie: Influences of Fire Frequency, Topography, and Vegetation." *Canadian Journal of Zoology* 66:1495–1501.

Evans, T. 1988. *Prairie: Images of Ground and Sky.* Lawrence: University Press of Kansas.

Evernden, N. 1983. "Beauty and Nothingness: Prairie as Failed Resource." *Landscape* 27:3–6.

Farrar, J. 1982. *The Rainwater Basin, Nebraska's Vanishing Wetlands.* Lincoln: Nebraska Game and Parks Commission.

Fenneman, N. 1931. *Phiography of the Western United States.* New York: McGraw-Hill.

Finch, D. M, et al. 1987. *Habitat Suitability Index Models: Lark Bunting.* Biological Report 82. Washington: U.S. Department of the Interior, Fish and Wildlife Service.

Finley, B. 1990. "Desert's Return Threatens Future of Great Plains." *Denver Post,* 22 July.

Fleharty, E. D., and K. W. Navo. 1983. "Irrigated Cornfields as Habitat for Small Mammals in the Sandsage Prairie Region of Western Kansas." *Journal of Mammalogy* 64:367–79.

Flickinger, E. L. 1981. "Wildlife Mortality at Petroleum Pits in Texas." *Journal of Wildlife Management* 45:560–64.

Flores, D. 1989. "Canyons of the Imagination." *Southwest Art* 18:70–76.

———. 1990. *Caprock Canyonlands: Journeys into the Heart of the Southern Plains.* Austin: University of Texas Press.

———. 1991a. "Bison Ecology and Bison Diplomacy: The Southern Plains from 1800 to 1850." *Journal of American History* 78:465–85.

———. 1991b. "The Plains and the Painters: Two Centuries of Landscape Art from the Llano Estacado." *Journal of American Culture* 14:19–28.

———. 1993. "The Grand Canyon of Texas: How Palo Duro Just Missed Becoming Texas's Third Great National Park." *Texas Parks and Wildlife Magazine* 51:4–14.

———. 1994. "Place: An Argument for Bioregional History." *Environmental History Review* 18:1–18.

Flores, D., and A. Winton. 1989. *Canyon Visions: Photographs and Pastels of the Texas Plains.* Lubbock: Texas Tech University Press.

Forman, R. T. T., and M. Godron. 1986. *Landscape Ecology.* New York: John Wiley and Sons.

Forrest, S. C., et al. 1988. "Population Attributes for the Black-footed Ferret (*Mustela nigripes*) at Meeteetse, Wyoming, 1981–1985." *Journal of Mammalogy* 69:261–73.

Frank, D. A., and S. J. McNaughton. 1993. "Evidence for the Promotion of Aboveground Grassland Production by Native Large Herbivores in Yellowstone National Park." *Oecologia* 96:157–61.

Franklin, I. R. 1986. "Evolutionary Change in Small Populations." In *Conservation Biology,* edited by M. E. Soulé and B. A. Wilcox, 135–50. Sunderland: Sinauer Associates.

Fredrickson, L. H., and F. R. Reid. 1990. "Impacts of Hydrologic Alteration on Management of Freshwater Wetlands." In *Management of Dynamic Ecosystems,* edited by J. M. Sweeney, 71–90. West Lafayette: North Central Section, Wildlife Society.

Freeman, A. M., III. 1993. *The Measurement of Environmental and Resource Values: Theory and Methods.* Washington, D.C.: Resources for the Future.

French, N. R. 1979. *Perspectives in Grassland Ecology.* Berlin: Springer-Verlag.

Frey, J. K. 1992. "Response of a Mammalian Faunal Element to Climatic Changes." *Journal of Mammalogy* 73:43–50.

———. 1993. "Modes of Peripheral Isolate Formation and Speciation." *Systematic Biology* 42:373–81.

Friend, M. 1982. *Playas and Wildlife Disease.* Fort Collins: U.S. Department of the Interior, Fish and Wildlife Service.

Friesen, G. 1985. *The Canadian Prairies: A History.* Toronto: University of Toronto Press.

Frost, D. R., and D. M. Hillis. 1990. "Species in Concept and Practice: Herpetological Applications." *Herpetologica* 46:87–104.

Gabig, P. J. 1985. "Mallards—Past and Prospects." *Nebraskaland* 63:18–33.

Galat, D. L., et al. In press. "Restoring Aquatic Resources to the Lower Missouri River: Issues and Initiatives." In *Overview of River-Floodplain Ecology in the Upper Mississippi River Basin.* Washington, D.C.: U.S. Government Printing Office.

Galatowitsch, S. M., and A. G. van der Valk. 1994. *Restoring Prairie Wetlands: An Ecological Approach.* Ames: Iowa State University Press.

Garcia, F. O., and C. H. Rice. 1994. "Microbial Biomass Dynamics in Tallgrass Prairie." *Soil Society of America Journal* 58:816–23.

Gartner, F. R., et al. 1978. "Prescribed Burning of Range Ecosystems in South Dakota." In *Proceedings of the First International Rangeland Congress,* edited by D. N. Hyder, 687–90. Denver: Society of Range Management.

———. 1986. "Vegetation Responses to Spring Burning in Western South Dakota." In *The Prairie—Past, Present, and Future: Proceedings of the Ninth North American Prairie Conference,* edited by G. K. Clambey and R. H. Pemble, 143–46. Fargo: Tri-College University Center for Environmental Studies.

Gaston, K. J. 1991. "The Magnitude of Global Insect Species Diversity." *Conservation Biology* 5:283–96.

Gaston, K. J., and B. H. McArdle. 1993. "All Else Is Not Equal: Temporal Population Variability and Insect Conservation." In *Perspectives on Insect Conservation,* edited by K. J. Gaston et al., 171–84. Andover, Hants: Intercept.

Genoways, H. H. 1985. "Badger, *Taxidea taxus.*" In *Species of Special Concern in Pennsylvania,* edited by H. H. Genoways and F. J. Brenner, 408–9. Pittsburgh: Carnegie Museum of Natural History.

———. 1986. "Causes for Species of Large Mammals to Become Threatened or Endangered." In *Endangered and Threatened Species Programs in Pennsylvania and Other States: Causes, Issues, and Management,* edited by S. K. Majumdar et al., 234–51. Philadelphia: Pennsylvania Academy of Science.

Genoways, H. H., and D. A. Schlitter. 1967. "Northward Dispersal of the Hispid Cotton Rat in Nebraska and Missouri." *Transactions of the Kansas Academy of Science* 69:356–57.

Genoways, H. H., et al. 1979. "Mammals of the Guadalupe Mountains National Park, Texas." In *Biological Investigations in the Guadalupe Mountains National Park, Texas,* edited by H. H. Genoways and R. J. Baker, 271–332. Washington, D.C.: U.S. Department of the Interior, National Park Service.

George, T. L., et al. 1992. "Impacts of a Severe Drought on Grassland Birds in Western North Dakota." *Ecological Applications* 2:275–84.

Gersib, R. A., et al. 1989. *Waterfowl Values by Wetland Type within Rainwater Basin Wetlands with Special Emphasis on Activity Time Budget and Census Data.* Lincoln: Nebraska Game and Parks Commission, and U.S. Department of the Interior, Fish and Wildlife Service.

———. 1992. *Looking to the Future: An Implementation Plan for the Rainwater Basin Joint Venture.* Lincoln: Nebraska Game and Parks Commission.

Gibbs. E. L., et al. 1971. "The Live Frog Is Almost Dead." *BioScience* 21:1027–34.

Gibson, D. J. 1988. "Regeneration and Fluctuation in Tallgrass Prairie Vegetation in Response to Burning Frequency." *Bulletin of the Torrey Botanical Club* 115:1–12.

Gibson, D. J., and L. C. Hulbert. 1987. "Effects of Fire, Topography, and Year-to-Year Climatic Variation on Species Composition in Tallgrass Prairie." *Vegetatio* 72:175–85.

Gibson, D. J., et al. 1990. "Effects of Small Mammal and Invertebrate Herbivory on Plant Species Richness and Abundance in Tallgrass Prairie." *Oecologia* 84:169–75.

Gilpin, M. E. 1987. "Spatial Structure and Population Vulnerability." In *Viable Populations for Conservation,* edited by M. E. Soulé, 125–39. New York: Cambridge University Press.

Ginsberg, M. H. 1985. "Nebraska's Sandhills Lakes: A Hydrogeologic Overview." *Water Resources Bulletin* 21:573–78.

Girt, J. 1990. *Common Ground: Recommendations for Policy Reform to Integrate Wildlife Habitat, Environmental and Agricultural Objectives on the Farm.* Ottawa: Wildlife Habitat Canada, 56 pp.

Glanz, M., and J. Ausubel. 1984. "The Ogallala Aquifer and Carbon Dioxide: Comparison and Convergence." *Environmental Conservation* 2:123–31.

Gleason, H. A. 1926. "The Individualistic Concept of the Plant Association." *Bulletin of the Torrey Botanical Club* 53:1–20.

Glenn, S. M., and S. L. Collins. 1990. "Patch Structure in Tallgrass Prairies: Dynamics of Satellite Species." *Oikos* 57:229–36.

————. 1992. "Effects of Scale and Disturbance on Rates of Immigration and Extinction of Species in Prairies." *Oikos* 63:273–80.

Glooschenko, W. A., et al. 1993. "The Wetlands of Canada." In *Wetlands of the World: Inventory, Ecology, and Management,* edited by D. F. Whiggams et al., 415–514. Dordrecht: Kluwer Academic Publishers.

Goldsmith, F. B. 1991. "The Selection of Protected Areas." In *The Scientific Management of Temperate Communities for Conservation,* edited by I. F. Spellerberg et al., 273–91. Oxford: Blackwell Scientific Publications.

Goldstein-Golding, E. L. 1991. "The Ecology and Structure of Urban Greenspaces." In *Habitat Structure: The Physical Arrangement of Objects in Space,* edited by S. S. Bell et al., 392–411. London: Chapman and Hall.

Gould, S. J. 1989. *Wonderful Life: The Burgess Shale and the Nature of History.* New York: W. W. Norton.

Government of Canada. 1991. *The State of Canada's Environment.* Ottawa: Minister of the Environment.

Graham, R. W. 1986. "Response of Mammalian Communities to Environmental Changes during the Late Quaternary." In *Community Ecology,* edited by J. W. Diamond and T. J. Case, 300–13. New York: Harper and Row.

Grant, W. E., et al. 1982. "Structure and Productivity of Grassland Small Mammal Communities Related to Grazing-induced Changes in Vegetative Cover." *Journal of Mammalogy* 63:248–60.

Graves B. M., and D. Duvall. 1993. "Reproduction, Rookery Use, and Thermoregulation in Free-ranging, Pregnant *Crotalus v. viridis.*" *Journal of Herpetology* 27:33–41.

Green, D. 1973. *Land of the Underground Rain: Irrigation on the Texas High Plains, 1910–1970.* Austin: University of Texas Press.

Greenwood, R. J. 1986. "Influence of Striped Skunk Removal on Upland Duck Nest Success in North Dakota." *Wildlife Society Bulletin* 14:6–11.

Greenwood, R. J., et al. 1995. "Factors Associated with Duck Nest Success in the Prairie Pothole Region of Canada." *Wildlife Society Monograph* 128.

Gregory P. T. 1982. "Reptilian Hibernation." In *Biology of the Reptilia.* Vol. 13, *Physiology,* edited by C. Gans and F. H. Pough, 53–154. London: Academic Press.

Grime, J. P. 1973. "Competitive Exclusion in Herbaceous Vegetation." *Nature* 242:344–47.

Grimm, J. W., and R. H. Yahner. 1987. "Small Mammal Responses to Roadside Habitat Management in South Central Minnesota." *Journal of the Minnesota Academy of Science* 53:16–21.

Grinnell, J. 1922. "The Trend of Avian Populations in California." *Science* 56:671–76.

Guthery, F. S., and F. C. Bryant. 1982. "Status of Playas in the Southern Great Plains." *Wildlife Society Bulletin* 10:309–17.

Guthery, F. S., et al. 1981. *Playa Assessment Study.* Amarillo: Water and Power Resources Service.

————. 1982. "Characterization of Playas in the North-central Llano Estacado in Texas." *Transactions of the North American Wildlife and Natural Resources Conference* 47:516–27.

Haferkamp, M. R., et al. 1993. "Effects of Mechanical Treatments and Climatic Factors on the Productivity of Northern Great Plains Rangelands." *Journal of Range Management* 46:346–50.

Hall, D. L., and M. R. Willig. 1994. "Mammalian Species Composition, Diversity, and Succession in Conservation Reserve Program Grasslands." *Southwestern Naturalist* 39:1–10.

Hall, E. R. 1962. "The Prairie National Park." *National Parks Magazine* 41:1–6.

Hammerson, G. A. 1986. *Amphibians and Reptiles in Colorado.* Denver: Colorado Division of Wildlife.

Hansen, H. A., and D. E. McKnight. 1964. "Emigration of Drought-displaced Waterfowl to the Arctic." *Transactions of the North American Wildlife and Natural Resource Conference* 29:119–27.

Hansen, R. M., and I. K. Gold. 1977. "Blacktail Prairie Dogs, Desert Cottontails, and Cattle Trophic Relations on Shortgrass Range." *Journal of Range Management* 30:210–13.

Hansen, J. R. 1984. "*Bison* Ecology in the Northern Great Plains and a Reconstruction of *Bison* Patterns for the North Dakota Region." *Plains Anthropologist* 29:93–113.

Hartley, W. 1950. "The Global Distribution of Tribes of Gramineae in Relation to Historical and Environmental Factors." *Australia Journal of Agricultural Research* 1:355–73.

Harty, F. M., et al. 1991. "Direct Mortality and Reappearance of Small Mammals in an Illinois Grassland after a Prescribed Burn." *Natural Areas Journal* 11:114–18.

Harwood Group. 1994. "Managing Policy Dialogues." Bethesda, Md.: Harwood Group.

————. 1995. Memo to the Great Plains Partnership. Denver: Western Governors' Association.

Heat-Moon, W. L. 1991. *Prairyerth.* Boston: Houghton Mifflin. Quoting Field, *Lost Horizon.*

Herkert, J. R. 1994a. "The Effects of Habitat Fragmentation on Midwestern Grassland Bird Communities." *Ecological Applications* 4:461–71.

————. 1994b. "Breeding Bird Communities of Midwestern Prairie Fragments: The Effects of Prescribed Burning and Habitat Area." *Natural Areas Journal* 14:128–35.

Heske, E. J., et al. 1993. "Effects of Kangaroo Rat Exclusion on Vegetation Structure and Plant Species Diversity in the Chihuahuan Desert." *Oecologia* 95:520–24.

Hesse, L. W. 1994. "The Status of Nebraska Fishes in the Missouri River, 5. Selected Chubs and Minnows (Cyprinidae)." *Transactions of the Nebraska Academy of Sciences* 21:99–108.

Hesse, L. W., et al. 1989. "Missouri River Fishery Resources in Relation to Past, Present, and Future Stresses." In *Proceedings of the International Large River Symposium*, edited by D. P. Dodge, 352–71. Canadian Special Publication of Fisheries and Aquatic Science 106. Ottawa: Department of Fisheries and Oceans.

Hester, T., ed. 1976. *The Texas Archaic: A Symposium*. San Antonio: Center for Archaeological Research, University of Texas at San Antonio.

Higgins, K. F. 1977. "Duck Nesting in Intensively Farmed Areas of North Dakota." *Journal of Wildlife Management* 41:232–42.

———. 1986. *Interpretation and Compendium of Historical Fire Accounts in the Northern Great Plains*. Resource Publication 161. Washington, D.C.: U.S. Department of the Interior, Fish and Wildlife Service.

Higgins, K. F., et al. 1984. "Breeding Bird Community Colonization of Sown Stands of Native Grasses in North Dakota." *Prairie Naturalist* 16:177–82.

———. 1989. *Prescribed Burning Guidelines in the Northern Great Plains*. U.S. Department of Agriculture Publication EC-760. Brookings: U.S. Department of the Interior, Fish and Wildlife Service, and Cooperative Extension Service, South Dakota State University.

High Plains Associates. 1982. *Six State High Plains Ogallala Aquifer Regional Resources Study*. Austin: High Plains Associates.

Hine, R. L., et al. 1981. *Leopard Frog Populations and Mortality in Wisconsin, 1974–1976*. Technical Bulletin 122. Madison: Wisconsin Department of Natural Resources.

Hofmann, L., and R. E. Ries. 1989. "Animal Performance and Plant Production from Continuously Grazed Cool-season Reclaimed and Native Pastures." *Journal of Range Management* 42:248–51.

Hoover, E. I., and T. B. Bragg. 1981. "Effects of Season of Burning and Mowing on an Eastern Nebraska Stipa-Andropogon Prairie." *American Midland Naturalist* 105:13–18.

Hopkins, H. H., et al. 1948. "Some Effects of Burning upon a Prairie in West-central Kansas." *Transactions of the Kansas Academy of Science* 51:131–41.

Hornaday, W. T. 1889. "The Extermination of the American Bison." *Report of the U.S. National Museum, 1886–1887*, 367–548. Washington, D.C.: U.S. National Museum.

Houtcooper, W. C. 1977. "Food Habits of Rodents in a Cultivated Ecosystem." *Journal of Mammalogy* 59:427–30.

———. 1978. "Distribution and Abundance of Rodents in Cultivated Ecosystems." *Proceedings of the Indiana Academy of Science* 87:434–37.

Howe, H. F. 1994a. "Response of Early- and Late-flowering Plants to Fire Season in Experimental Prairies." *Ecological Applications* 4:121–33.

———. 1994b. "Managing Species Diversity in Tallgrass Prairie: Assumptions and Implications." *Conservation Biology* 8:691–704.

Howarth, B. 1993. "Literature of Place: Environmental Writers." *Isle: Interdisciplinary Studies in Literature and Environment* 1:167–73.

Hulbert, L. C. 1969. "Fire and Litter Effects in Undisturbed Bluestem Prairie." *Ecology* 50:874–77.

———. 1988. "Causes of Fire Effects in Tallgrass Prairie." *Ecology* 69:46–58.

Huntington, M. H., and J. D. Echeverria. 1990. *The American Rivers Outstanding Rivers List.* Washington D.C.: American Rivers.

Huntly, N., and R. Inouye. 1988. "Pocket Gophers in Ecosystems: Patterns and Mechanisms." *BioScience* 38:786–93.

Hurt, D. 1985. "The National Grasslands: Origin and Development in the Dust Bowl." In *The History of Soil and Water Conservation,* edited by D. Helms and S. L. Flader, 144–57. Washington, D.C.: Agricultural History Society.

Imboden, C. 1988. Foreword to *Grassland Birds,* edited by P. D. Goriup. Cambridge: International Council for Bird Preservation.

Ingham, R. E., and J. K. Detling. 1990. "Effects of Root-feeding Nematodes on Aboveground Net Primary Production in a North American Grassland." *Plant and Soil* 121:279–81.

Inouye, R. S., et al. 1987. "Pocket Gophers *(Geomys bursarius),* Vegetation, and Soil Nitrogen along a Successional Sere in East Central Minnesota." *Oecologia* 72:178–84.

Iverson, C. G., et al. 1985. "Distribution and Abundance of Sandhill Cranes Wintering in Western Texas." *Journal of Wildlife Management* 49: 250–55.

Jackson, W. 1980. *New Roots for Agriculture.* Lincoln: University of Nebraska Press.

James, E. 1906. *Early Western Travels, 1748–1846,* edited by R. G. Thwaites. Vol. 17, *Account of an Expedition from Pittsburgh to the Rocky Mountains Performed in the Years 1819, 1829.* Cleveland: Arthur C. Clark.

James, S. W. 1988. "The Postfire Environment and Earthworm Populations in Tallgrass Prairie." *Ecology* 69:476–83.

Janzen, D. H. 1984. "Dispersal of Small Seeds by Big Herbivores: Foliage Is the Fruit." *American Naturalist* 123:338–53.

Jaramillo, V. J., and J. K. Detling. 1992a. "Small-scale Grazing in a Semi-arid North American Grassland. 1. Tillering, N Uptake, and Retranslocation in Simulated Urine Patches." *Journal of Applied Ecology* 29:1–8.

———. 1992b. "Small-scale Grazing in a Semi-arid North American Grassland. 2. Cattle Grazing of Simulated Urine Patches." *Journal of Applied Ecology* 29:9–13.

Jenny, H. 1941. *Factors in Soil Formation.* New York: McGraw-Hill.

Joern, A. 1992. "Variable Impact of Avian Predation on Grasshopper Assemblies." *Oikos* 64:458–63.

Johnson, D. H., and J. W. Grier. 1988. "Determinants of Breeding Distributions of Ducks." *Wildlife Monographs* 100.

Johnson, D. H., and M. D. Schwartz. 1993. "The Conservation Reserve Program and Grassland Birds." *Conservation Biology* 7:934–37.

Johnson, E., ed. 1987. *Lubbock Lake: Late Quaternary Studies on the Southern High Plains.* College Station: Texas A&M University Press.

Johnson, S. R. 1995. "The Future Direction of U.S. Farm Policy." Paper presented at the Grain World 1995 Conference. Winnipeg: Canadian Wheat Board.

Johnson, W. C. 1994. "Woodland Expansion in the Platte River, Nebraska: Pattern and Causes." *Ecological Monographs* 64:45–84.

Jones, J. K., Jr. 1964. *Distribution and Taxonomy of Mammals of Nebraska.* Lawrence: University of Kansas Publications, Museum of Natural History.

Jones, J. K., Jr., et al. 1983. *Mammals of the Northern Great Plains.* Lincoln: University of Nebraska Press.

Jones, K. B. 1981. "Effects of Grazing on Lizard Abundance and Diversity in Western Arizona." *Southwestern Naturalist* 26:107–15.

Jordan, T. 1994. *North American Cattle Ranching Frontiers.* Lincoln: University of Nebraska Press.

Josephson, R. A. 1992. "An Economic Evaluation of Land Use Changes in Southwest Manitoba." Winnipeg: Manitoba Habitat Heritage Corporation. Personal communication.

Judson, S. 1950. "Depressions of the Northern Portions of the Southern High Plains of Eastern New Mexico." *Bulletin of the Geological Society of America.* 61:253–73.

Junk, W. J., et al. 1989. "The Flood-pulse Concept in River-Floodplain Systems." In *Proceedings of the International Large River Symposium,* edited by D. P. Dodge, 110–27. Canadian Special Publication of Fisheries and Aquatic Science 106. Ottawa: Department of Fisheries and Oceans.

Kahn, J. R. 1995. *The Economic Approach to Environmental and Natural Resources.* Orlando: Dryden Press.

Kantrud, H. A. 1981. "Grazing Intensity Effects on the Breeding Avifauna of North Dakota Native Grasslands." *Canadian Field-Naturalist* 95:404–17.

Kantrud, H. A., and R. L. Kologiski. 1982. *Effects of Soils and Grazing on Breeding Birds of Uncultivated Upland Grasslands of the Northern Great Plains.* Wildlife Research Report 15. Washington, D.C.: U.S. Department of the Interior, Fish and Wildlife Service.

Kantrud, H. A., et al. 1989a. *Prairie Basin Wetlands of Dakotas: A Community Profile.* Biological Report 85. Washington, D.C.: U.S. Department of the Interior, Fish and Wildlife Service.

———. 1989b. "Vegetation of Wetlands of the Prairie Pothole Region." In *Northern Prairie Wetlands,* edited by A. G. van der Valk, 132–87. Ames: Iowa State University Press.

Karr, J. R., and D. R. Dudley. 1981. "Ecological Perspective on Water Quality Goals." *Environmental Management* 5:55–68.

Karr, J. R., et al. 1985. "Fish Communities and Midwestern Rivers: A History and Degradation." *BioScience* 35:90–95.

Kaufman, D. W., et al. 1983. "Effects of Fire on Rodents in Tallgrass Prairie of the Flint Hills Region of Eastern Kansas." *Prairie Naturalist* 15:49–56.

————. 1990. "Small Mammals and Grassland Fires." In *Fire in the North American Tallgrass Prairie,* edited by S. L. Collins and L. L. Wallace, 46–80. Norman: University of Oklahoma Press.

Kaufman, G. A., et al. 1988. "Influence of Fire and Topography on Habitat Selection by *Peromyscus maniculatus* and *Reithrodontomys megalotis* in Ungrazed Tallgrass Prairie." *Journal of Mammalogy* 69:342–52.

Kaul, R. B. 1990. "Plants." In *An Atlas of the Sand Hills,* edited by A. Bleed and C. Flowerday, 127–42. Lincoln: University of Nebraska Conservation and Survey Division.

Kaul, R. B., and S. B. Rolfsmeier. 1993. *Native Vegetation of Nebraska.* Map. Lincoln: University of Nebraska Conservation and Survey Division.

Kaul, R. B., et al. 1988. "The Niobrara River Valley, a Postglacial Migration Corridor and Refugium of Forest Plants and Animals in the Grasslands on Central North America." *Botanical Review* 54:44–81.

Kemmis, T. J. 1991. "Glacial Landforms, Sedimentology, and Depositional Environments of the Des Moines Lobe, Northern Iowa." Ph.D. diss., University of Iowa, Iowa City.

Kendeigh, S. C. 1941. "Birds of a Prairie Community." *Condor* 43:165–74.

Kiel, W. H., et al. 1972. *Waterfowl Habitat Trends in the Aspen Parkland of Manitoba.* Report Series 18. Ottawa: Canadian Wildlife Service.

Kiester, A. R. 1971. "Species Density of North American Amphibians and Reptiles." *Systematic Zoology* 20:127–37.

King, M. B., and D. Duvall. 1990. "Prairie Rattlesnake Seasonal Migrations: Episodes of Movement, Vernal Foraging, and Sex Differences." *Animal Behavior* 39:924–35.

King, P. 1967. "Tectonic Features of the United States" In *The National Atlas of the United States,* edited by A. Gerlach, 69–71. Washington, D.C.: U.S. Department of the Interior, Geological Survey.

Kirsch, L. M., and A. D. Kruse. 1973. "Prairie Fires and Wildlife." *Proceedings of the Tall Timbers Fire Ecology Conference* 12:289–303.

Kittredge, W., and A. Smith, eds. 1988. *The Last Best Place: A Montana Anthology.* Seattle: University of Washington Press.

Klauber, L. M. 1972. *Rattlesnakes: Their Habits, Life Histories, and Influence on Mankind.* Berkeley: University of California Press.

Klett, A. T., et al. 1988. "Duck Nest Success in the Prairie Pothole Region." *Journal of Wildlife Management* 52:431–40.

Klopatek, J. M., et al. 1979. "Land-use Conflicts with Natural Vegetation in the United States." *Environmental Conservation* 6:191–99.

Knapp, A. K. 1984. "Post-burn Differences in Solar Radiation, Leaf Temperature, and Water Stress Influencing Production in a Lowland Tallgrass Prairie." *American Journal of Botany* 71:220–27.

Knapp, A. K., and T. R. Seastedt. 1986. "Detritus Accumulation Limits Productivity of Tallgrass Prairie." *BioScience* 36:662–69.

Knopf, F. L. 1986. "Changing Landscapes and the Cosmopolitanism of Eastern Colorado Avifauna." *Wildlife Society Bulletin* 14:132–42.

———. 1992. "Faunal Mixing, Faunal Integrity, and the Biopolitical Template for Diversity Conservation." *Transactions of the North American Wildlife and Natural Resources Conference* 57:330–42.

———. 1994. "Avian Assemblages on Altered Grasslands." *Studies in Avian Biology* 15:247–57.

———. 1996. "Perspectives on Grazing Nongame Bird Habitats." In *Rangeland Wildlife,* edited by P. R. Krausman, 51–58. Denver: Society for Range Management.

Knopf, F. L., and B. J. Miller. 1994. "*Charadrius montanus*—Montane, Grassland, or Bare-Ground Plover?" *Auk* 111:505–6.

Knopf, F. L., and J. R. Rupert. 1996. "Reproduction and Movements of Mountain Plovers Breeding in Colorado." *Wilson Bulletin* 108:28–35.

Knopf, F. L., and F. B. Samson. 1995. "Conserving the Biotic Integrity of the Great Plains." In *Conservation of Great Plains Ecosystems: Current Science, Future Options,* edited by S. R. Johnson and A. Bouzaher, 121–133. Dordrecht: Kluwer Academic Press.

Knopf, F. L., and M. L. Scott. 1990. "Altered Flows and Created Landscapes in the Platte River Headwaters, 1840–1990." In *Management of Dynamic Ecosystems,* edited by J. M. Sweeney, 47–70. West Lafayette: North Central Section, Wildlife Society.

Koonz, W. 1992. "Amphibians in Manitoba." *Canadian Wildlife Service Occasional Paper* 76:19–20.

Kosztarab, M., and C. W. Schaefer, eds. 1990. *Systematics of the North American Insects and Arachnids: Status and Needs.* Virginia Agricultural Experiment Station Information Series 90–91.

Kremen, C., et al. 1993. "Terrestrial Arthropod Assemblages: Their Use in Conservation Planning." *Conservation Biology* 7:796–808.

Kromm, D., and S. White. 1985. *Conserving the Ogallala: What Next?* Manhattan: Kansas State University.

———, eds. 1994. *Groundwater Exploitation in the High Plains.* Lawrence: University Press of Kansas.

Krueger, K. 1986. "Feeding Relationships among Bison, Pronghorn, and Prairie Dogs: An Experimental Analysis." *Ecology* 67:760–70.

Kucera, C. L. 1992. "Tall-grass Prairie." In *Ecosystems of the World.* Vol. 8A, *Natural Grasslands, Introduction and Western Hemisphere,* edited by R. T. Coupland, 227–68. Amsterdam: Elsevier Scientific Publishing.

Küchler, A. W. 1985. "Potential National Vegetation." *National Atlas of the United States of America.* Map. Reston: U.S. Department of the Interior, Geological Survey.

Labedz, T. E. 1990. "Birds." In *An Atlas of the Sand Hills,* edited by A. Bleed and C. Flowerday, 161–80. Lincoln: University of Nebraska Conservation and Survey Division.

Landres, P. B., et al. 1988. "Ecological Uses of Vertebrate Indicator Species: A Critique." *Conservation Biology* 2:316–29.

Lanoo, M. J., et al. 1994. "An Altered Amphibian Assemblage: Dickinson County, Iowa, Seventy Years after Frank Blanchard's Survey." *American Midland Naturalist* 131:311–19.

Lant, C. L., et al. 1995. "The 1990 Farm Bill and Water Quality in Corn Belt Watersheds: Conserving Remaining Wetlands and Restoring Farmed Wetlands." *Journal of Soil and Water Conservation* 50:201–5.

Lauenroth, W. K. 1979. "Grassland Primary Production: North American Grasslands in Perspective." In *Perspectives in Grassland Ecology,* edited by N. R. French, 3–24. New York: Springer-Verlag.

Lauenroth, W. K., and D. G. Milchunas. 1992. "Short-grass Steppe." In *Ecosystems of the World.* Vol. 8A, *Natural Grasslands, Introduction and Western Hemisphere,* edited by R. T. Coupland, 183–226. Amsterdam: Elsevier Scientific Publishing.

Lauenroth, W. K., and E. O. Sala. 1992. "Long-term Forage Production on North American Shortgrass Steppe." *Ecological Applications* 2:397–403.

Launchbaugh, J. L. 1964. "Effects of Early Spring Burning on Yields of Native Vegetation." *Journal of Range Management* 17:5–6.

———. 1973. "Effects of Fire on Shortgrass and Mixed Prairie Species." *Proceedings of the Tall Timbers Fire Ecology Conference* 12:129–51.

Lawton, D. 1984. "Physical Characteristics of the Sandhills: Groundwater Hydrogeology and Stream Hydrology." In *The Sandhills of Nebraska: Yesterday, Today, and Tomorrow.* Proceedings of the 1984 Water Resources Seminar Series, 44–53. Lincoln: University of Nebraska.

Lawton, J. H. 1983. "Plant Architecture and the Diversity of Phytophagous Insects." *Annual Review Entomology* 28:23–39.

Lawton, J. H., and M. MacGarvin. 1986. "The Organization of Herbivore Communities." In *Community Ecology: Pattern and Process,* edited by J. Kikkawa and D. J. Anderson, 163–86. Oxford: Blackwell Scientific Publications.

Lerohl, M. L. 1990. "The Western Canadian Farm Policy Dilemma." *Canadian Journal of Agriculture Economics* 38:557.

Lesica, P. 1995. "Endless Sea of Grass—No Longer." *Kelseya* 8:1–9.

Levell, J. P. 1995. *A Field Guide to Reptiles and the Law.* Excelsior: Serpent's Tale.

Lewis, G. L., and R. J. Bockelman. 1988. "Wetland Impacts of Large Scale Center-pivot Irrigation in Nebraska's Sandhills: Hydrologic and Habitat Issues." In *Proceedings of the National Symposium on Protection of Wetlands from Agricultural Impacts,* edited by P. J. Stuber, 30–37. Biological Report 88. Washington, D.C.: U.S. Department of the Interior.

Lipnack, J., and J. Stamps. 1994. *The Age of the Network.* Essex Junction, Vt.: Oliver Wight Publications.

Lissey, A. 1971. "Depression-focused Transient Groundwater Flow Patterns in Manitoba." *Geological Association of Canada Special Paper* 9:333–41.

Loope, D. B. 1986. "Recognizing and Utilizing Vertebrate Tracks in Cross Section: Cenozoic Hoofprints from Nebraska." *Palaios* 1:141–51.

Loucks, O. L., et al. 1985. "Gap Processes and Large-scale Disturbances in Sand Prairies." In *The Ecology of Natural Disturbance and Patch Dynamics,* edited by S. T. A. Pickett and P. S. White, 71–83. New York: Academic Press.

Loveland, T. R., et al. 1991. "Development of a Land-cover Characteristics Data Base for the Conterminous U.S." *Photogrammetric Engineering and Remote Sensing* 57:1453–63.

Lovell, D. C., et al. 1985. "Succession of Mammals in a Disturbed Area of the Great Plains." *Southwestern Naturalist* 30:335–42.

Lundelius, E. L., et al. 1983. "Terrestrial Vertebrate Faunas." In *Late Quaternary Environments of the United States.* Vol. 1., edited by S. C. Porter, 311–53. Minneapolis: University of Minnesota Press.

Lura, C. L., et al. 1988. "Range Plant Communities of the Central Grasslands Research Station in South Central North Dakota." *Prairie Naturalist* 20:177–92.

Lynch-Stewart, P. 1983. *Land Use Change in Wetlands of Southern Canada: Review and Bibliography.* Working Paper 26. Ottawa: Environment Canada, Lands Directorate.

Mack, R. N., and J. N. Thompson. 1982. "Evolution in Steppe with Few Large Hooved Mammals." *American Naturalist* 119:757–73.

Madole, R. 1994. "Stratigraphic Evidence of Desertification in the West-Central Great Plains within the Past One Thousand Yr." *Geology* 22:483–86.

Malin, J. 1984. *History and Ecology: Studies of the Grasslands,* edited by Robert Swierenga. Lincoln: University of Nebraska Press.

Malone, M., and R. Roeder. 1991. *Montana: A History of Two Centuries.* Seattle: University of Washington Press.

Marshall, L. G. 1984. "Who Killed Cock Robin? An Investigation of the Extinction Controversy." In *Quaternary Extinctions: A Prehistoric Revolution,* edited by P. S. Martin and R. G. Klein, 785–806. Tucson: University of Arizona Press.

Martin, L. D., and A. M. Neuner. 1978. "The End of the Pleistocene in North America." *Transactions of the Nebraska Academy of Sciences* 6:117–26.

Martin, P. S. 1984. "Prehistoric Overkill: The Global Model." In *Quaternary Extinc-*

tions: A Prehistoric Revolution, edited by P. S. Martin and R. G. Klein, 354–403. Tucson: University of Arizona Press.

Martin, P. S., and R. G. Klein, eds. 1984. *Quaternary Extinctions: A Prehistoric Revolution.* Tucson: University of Arizona Press.

Martin, P. S., and H. Wright Jr., eds. 1967. *Pleistocene Extinctions: The Search for a Cause.* New Haven: Yale University Press.

Martin, T. E. 1981. "Limitation in Small Habitat Islands: Chance or Competition?" *Auk* 98:715–34.

Maschinski, J., and T. Whitman. 1989. "The Continuum of Plant Responses to Herbivory: The Influences of Plant Association, Nutrient Availability, and Timing." *American Naturalist* 134:1–19.

Master, L. L. 1991. "Assessing Threats and Setting Priorities for Conservation." *Conservation Biology* 5:559–63.

Masters, R. A., et al. 1990. *Conducting a Prescribed Burn and Prescribed Burn Checklist.* Extension Publication EC-90-121. Lincoln: Nebraska Cooperative Extension Service.

Matthews, A. 1992. *Where the Buffalo Roam: The Storm over the Revolutionary Plan to Restore America's Great Plains.* New York: Grove Press.

Matthews, W. J. 1988. "North American Prairie Streams as Systems for Ecological Study." *Journal of the North American Benthological Society* 7:387–409.

Mayewski, P. A., et al. 1981. "The Last Wisconsin Ice Sheets in North America." In *The Last Great Ice Sheets,* edited by G. H. Denton and T. J. Hughes, 67–178. New York: John Wiley and Sons.

Mayr, E., and P. D. Ashlock. 1991. *Principles of Systematic Zoology.* New York: McGraw-Hill.

McBrien, H., et al. 1983. "A Case of Insect Grazing Affecting Plant Succession." *Ecology* 64:1035–39.

McCoy, E. D., and H. R. Mushinsky. 1992. "Rarity of Organisms in the Sand Pine Scrub Habitat of Florida." *Conservation Biology* 6:537–48.

McCullough, D. R. 1971. "The Tule Elk: Its History, Behaviour, and Ecology." *University of California Publications in Zoology* 88:1–191.

McDonald, J. N. 1981. *North American Bison: Their Classification and Evolution.* Berkeley: University of California Press.

McNaughton, S. J. 1983. "Serengeti Grassland Ecology: The Role of Composite Environment Factors and Contingency in Community Organization." *Ecological Monographs* 53:291–320.

———. 1984. "Grazing Lawns: Animals in Herds, Plant Form, and Coevolution." *American Naturalist* 124:863–86.

———. 1985. "Ecology of a Grazing Ecosystem: The Serengeti." *Ecological Monographs* 55:259–94.

————. 1992. "Grasses and Grazing, Science and Management." *Ecological Applications* 3:17–20.

McNaughton, S. J., et al. 1988. "Large Mammals and Process Dynamics in African Ecosystems: Herbivorous Mammals Affect Primary Productivity and Regulate Recycling Balances." *BioScience* 38:794–800.

McNicholl, M. K. 1988. "Ecological and Human Influences on Canadian Populations of Grassland Birds." In *Grassland Birds,* edited by P. D. Goriup, 1–25. Cambridge: International Council for Bird Preservation.

Mendelsohn, R. 1993. "Assessing Natural Resource Damages with Indirect Methods: Comments on Chapter 8." In *Valuing Natural Assets: The Economics of Natural Resource Damage Assessment,* edited by R. J. Kopp and V. K. Smith. Washington, D.C.: Resources for the Future.

Mengel, R. M. 1970. "The North American Central Plains as an Isolating Agent in Bird Speciation." In *Pleistocene and Recent Environments of the Central Great Plains,* edited by W. Dort and J. K. Jones Jr., 279–340. Lawrence: University Press of Kansas.

Menzel, R. W., et al. 1984. "Ecological Alterations of Iowa Prairie-Agricultural Streams." *Iowa State Journal of Research* 59.5–30.

Merrell, D. J. 1977. "Life History of the Leopard Frog, *Rana pipiens,* in Minnesota." *Occasional Papers of the Bell Museum of Natural History* 15:1–23.

Messier, F. 1994. "Ungulate Population Models with Predation: A Case Study with the North American Moose." *Ecology* 75:478–88.

Michler, N., Jr. 1850. "Routes from the Western Boundary of Arkansas to Santa Fe and Valley of the Rio Grande." *House Executive Document 67,* 31st Congress, 1st Session, Public Document 577. Washington, D.C.

Mihlbacher, B. S., et al. 1989. "Long-term Grass Dynamics within a Mixed-grass Prairie." In *Proceedings of the Eleventh North American Prairie Conference,* edited by T. B. Bragg and J. Stubbendieck, 25–28. Lincoln: University of Nebraska Press.

Milchunas, D. G., and W. K. Lauenroth. 1993. "Quantitative Effects of Grazing on Vegetation and Soils over a Global Range of Environments." *Ecological Monographs* 63:327–66.

Milchunas, D. G., et al. 1988. "A Generalized Model of the Effects of Grazing by Large Herbivores on Grassland Community Structure." *American Naturalist* 132:87–106.

Miller, B., et al. 1994. "The Prairie Dog and Biotic Diversity." *Conservation Biology* 8:677–81.

Miller, K. 1984. "The Natural Protected Areas of the World." In *National Parks: Conservation and Development, The Role of Protected Areas in Sustaining Society,* edited by J. McNeely and K. Miller, 17–22. Washington, D.C.: Smithsonian Institution Press.

Miller, R. R., et al. 1989. "Extinctions of North American Fishes during the Past Century." *Fisheries* 14:22–38.

Minnesota Department of Natural Resources. 1995. "Draft Internal Document: Directions for Natural Resources—An Ecosystem Based Framework for Natural Resource Management." St. Paul: Minnesota Department of Natural Resources.

Mitchell, J. E. 1988. *Conservation Reserve Program in the Great Plains.* Fort Collins: U.S. Department of Agriculture, Forest Service.

Mitchell, R. C., and R. T. Carson. 1989. *Using Surveys to Value Public Goods.* Washington, D.C.: Resources for the Future.

Moore, C. T. 1972. "Man and Fire in the Central North American Grassland 1535–1890: A Documentary Historical Geography." Ph.D. diss., University of California, Los Angeles.

Morrison, M. L. 1986. "Bird Populations as Indicators of Environmental Change." *Current Ornithology* 3:429–51.

Morse, L. E. 1993. "Standard and Alternative Taxonomic Data in the Multi-institutional Natural Heritage Data Center Network." In *Designs for a Global Plant Species Information System,* edited by F. A. Bisby et al., 69–79. Oxford: Clarendon Press.

Moulton, M. P., et al. 1981. "Associations of Small Mammals on the Central High Plains of Eastern Colorado." *Southwestern Naturalist* 26:53–59.

Munn, L. C. 1993. "Effects of Prairie Dogs on Physical and Chemical Properties of Soils." In *Management of Prairie Dog Complexes for the Reintroduction of the Black-footed Ferret,* edited by J. L. Oldemeyer et al., 11–17. Washington, D.C.: U.S. Department of the Interior, Fish and Wildlife Service.

Munro, R. E., and C. F. Kimball. 1982. *Population Ecology of the Mallard* Vol. 7, *Distribution and Derivation of the Harvest.* Resource Publication 147. Washington, D.C.: U.S. Department of the Interior, Fish and Wildlife Service.

Murphy, D. D., and B. A. Wilcox. 1986. "Butterfly Diversity in Natural Habitat Fragments: A Test of the Validity of Vertebrate-Based Management." In *Wildlife 2000: Modeling Habitat Relationships of Terrestrial Vertebrates,* edited by J. Verner et al., 287–92. Madison: University of Wisconsin Press.

Mushinsky, H. R. 1985. "Fire and the Florida Sandhill Herpetological Community: With Special Attention to Responses of *Cnemidophorus sexlineatus.*" *Herpetologica* 41:333–42.

Naeem, S., et al. 1994. "Declining Biodiversity Can Alter the Performance of Ecosystems." *Nature* 368:734–37.

Nagel, H. G. 1979. "Analysis of Invertebrate Diversity in a Mixed Prairie Ecosystem." *Journal of the Kansas Entomological Society* 52:777–86.

———. 1983. "Effect of Spring Burning Date on Mixed-prairie Soil Moisture, Productivity, and Plant Species Composition." In *Proceedings of the Seventh North American Prairie Conference,* edited by C. L. Kucera, 259–63. Springfield: Southwest Missouri State University.

———. 1994. "Willa Cather Memorial Prairie: Seventeen Years of Vegetative Change with Limited Grazing and Fire." Abstract. *Fourteenth North American Prairie Conference.* Manhattan: Kansas State University.

Nash, R. 1983. *Wilderness and the American Mind.* New Haven: Yale University Press.

Natural Heritage Data Center Network. 1993. *Perspectives on Species Imperilment.* Arlington: Nature Conservancy.

Nature Conservancy. 1987. *Preserve Selection and Design Operations Manual.* Arlington: Nature Conservancy.

———. 1994. *Standardized National Vegetation Classification System.* Arlington: Nature Conservancy.

Navo, K. W., and E. D. Fleharty. 1983. "Small Mammals of Winter Wheat and Grain Sorghum Croplands in West-Central Kansas." *Prairie Naturalist* 15:159–72.

Nebraska Game and Parks Commission. 1984. *Survey of Habitat Work Plan K-83.* W-15-R-40. Lincoln: Nebraska Game and Parks Commission.

Nebraska Natural Resources Commission. 1993. *Report on the Sandhills Area Study.* Lincoln: Nebraska Natural Resources Commission.

Nelson, J. W., and M. B. Connolly. In press. "Techniques for Increasing the Quality of the Breeding Areas for Anatidae in North America." In *Proceedings IWRB Symposium,* edited by P. Havet and M. Birkan. Slimbridge, Eng.: IWRB.

Nelson, R. W., et al. 1983. *Playa Wetlands and Wildlife of the Southern Great Plains: A Characterization of Habitat.* Washington, D.C.: U.S. Department of the Interior, Fish and Wildlife Service.

Neuenschwander, L. F., et al. 1978. "The Effect of Fire on a Tobosagrass-Mesquite Community in the Rolling Plains of Texas." *Southwestern Naturalist* 23:315–38.

Nevins, A., ed. 1956. *Narratives of Exploration and Adventure by John Charles Frémont.* New York: Longmans, Green.

New, L. 1979. *High Plains Irrigation Survey.* College Station: Texas Agriculture Extension Service.

New, T. R. 1984. *Insect Conservation: An Australian Perspective.* Dordrecht: Junk.

Nickles, S. H. 1987. "Setting Farmers Free: Righting the Unintended Anomaly of UCC Section 9-312(2)." *Minnesota Law Review* 71:1135–218.

Norman, M. J. T., and J. O. Green. 1958. "The Local Influence of Cattle Dung and Urine upon the Yield and Botanical Composition of Permanent Pasture." *Journal of the British Grassland Society* 13:39–45.

Noss, R. F. 1987. "Protecting Natural Areas in Fragmented Landscapes." *Natural Areas Journal* 7:2–13.

———. 1990. "Indicators for Monitoring Biodiversity: A Hierarchial Approach." *Conservation Biology* 4:355–64.

Noss, R. F., et al. 1995. *Endangered Ecosystems of the United States: A Preliminary Assessment of Loss and Degradation.* Biological Report 28. Washington, D.C.: U.S. Department of the Interior, National Biological Service.

Novacek, J. M. 1989. "The Water and Wetland Resources of the Nebraska Sandhills." In *Northern Prairie Wetlands,* edited by A. G. van der Valk, 340–84. Ames: Iowa State University. Quoting McCarraher.

Novak, B. 1979. *Nature and Culture: American Landscape and Painting, 1825–1875.* New York: Oxford University Press.

Ode, D. J., et al. 1980. "The Seasonal Contribution of C3 and C4 Plant Species to Primary Production in a Mixed Prairie." *Ecology* 61:1304–11.

Oldemeyer, J. L., et al. 1993. *Management of Prairie Dog Complexes for the Reintroduction of the Black-footed Ferret.* Washington, D.C.: U.S. Department of the Interior, Fish and Wildlife Service.

O'Meilia, M. E., et al. 1982. "Some Consequences of Competition between Prairie Dogs and Beef Cattle." *Journal of Range Management* 35:580–85.

Omernick, J. 1987. "Ecoregions of the Conterminous United States." *Annals of Association of American Geographers* 77:88–125.

Opler, P. A. 1981. "Management of Prairie Habitats for Insect Conservation." *Journal of the Natural Areas Association* 1:3–6.

———. 1991. "North American Problems and Perspectives in Insect Conservation." In *The Conservation of Insects and Their Habitats,* edited by N. M. Collins and J. A. Thomas, 9–32. Fifteenth Symposium of the Royal Entomological Society of London. San Diego: Academic Press.

Orians, G. H. 1986. "Site Characteristics Favoring Invasions." In *Ecology of Biological Invasions of North America and Hawaii,* edited by H. A. Mooney and J. A. Drake, 133–48. New York: Springer-Verlag.

Orr, C. C., and O. J. Dickerson. 1966. "Nematodes in True Prairie Soils of Kansas." *Transactions of the Kansas Academy of Science* 69:317 34.

Otte, D. 1981. *The North American Grasshoppers: Gomphocerinae and Acridinae.* Vol. 1. Cambridge: Harvard University Press.

———. 1990. "Orthoptera (Saltatoria) of the United States and Canada." In *Systematics of the North American Insects and Arachnids: Status and Needs,* edited by M. Kosztarab and C. W. Schaefer, 63–75. Virginia Agricultural Experiment Station Information Series 90–91.

Page, J. C., et al. 1938. "What Is or Should Be the Status of Wildlife as a Factor in Drainage and Reclamation Planning?" *Transactions of the North American Wildlife Conference* 3:109–25.

Panzer, R. 1988. "Managing Prairie Remnants for Insect Conservation." *Natural Areas Journal* 8:83–90.

Parker, I. M., et al. 1993. "Distribution of Seven Native and Two Exotic Plants in a Tallgrass Prairie in Southeastern Wisconsin: The Importance of Human Disturbance." *American Midland Naturalist* 130:43–55.

Parker, W. S., and W. S. Brown. 1974. "Mortality and Weight Changes of Great Basin Rattlesnakes (*Crotalus viridis*) at a Hibernaculum in Northern Utah." *Herpetologica* 30:234–39.

Patterson, J. H. 1993. "Trade Liberalization, Agricultural Policy, and Wildlife: Reforming the Landscape." In *NAFTA and the Environment,* edited by T. L. Anderson, 61–68. San Francisco: Pacific Research Institute for Public Policy.

Pearce, D. W., and R. K. Turner. 1990. *Economics of Natural Resources and the Environment.* Baltimore: Johns Hopkins University Press.

Pearsall, S. H., et al. 1986. "Evaluation Methods in the United States." In *Wildlife Conservation Evaluation,* edited by M. B. Usher, 111–33. London: Chapman and Hall.

Pechmann, J. H. K., et al. 1991. "Declining Amphibian Populations: The Problem of Separating Human Impacts from Natural Fluctuations." *Science* 253:892–95.

Peden, D. G., et al. 1974. "The Trophic Ecology of *Bison Bison L.* on Shortgrass Plains." *Journal of Applied Ecology* 11:489–98.

Peek, J. M., and P. G. Risser. 1979. *Wildlife and Range Management on the National Grasslands.* National Audubon Society Report. Lakewood: National Audubon Society.

Poirot, E. M. 1964. *Our Margin of Life.* New York: Vantage Press.

Peterjohn, B. G., and J. R. Sauer. 1993. "North American Breeding Bird Survey Annual Summary 1990–1991." *Bird Populations* 1:1–15.

Peterson, G. A., and C. V. Cole. In press. "Productivity of Great Plains Soils: Past, Present, and Future." In *Conservation of Great Plains Ecosystems: Current Science, Future Options,* edited by S. Johnson and A. Bouzaher. Dordrecht: Kluwer Academic Press.

Pfeiffer, K. E., and A. A. Steuter. 1994. "Preliminary Response of Sandhills Prairie to Fire and Bison Grazing." *Journal of Range Management* 47:395–97.

Pflieger, W. L., and T. B. Grace. 1987. "Changes in the Fish Fauna of the Lower Missouri River, 1940–1983." In *Community and Evolutionary Ecology of North American Stream Fishes,* edited by W. J. Matthews and D. C. Heins, 166–77. Norman: University of Oklahoma Press.

Pickett, S. T. A., and P. S. White, eds. 1985. *The Ecology of Natural Disturbance and Patch Dynamics.* New York: Academic Press.

Pielou, E. C. 1992. *After the Ice Age: The Return of Life to Glaciated North America.* Chicago: University of Chicago Press.

Pike, A. 1969. *Albert Pike's Journeys in the Prairie, 1831–1832,* edited by J. E. Haley. Canyon: Panhandle-Plains Historical Society.

Pike, Z. 1966. *The Journals of Zebulon Montgomery Pike,* edited by D. Jackson. Norman: University of Oklahoma Press.

Pimentel, D. 1986. "Agroecology and Economics." In *Ecological Theory and Integrated Pest Management Practice,* edited by M. Kogan, 299–319. New York: John Wiley and Sons.

Pizzimenti, J. J. 1981. "Increasing Sexual Dimorphism in Prairie Dogs: Evidence for Changes During the Past Century." *Southwestern Naturalist* 26:43–48.

Platt, W. J. 1975. "The Colonization and Formation of Equilibrium Plant Associations on Badger Disturbances in Tall-grass Prairie." *Ecological Monographs* 45:285–305.

Platt, W. J., and I. M. Weiss. 1985. "An Experimental Study of Competition among Fugitive Prairie Plants." *Ecology* 66:708–20.

Plumb, G. E., and J. L. Dodd. 1993 "Foraging Ecology of Bison and Cattle on a Northern Mixed Prairie: Implications for Natural Area Management." *Ecological Applications* 3:631–43.

Pollard, E. 1982. "Monitoring Butterfly Abundance in Relation to the Management of a Nature Reserve." *Biological Conservation* 24:317–28.

Pollard, E., and T. J. Yates. 1993. *Monitoring Butterflies for Ecology and Conservation.* New York: Chapman and Hall.

Polley, H. W., and S. L. Collins. 1984. "Relationships of Vegetation and Environment in Buffalo Wallows." *American Midland Naturalist* 112:179–86.

Ponting, C. 1992. *A Green History of the World.* New York: St. Martin's Press.

Popper, D. E., and F. Popper. 1987. "A Daring Proposal for Dealing with an Inevitable Disaster." *Planning* (December): 12–18.

———. 1988. "The Fate of the Plains." *High Country News* 20:15–19.

Potvin, M. A., and A. T. Harrison. 1984. "Vegetation and Litter: Changes of a Nebraska Sandhills Prairie Protected from Grazing." *Journal of Range Management* 31:55–58.

Pough, F. H. 1983. "Amphibians and Reptiles as Low Energy Systems." In *Behavioral Energetics,* edited by W. P. Aspey and S. Lustick, 141–88. Columbus: Ohio State University Press.

Prairie Conservation Coordinating Committee. 1994. *Assessment of the Prairie Conservation Action Plan, 1989–1994: What Has Been Accomplished in Alberta?* Lethbridge: Alberta Environmental Protection, 31 pp.

Prairie Conservation Forum. 1996. *Alberta Prairie Conservation Action Plan, 1996–2000,* 17 pp.

Prairie Habitat Joint Venture Advisory Board. 1990. *North American Waterfowl Management Plan, Prairie Habitat: Prospectus.* Edmonton: Canadian Wildlife Service, 32 pp.

Pulliam, H. R., and B. J. Danielson. 1991. "Sources, Sinks, and Habitat Selection: A Landscape Perspective on Population Dynamics." *American Naturalist* 137:50–66.

Pyne, S. 1982. *Fire in America: A Cultural History of Wildland and Rural Fire.* Princeton: Princeton University Press.

Quortrup, E. R., et al. 1946. "An Outbreak of Pasteurellosis in Wild Ducks." *Journal of the American Veterinary Medicine Association* 108:94–100.

Rauzi, F., and M. L. Fairbourn. 1983. "Effects of Annual Applications of Low N Fertilizer Rates on a Mixed Grass Prairie." *Journal of Range Management* 36:359–62.

Reading, R. P., et al. 1989. "Attributes of Black-tailed Prairie Dog Colonies, Associated Species, and Management Implications." In *The Prairie Dog Ecosystem: Managing for Biological Diversity,* 13–27. Billings: Bureau of Land Management.

Redmann, R. E., et al. 1993. "Impacts of Burning on Primary Productivity of *Festuca* and *Stipa-Agropyron* Grasslands in Central Saskatchewan." *American Midland Naturalist* 130:262–73.

Reed, K. M., and J. R. Choate. 1986. "Natural History of the Plains Pocket Mouse in Agriculturally Disturbed Sandsage Prairie." *Prairie Naturalist* 18:79–90.

Reed, T. M. 1983. "The Role of Species-Area Relationships in Reserve Choice: A British Example." *Biological Conservation* 25:263–71.

Reeves, C. C., Jr. 1966. "Pluvial Lake Basins of West Texas." *Journal of Geology* 74:269–71.

Reichman, O. J. 1987. *Konza Prairie: A Tallgrass Natural History.* Lawrence: University Press of Kansas.

Reichman, O. J., and S. C. Smith. 1985. "Impact of Pocket Gopher Burrows on Overlying Vegetation." *Journal of Mammalogy* 66:720–25.

Reichman, O. J., et al. 1982. "Adaptive Geometry of Burrow Spacing in Two Pocket Gopher Populations." *Ecology* 63:687–95.

Remmert, H. ed. 1991. *The Mosaic-Cycle Concept of Ecosystems.* New York: Springer-Verlag.

Reynolds, R. E., et al. 1994. "Conservation Reserve Program: Benefit for Grassland Birds in the Northern Plains." *Transactions of the North American Wildlife and National Resources Conference* 59:328–36.

Reynolds, T. D. 1979. "Responses of Reptile Populations to Different Land Management Practices on the Idaho National Engineering Laboratory Site." *Great Basin Naturalist* 39:255–62.

Rhodes, M. J., and J. D. Garcia. 1981. "Characteristics of Playa Lakes Related to Summer Waterfowl Use." *Southwestern Naturalist* 26:231–35.

Rhodes, R. S., and H. A. Semken Jr. 1986. "Quaternary Biostratigraphy and Paleoecology of Fossil Mammals from the Loess Hills Region of Western Iowa." *Proceedings of the Iowa Academy of Science* 93:94–130.

Ribaudo, M.O., et al. 1990. *Natural Resources and User Benefits from the Conservation Reserve Program.* AER Report 627. Washington, D.C.: U.S. Department of Agriculture.

Richardson, N. 1989. *Land Use Planning and Sustainable Development in Canada.* Canadien Environmental Advisory Council. 49 pp.

Richardson, N. 1989. "Land Use Planning and Sustainable Development in Prairie." *Soil Science Society of America Journal* 58:816–23.

Ricklefs, R. E., and D. Schluter. 1993. *Species Diversity in Ecological Communities.* Chicago: University of Chicago Press.

Riebsame, W. 1986. "The Dust Bowl: Historical Image, Psychological Anchor, and Ecological Taboo." *Great Plains Quarterly* 6:127–36.

———. 1990. "The United States Great Plains." In *The Earth as Transformed by Human Action,* edited by B. L. Turner II et al., 417–34. New York: Cambridge University Press.

———. 1991. "Sustainability of the Great Plains in an Uncertain Climate." *Great Plains Research* 1:133–51.

———. 1993. "Sustainable Development Questioned: The Historical Debate between Adaptationism and Catastrophism in Great Plains Studies." Paper presented at the American Society for Environmental History Conference, Pittsburgh.

Ring, C. B., II, et al. 1985. "Vegetational Traits of Patch-grazed Rangeland in West-central Kansas." *Journal of Range Management* 38:51–55.

Risser, P. G. 1985. "Grasslands." In *Physiological Ecology of North American Plant Communities,* edited by B. F. Chabot and H. A. Mooney, 232–56. New York: Chapman and Hall.

———. 1988. "Diversity in and among Grasslands." In *Biodiversity,* edited by E. O. Wilson, 176–80. Washington, D.C.: National Academy Press.

Risser, P. G., et al. 1981. *The True Prairie Ecosystem.* Stroudsburg: Hutchinson Ross.

Robbins, C. S., and W. T. Van Velzen. 1969. *Breeding Bird Survey 1967 and 1968.* Special Scientific Report 124. Washington, D.C.: U.S. Department of the Interior, Bureau of Sport Fisheries and Wildlife.

Robbins, C. S., et al. 1986. *The Breeding Bird Survey: Its First Fifteen Years, 1965–1979.* Resource Publication 157. Washington, D.C.: U.S. Department of the Interior, Fish and Wildlife Service.

Robinson, G. R., et al. 1992. "Diverse and Contrasting Effects of Habitat Fragmentation." *Science* 257:524–26.

Robinson, S. K., et al. 1995. "Regional Forest Fragmentation and the Nesting Success of Migratory Birds." *Science* 267:1987–90.

Roe, F. 1970. *The North American Buffalo: A Critical Study of the Species in its Wild State.* Toronto: University of Toronto Press.

Rosaasen, K. A., and J. A. Lokken. 1996. "Canadian Agricultural Policies and Other 'Land Mining' Initiatives for Prairie Agriculture." In *Proceedings of the Fourth Prairie Conservation and Endangered Species Workshop.* Natural History Occasional Paper No. 23. Edmonton: Provincial Museum of Alberta.

Rowe, S. 1990. *Home Place: Essays on Ecology.* Edmonton: NeWest Publishers, 253 pp.

Rubec, C. D. A., et al. 1995. "NAFTA Opportunities for Conserving Continental Biodiversity." Paper presented at the North American Commission on Environmental Cooperation, Montreal, Canada.

Rundquist, D. C., et al. 1981. *Wetland Atlas of the Omaha District.* Contract no. DACW-80-C-0220. Omaha: Army Corps of Engineers.

Runte, A. 1987. *National Parks: The American Experience.* Lincoln: University of Nebraska Press.

Rzedowski, J. 1978. *Vegetation de Mexico.* Mexico City: Editorial Limusa.

Samson, F. B. 1980. "Island Biogeography and the Conservation of Prairie Birds." In *Seventh North American Prairie Conference,* edited by C. L. Kucera, 293–305. Springfield: Southwest Missouri State University.

Samson, F. B., and F. L. Knopf. 1994. "Prairie Conservation in North America." *BioScience* 44:418–21.

Samways, M. J. 1993a. "A Spatial and Process Sub-regional Framework for Insect and

Biodiversity Conservation Research and Management." In *Perspectives on Insect Conservation,* edited by K. J. Gaston et al., 1–27. Andover, Hants: Intercept.

————. 1993b. "Dragonflies (Odonata) in Taxic Overlays and Biodiversity Conservation." In *Perspectives on Insect Conservation,* edited by K. J. Gaston et al., 111–23. Andover, Hants: Intercept.

————. 1994. *Insect Conservation Biology.* Conservation Biology Series. New York: Chapman and Hall.

Saskatchewan PCAP Evaluation Committee. 1995. *Prairie Conservation Action Plan: Review and Evaluation for Saskatchewan.* A Report on Conservation Progress. Saskatoon: Saskatchewan Environment and Resource Management, 59.

Sauer, J. R., and P. H. Geissler. 1990. "Estimation of Annual Indices from Roadside Surveys." In *Survey Designs and Statistical Methods for the Estimation of Avian Population Trends.* Biological Report 90, edited by J. R. Sauer and S. Droege, 58–62. Washington, D.C.: U.S. Department of the Interior, Fish and Wildlife Service.

Salyer, J. C., and F. G. Gillett. 1964. "Federal Refuges." In *Waterfowl Tomorrow,* edited by J. P. Linducka, 497–508. Washington, D.C.: U.S. Department of the Interior.

Schacht, W., and J. Stubbendieck. 1985. "Prescribed Burning in the Loess Hills Mixed Prairie of Southern Nebraska." *Journal of Range Management* 38:47–51.

Schlesier, Karl, ed. 1994. *Plains Indians, a.d. 500–1500: The Archaeological Past of Historic Groups.* Norman: University of Oklahoma Press.

Schmidt, N. D., and C. L. Kucera. 1975. "Arthropod Food Chain Energetics in Missouri Tall Grass Prairie." In *Prairie: A Multiple View,* edited by M. K. Wali, 143–54. Grand Forks: University of North Dakota Press.

Scott, J. A., et al. 1979. "Patterns of Consumption in Grasslands." In *Perspectives in Grassland Ecology,* edited by N. French, 89–105. Berlin: Springer-Verlag.

Scott, N. J., Jr. In press. "Evolution and Management of the North American Grassland Herpetofauna." *Conservation Biology.*

Seastedt, T. R. 1985. "Maximization of Primary and Secondary Productivity by Grazers." *American Naturalist* 126:559–64.

Seastedt, T. R., and A. K. Knapp. 1993. "Consequences of Nonequilibrium Resource Availability across Multiple Time Scales: The Transient Maxima Hypothesis." *American Naturalist* 141:621–33.

Seastedt, T. R., and R. A. Ramundo. 1990. "The Influence of Fire on Belowground Processes of Tallgrass Prairie." In *Fire in North American Tallgrass Prairies,* edited by S. L. Collins and L. L. Wallace, 99–117. Norman: University of Oklahoma Press.

Seastedt, T. R., et al. 1991. "Controls of Nitrogen Limitation in Tallgrass Prairie." *Oecologia* 87:72–79.

Seburn, C. N. L. 1992. "The Status of Amphibian Populations in Saskatchewan." *Canadian Wildlife Service Occasional Paper* 76:17–18.

Sedjo, R. A. 1992. "Preserving Biodiversity as a Resource." *Resources* 106:26–29.

Seigel, R. A. 1986. "Ecology and Conservation of an Endangered Rattlesnake, *Sistrurus catenatus,* in Missouri, USA." *Biological Conservation* 35:333–46.

Senner, S. E., and B. Ladd. 1995. *Grassland Birds and Ecosystem Management on the National Grasslands.* National Audubon Society Report. Boulder: National Audubon Society.

Sexton, M. L. 1980. "Destruction of Sandsage Prairie in Southwest Kansas." In *Proceedings of the Seventh North American Prairie Conference,* edited by C. L. Kucera, 113–15. Springfield: Southwest Missouri State University.

Shackford, J. S. 1991. "Breeding Ecology of the Mountain Plover in Oklahoma." *Bulletin Oklahoma Ornithological Society* 24:9–13.

Shafer, C. L. 1990. *Nature Reserves: Island Theory and Conservation Practice.* Washington, D.C.: Smithsonian Institution Press.

———. 1995. "Values and Shortcomings of Small Reserves." *BioScience* 45:80–88.

Shariff, A. R., et al. 1994. "Grazing Intensity Effects on Litter Decomposition and Soil Nitrogen Mineralization." *Journal of Range Management* 47:444–49.

Sharps J. C., and D. W. Uresk. 1990. "Ecological Review of Black-tailed Prairie Dogs and Associated Species in Western South Dakota." *Great Basin Naturalist* 50:339–45.

Sharrow, S. H., and H. A. Wright. 1977. "Proper Burning Intervals for Tobosagrass in West Texas Based on Nitrogen Dynamics." *Journal of Range Management* 30:343–46.

Shaw, J. H., and T. S. Carter. 1990. "Bison Movements in Relation to Fire and Seasonality." *Wildlife Society Bulletin* 18:426–30.

Shelford, V. 1933. "Preservation of Natural Biotic Communities." *Ecology* 14:240–45.

Shepard, J. 1995. "Singing Out of Tune: Historical Perceptions and National Parks on the Great Plains." Ph.D. diss., Texas Tech University, Lubbock.

Shepard, J., et al. 1994. *A New Vision of the Heartland: The Great Plains in Transition.* Denver: Center for the New West.

Shmida, A., and M. V. Wilson. 1985. "Biological Determinants of Species Diversity." *Journal of Biogeography* 12:1–20.

Simpson, C. D., et al. 1981. "Significance of Playas to Migratory Wildlife." *Proceedings of the Playa Lakes Symposium,* edited by J. S. Barclay and W. V. White, 35–45. Washington, D.C.: U.S. Department of the Interior, Fish and Wildlife Service.

Sims, P., and J. Singh. 1978. "The Structure and Function of Ten Western North American Grasslands. 3. Net Primary Production, Turnover, and Efficiencies of Energy Capture and Water Use." *Journal of Ecology* 66:573–97.

Sims, P. L. 1988. "Grasslands." In *North American Terrestrial Vegetation,* edited by M. G. Barbour and W. D. Billings, 265–86. New York: Cambridge University Press.

Sims, P. L., et al. 1978. "The Structure and Function of Ten Western North American Grasslands. 1. Abiotic and Vegetational Characteristics." *Journal of Ecology* 66:251–85.

Singh, J. S., et al. 1983. "Structural and Functional Attributes of the Vegetation of Northern Mixed Prairie of North America." *Botanical Review* 49:117–49.

Smith, B. J., and K. F. Higgins. 1990. "Avian Cholera and Temporal Changes in Wetland Numbers and Densities in Nebraska's Rainwater Basin Area." *Wetlands* 10:1–5.

Smith, B. J., et al. 1989. "Land Use Relationships to Avian Cholera Outbreaks in the Nebraska Rainwater Basin Area." *Prairie Naturalist* 21:125–38.

Smith, H. N. 1950. *Virgin Land: The American West in Symbol and Myth.* Cambridge: Harvard University Press.

Smith, R. I. 1970. "Response of Pintail Breeding Populations to Drought." *Journal of Wildlife Management* 34:943–46.

Smoliak, S. 1956. "Influence of Climatic Conditions on Forage Production of Shortgrass Rangeland." *Journal of Range Management* 9:89–91.

Smolik, J. D. 1974. "Nematode Studies at the Cottonwood Site." US/IBP Grassland Biome Technical Report, no. 251. Fort Collins: Colorado State University.

Smolik, J. D., and J. K. Lewis. 1982. "Effect of Range Condition on Density and Biomass of Nematodes in Terrestrial Ecosystems." *Journal of Range Management* 35:657–63.

Sneft, R. L., et al. 1987. "Large Herbivore Foraging and Ecological Hierarchies." *BioScience* 37:790–99.

Sohlenius, B. 1980. "Abundance, Biomass, and Contribution to Energy Flow by Soil Nematodes in Terrestrial Ecosystems." *Oikos* 34:186–94.

Sopuck, R. 1993. *Canada's Agricultural and Trade Policies: Implications for Rural Renewal and Biodiversity.* Ottawa: National Round Table on Environment and Economy, 51 pp.

Sparrow, H. O. 1984. *Soils at Risk: Canada's Eroding Future.* Standing Committee on Agriculture, Fisheries and Forestry. Ottawa: Senate of Canada, 129 pp.

Stanford, J. A., and J. V. Ward. 1979. "Stream Regulation in North America." In *The Ecology of Regulated Streams,* edited by J. V. Wasr and J. A. Stanford, 215–36. New York: Plenum Press.

Stanton, N. L. 1988. "The Underground in Grasslands." *Annual Review of Ecology and Systematics* 19:573–89.

Statistics Canada. 1992. *1991 Census of Agriculture.* Ottawa: Agriculture Canada.

Stebbins, R. C. 1985. *A Field Guide to Western Reptiles and Amphibians.* Boston: Houghton Mifflin.

Stegner, W. 1962. *Wolf Willow.* New York: Viking Press.

Steiger, T. L. 1930. "Structure of Prairie Vegetation." *Ecology* 11:170–217.

Steinauer, E. M. 1994. "Effects of Urine Deposition on Small-scale Patch Structure and Vegetative Patterns in Prairie Vegetation." Ph.D. diss., University of Oklahoma, Norman.

Steinauer, E. M., and T. B. Bragg. 1987. "Ponderosa Pine (*Pinus ponderosa*) Invasion of Nebraska Sandhills Prairie." *American Midland Naturalist* 118:358–65.

Steinauer, E. M., and S. L. Collins. 1995. "Effects of Urine Deposition on Small-scale Patch Structure in Prairie Vegetation." *Ecology* 76:1195–205.

Steuter, A. A. 1987. "C3/C4 Production Shift on Seasonal Burns—Northern Mixed Prairie." *Journal of Range Management* 40:27–31.

———. 1991. "Human Impacts on Biodiversity in America: Thoughts from the Grassland Biome." *Conservation Biology* 5:136–37.

Steuter, A. A., and H. A. Wright. 1983. "Spring Burning Effects on Redberry Juniper-mixed Grass Habitats." *Journal of Range Management* 36:161–64.

Steuter, A. A., et al. 1990a. "Woodland/Grassland Boundary Changes in the Middle Niobrara Valley of Nebraska Identified by d13C Values of Soil Organic Matter." *American Midland Naturalist* 124:301–8.

———. 1990b. "A Synthetic Approach to Research and Management Planning: The Conceptual Development and Implementation." *Natural Areas Journal* 10:61–68.

———. 1995. "Distribution and Diet of Bison and Pocket Gophers in a Sandhills Prairie." *Ecological Applications* 5:756–66.

Stewart, R. B., and H. A. Kantrud. 1971. *Classification of Natural Ponds and Lakes in the Glaciated Prairie Region.* Resource Publication 92. Washington, D.C.: U.S. Department of the Interior, Bureau of Sport Fisheries and Wildlife.

Strong, D. R., et al. 1984. *Insects on Plants: Community Patterns and Mechanisms.* Oxford: Blackwell Scientific Publications.

Stubbendieck, J., et al. 1993. "Establishment and Survival of the Endangered Blowout Penstemon." *Great Plains Research* 3.3–19.

Summers, C. A., and R. L. Linder. 1978. "Food Habits of the Black-tailed Prairie Dog in Western South Dakota." *Journal of Range Management* 31:134–36.

Swanson, G. A., and H. F. Duebbert. 1989. "Wetland Habitat of Waterfowl in the Prairie Pothole Region." In *Northern Prairie Wetlands,* edited by A. G. van der Valk, 228–67. Ames: Iowa State University Press.

Swineheart, J. B. 1984. "Physical Characteristics of the Sandhills Geology." In *The Sandhills of Nebraska: Yesterday, Today, and Tomorrow,* 151–77. Proceedings of the 1984 Water Resources Seminar Series. Lincoln: University of Nebraska.

Taff, S. J. 1992. "What Is a Wetland Worth? Concepts and Issues in Economic Valuation." Staff Paper P92-1. Department of Agriculture and Applied Economics. St. Paul: University of Minnesota, Institute of Agriculture, Forestry, and Home Economics.

Temple, S. A., and J. A. Wiens. 1989. "Bird Populations and Environmental Changes: Can Birds Be Bio-indicators?" *American Birds* 43:260–70.

Terri, J., and L. Stowe. 1976. "Climate Patterns and the Distribution of C4 Grasses in North America." *Oecologia* 23:1–12.

Thorne, E. T., and E. S. Williams. 1988. "Disease and Endangered Species: The Black-footed Ferret as a Recent Example." *Conservation Biology* 2:66–74.

Thwaites, R. G. 1905. *Early Western Travels, 1748–1846,* Vol. 15, *James's Account of S. H. Long's Expedition.* Cleveland: Arthur H. Clark. Quoting Stephen Long.

Tilman, D. 1985. "The Resource-ratio Hypothesis of Plant Succession." *American Naturalist* 125:827–52.

Tilman, D., and J. A. Downing. 1994. "Biodiversity and Stability in Grasslands. *Nature* 367:363–65.

Tilman, D., and A. El Haddi. 1992. "Drought and Biodiversity in Grasslands." *Oecologia* 89:257–64.

Tiner, R. W., Jr. 1984. *Wetlands of the United States: Current Status and Recent Trends.* Washington, D.C.: U.S. Department of the Interior, Fish and Wildlife Service.

Tomanek, G. W., and F. W. Albertson. 1953. "Some Effects of Different Intensities of Grazing on Mixed Prairies near Hays, Kansas." *Journal of Range Management* 6:299–306.

———. 1957. "Variations in Cover, Composition, Production, and Roots of Vegetation in Two Prairies in Western Kansas." *Ecological Monographs* 27:267–81.

Tomanek, G. W., and G. K. Hulett. 1970. "Effects of Historical Droughts on Grassland Vegetation in the Central Great Plains." In *Pleistocene and Recent Environments of the Central Great Plains,* edited by W. Dort Jr. and J. K. Jones Jr., 203–10. Lawrence: University Press of Kansas.

Traweek, M. S. 1978. *Texas Waterfowl Production Survey.* Federal Aid Project W-106-R-5. Austin: Texas Parks and Wildlife Department.

Trottier, G. C. 1992. *A Landowner's Guide: Conservation of Canadian Prairie Grasslands.* Canadian Wildlife Service, 92 pp.

Turner, F. 1989. *Spirit of Place: The Making of an American Literary Landscape.* San Francisco: Sierra Club Books.

Turner, R. W., and H. J. Stains. 1967. "Effects of a Cornfield upon the Movement of *Peromyscus leucopus.*" *Transactions of the Illinois State Academy of Science* 60:282–98.

Tyrchniewicz, A., and A. Wilson. 1994. *Sustainable Development for the Great Plains: Policy Analysis.* Winnipeg: International Institute for Sustainable Development, 35 pp.

Udvardy, M. D. F. 1958. "Ecological and Distributional Analysis of North American Birds." *Condor* 60:50–66.

Umbanhower, C. E. 1992. "Abundance, Vegetation, and Environment of Four Patch Types in a Northern Mixed Prairie." *Canadian Journal of Botany* 70:277–84.

United Nations Educational, Scientific, and Cultural Organization. 1973. *International Classification and Mapping of Vegetation. Series 6, Ecology and Conservation.* Paris: United Nations Educational, Scientific, and Cultural Organization.

Uresk, D. W. 1993. "Relation of Black-tailed Prairie Dogs and Control Programs to Vegetation, Livestock, and Wildlife." In *Management of Prairie Dog Complexes for the Reintroduction of the Black-footed Ferret,* edited by J. L. Oldemeyer et al., 8. Washington, D.C.: U.S. Department of the Interior, Fish and Wildlife Service.

U.S. Bureau of the Census. 1993. *Population and Housing Unit Counts: Nebraska.* Washington, D.C.: U.S. Government Printing Office.

U.S. Department of Agriculture. 1985. *Medicine Bow National Forest and Thunder Basin National Grassland Land and Resource Management Plan.* Denver: U.S. Department of Agriculture, Forest Service.

U.S. Environmental Protection Agency. 1986. *Fact Sheet: Rainwater Basin Wetlands Advance Identification Project.* Washington, D.C.: Environmental Protection Agency.

U.S. Fish and Wildlife Service. 1986. *North American Waterfowl Management Plan.* Washington, D.C.: U.S. Department of the Interior, Fish and Wildlife Service.

————. 1989a. *Playa Lakes Region Waterfowl Habitat Concept Plan: Category 24 of the North American Waterfowl Management Plan.* Denver: U.S. Department of the Interior, Fish and Wildlife Service.

————. 1989b. *Plan of Action: The Chase Lake Prairie Project.* Woodworth, N.D.: U.S. Department of the Interior, Fish and Wildlife Service.

————. 1993a. *Chase Lake Prairie Project Accomplishment Report.* Woodworth, N.D.: U.S. Department of the Interior, Fish and Wildlife Service.

————. 1993b. "Grassland Birds Declining, Research Finds." U.S. Department of the Interior News Release, 24 June.

————. 1994a. *Draft Ecosystem Planning for the Prairie, Wetlands, and Missouri River Main Stem.* Bismarck: U.S. Department of the Interior, Fish and Wildlife Service.

————. 1994b. *1994 Update to the North American Waterfowl Management Plan: Expanding the Commitment.* Washington, D.C.: U.S. Department of the Interior, Fish and Wildlife Service.

————. 1994c. "Endangered and Threatened Wildlife and Plants: Animal Candidate Review for Listing as Endangered or Threatened Species; Proposed Rule." *Federal Register* 59:58982–9028.

U.S. Geological Survey. 1993. *Prototype 1990 Conterminous United States Land Cover Characteristics Data Set CD-ROM.* Sioux Falls: EROS Data Center.

U.S. Prairie Pothole Joint Venture. 1994a. *Partnerships in Progress: U.S. Prairie Pothole Joint Venture Accomplishments.* Washington, D.C.: U.S. Department of the Interior, Fish and Wildlife Service.

————. 1994b. *Implementation Plan Update.* Denver: U.S. Prairie Pothole Joint Venture.

Usher, M. B. 1985. "Implications of Species-Area Relationships for Wildlife Conservation." *Journal of Environmental Management* 21:181–91.

————. 1991. "Habitat Structure and the Design of Nature Reserves." In *Habitat Structure: The Physical Arrangement of Objects in Space,* edited by S. S. Bell et al., 373–91. New York: Chapman and Hall.

Usher, M. B., and R. G. Jefferson. 1991. "Creating New and Successional Habitats for Arthropods." In *The Conservation of Insects and Their Habitats,* edited by N. M. Collins and J. A. Thomas, 263–91. Fifteenth Symposium of the Royal Entomological Society of London. San Diego: Academic Press.

Vacanti, P. L., and K. N. Geluso. 1985. "Recolonization of a Burned Prairie by Meadow Voles (*Microtus pennsylvanicus*)." *Prairie Naturalist* 17:15–22.

Vallentine, J. F. 1980. *Range Development and Improvements.* Provo: Brigham Young University Press.

Vankat, J. L. 1979. *The Natural Vegetation of North America.* New York: John Wiley and Sons.

van der Valk, A. G. 1989a. Introduction to *Northern Prairie Wetlands,* edited by A. G. van der Valk, x–xiv. Ames: Iowa State University Press.

———, ed. 1989b. *Northern Prairie Wetlands.* Ames: Iowa State University Press.

van der Valk, A. G., and C. B. Davis. 1978. "The Role of Seed Banks in the Vegetation Dynamics of Prairie Glacial Marshes." *Ecology* 59:322–35.

———. 1979. "A Reconstruction of the Recent Vegetational History of a Prairie Marsh, Eagle Lake, Iowa, from Its Seed Bank." *Aquatic Botany* 6:29–51.

Vasadava, U. 1991. "Trade Policy Implications of Sustainable Agriculture." *Canadian Journal of Agriculture Economics* 39:595.

Vial, J. L., and L. Saylor. 1994. "Preliminary Report on the Status of Amphibian Populations." Manuscript. Chicago: World Conservation Union Species Survival Commission.

Vickery, H. 1994. "Breeding Duck Populations Rebound in Response to Wet Weather, Improved Habitat Conditions." U.S. Department of the Interior News Release, 22 July.

Vinton, M. A., et al. 1993. "Interactive Effects of Fire, Bison *(Bison bison)* Grazing, and Plant Community Composition in Tallgrass Prairie." *American Midland Naturalist* 129:10–18.

Vogl, R. J. 1974. "Effects of Fire on Grasslands." In *Fire and Ecosystems,* edited by T. T. Kozlowski and C. E. Ahlgren, 139–94. New York: Academic Press.

Voorhies, M. R., and R. G. Corner. 1985. "Small Mammals with Boreal Affinities in Late Pleistocene (Rancholabrean) Deposits of Eastern and Central Nebraska." *Institute for Tertiary-Quaternary Studies-TER-QUA Symposium Series* 1:125–42.

Wakeley, J. S. 1978. "Factors Affecting the Use of Hunting Sites by Ferruginous Hawks." *Condor* 80:316–26.

Wallace, L. L. 1987. "Effects of Clipping and Soil Compaction on Growth, Morphology, and Mycorrhizal Colonization of *Schizachyrium scoparium,* a C4 Bunch Grass." *Oecologia* 72:423–28.

Wallach, B. 1985. "The Return of the Prairie." *Landscape* 28:1–5.

Walter, H. 1975. *Climate Diagram Maps of the Individual Continents of the World and the Ecological Climatic Regions of the Earth.* New York: Springer-Verlag.

Warner, R. E. 1994. "Agriculture Land Use and Grassland Habitat in Illinois: Future Shock for Midwestern Birds?" *Conservation Biology* 8:147–56.

Warren, A. 1976. "Morphology and Sediments of the Nebraska Sandhills in Relation to Pleistocene Winds the Development of Aeolian Bedforms." *Journal of Geology* 84:685–700.

Warren, S. D., et al. 1987. "Response of Grassland Arthropods to Burning: A Review." *Agriculture, Ecosystems, and Environment* 19:105–30.

Weaver, J. E. 1954. *North American Prairie.* Lincoln: Johnsen Publishing.

———. 1965. *Native Vegetation of Nebraska.* Lincoln: University of Nebraska Press.

———. 1968. *Prairie Plants and Their Environment: A Fifty-year Study in the Midwest.* Lincoln: University of Nebraska Press.

Weaver, J. E., and F. W. Albertson. 1956. *Grasslands of the Great Plains: Their Nature and Use.* Lincoln: Johnsen Publishing.

Weaver, J. E., and W. E. Brunner. 1954. "Nature and Place of Transition from True Prairie to Mixed Prairie." *Ecology* 35:117–26.

Weaver, J. E., and N. W. Roland. 1952. "Effects of Excessive Natural Mulch on Development, Yield, and Structure of Native Grassland." *Botanical Gazette* 114:1–19.

Weaver, T. 1983. "The Yeild Response of High Plains Grasslands to Water Enrichment, Three Phases." In *The High Plains Cooperative Program: 1981–1983,* edited by L. Holman and G. Knudsen, 77–94. Helena: Montana Department of Natural Resources.

Weaver, T., et al. 1993. "Plants Colonizing Disturbed Areas in Fifteen Rocky Mountain Environments: Weeds and Reclamation Candidates." *National Park Service Annual Report* 17:20–28.

Webb, W. 1931. *The Great Plains: A Study in Institutions and Environment.* Boston: Ginn.

Wedel, W. 1953. "Some Aspects of Human Ecology on the Central Plains." *American Anthropologist* 55:409–513.

Weir, J. 1992. "The Sweetwater Rattlesnake Round-up: A Case Study in Environmental Ethics." *Conservation Biology* 6:116–27.

Weller, M. W., and L. W. Fredrickson. 1974. "Avian Ecology of a Managed Glacial Marsh." *Living Bird* 12:269–91.

Weller, M. W., and C. E. Spatcher. 1965. *The Role of Habitat in the Distribution and Abundance of Marsh Birds.* Special Report 43. Ames: Iowa State University Agriculture and Home Economics Experiment Station.

Wells, P. V. 1965. "Scarp Woodlands, Transported Grassland Soils, and the Concept of Grassland Climate in the Great Plains Region." *Science* 148:246–49.

———. 1970. "Historical Factors Controlling Vegetation Patterns and Floristic Distributions in the Central Plains Region of North America." In *Pleistocene and Recent Environments of the Central Great Plains,* edited by W. Dort Jr. and J. K. Jones Jr., 211–21. Lawrence: University Press of Kansas.

West, T. 1990. "USDA Forest Service Management of the National Grasslands." *Agricultural History* 64:86–98.

Whicker, A. D., and J. K. Detling. 1988. "Ecological Consequences of Prairie Dog Disturbances." *BioScience* 38:778–85.

————. 1993. "Control of Grassland Ecosystem Processes by Prairie Dogs." In *Management of Prairie Dog Complexes for the Reintroduction of the Black-footed Ferret,* edited by J. L. Oldemeyer et al., 18–27. Washington, D.C.: U.S. Department of the Interior, Fish and Wildlife Service.

Whisenant, S. G. 1990. "Postfire Population Dynamics of *Bromus japonicus.*" *American Midland Naturalist* 123:301–8.

Whisenant, S. G., and D. W. Uresk. 1990. "Spring Burning Japanese Brome in a Western Wheatgrass Community." *Journal of Range Management* 43:205–8.

Whisenant, S. G., et al. 1984. "Effects of Fire on Texas Wintergrass Communities." *Journal of Range Management* 37:387–91.

Whitaker, J. O. 1966. " Food of *Mus musculus, Peromyscus maniculatus bairdi,* and *Peromyscus leucopus* in Vigo County, Indiana." *Journal of Mammalogy* 47:473–86.

Whitcomb, R. F., et al. 1984. "Biogeography of Leafhopper Specialists of the Shortgrass Prairie." *American Entomologist* (spring):19–35.

White, J. 1986. "Why Bother to Protect Prairies along Railroads?" In *The Prairie—Past, Present, and Future: Proceedings of the Ninth North American Prairie Conference,* edited by G. K. Clamby and R. H. Pemble, 172–73. Fargo: Tri College University Center for Environmental Studies.

Whitman, W. 1983. *Leaves of Grass.* New York: Bantam Books.

Whitman, W. C. 1974. "Influence of Grazing on the Microclimate of Mixed Grass Prairie." In *Plant Morphogenesis as the Basis for Scientific Management of Range Resources,* 207–18. Miscellaneous Publication 1271. Washington, D.C.: U.S. Department of Agriculture.

Wiens, J. A. 1977. "Assessing the Potential Impact of Granivorous Birds in Ecosystems." In *Granivorous Birds in Ecosystems,* edited by J. Pinowski and S. C. Kendeigh, 205–66. New York: Cambridge University Press.

Wilcove, D. S., et al. 1986. "Habitat Fragmentation in the Temperate Zone." In *Conservation Biology: The Science of Scarcity and Diversity,* edited by M. E. Soulé and B. A. Wilcox, 237–56. Sunderland: Sinauer Associates.

Wilcox, B. A. 1986. "Insular Ecology and Conservation." In *Conservation Biology: The Science of Scarcity and Diversity,* edited by M. E. Soulé and B. A. Wilcox, 95–118. Sunderland: Sinauer Associates.

Wilcox, B. A., and D. D. Murphy. 1985. "Conservation Strategy: The Effects of Fragmentation on Extinction. *American Naturalist* 125:879–87.

Wildlife Habitat Canada. 1991. *The Status of Wildlife Habitat in Canada: Realities and Visions.* Ottawa: Wildlife Habitat Canada, 102 pp.

Wilen, W. O., and R. W. Tiner. 1993. "Wetlands of the United States." *Wetlands of the World: Inventory, Ecology, and Management,* edited by D. F. Whiggams et al., 515–635. Dordrecht: Kluwer Academic Publishers.

Wiley, M. J., et al. 1990. "Longitudinal Structure of an Agricultural Prairie River System and Its Relationship to Current Stream Ecosystem Theory." *Canadian Journal of Fisheries and Aquatic Sciences* 47:373–84.

Willard, J. R. 1974. "Soil Invertebrates. 7. A Summary of Populations and Biomass." Technical Report 56. Saskatoon: Matador Project.

Williams, J. E., et al. 1989. "Fishes of North America Endangered, Threatened, or of Special Concern." *Fisheries* 14:2–20.

Willms, W. D., et al. 1980. "Effects of Clipping or Burning on Some Morphological Characteristics of *Agropyron spicatum.*" *Canadian Journal of Botany* 58:2309–312.

————. et al. 1986. "Herbage Production Following Litter Removal on Alberta Native Grasslands." *Journal of Range Management* 39:536–40.

————. 1993. "Influence of Litter on Herbage Production in the Mixed Prairie." *Journal of Range Management* 46:320–24.

Windingstad, R. M., et al. 1984. "Avian Cholera in Nebraska's Rainwater Basin." *Transactions of the North American Wildlife and Natural Resources Conference* 49:576–83.

Wink, R. L., and H. A. Wright. 1973. "Effects of Fire on an Ashe Juniper Community." *Journal of Range Management* 26:236–329.

Winter, T. C. 1986. "Effect of Ground-water Recharge on Configuration of the Water Table beneath Sand Dunes and on Seepage in Lakes in the Sandhills of Nebraska." *Journal of Hydrology* 86:221–37.

————. 1988. "A Conceptual Framework for Assessing Cumulative Impacts on the Hydrology of Nontidal Wetlands." *Environmental Management* 12:605–20.

————. 1989. "Hydrologic Studies of Wetlands in the Northern Prairie." In *Northern Prairie Wetlands,* edited by A. G. van der Valk, 16–54. Ames: Iowa State University Press.

Wolfe, C. 1984. "Physical Characteristics of the Sandhills: Wetlands, Fisheries, and Wildlife." In *The Sandhills of Nebraska: Yesterday, Today, and Tomorrow,* 54–61. Proceedings of the 1984 Water Resources Seminar Series. Lincoln: University of Nebraska.

Woodmansee, R. G. 1978. "Additions and Losses of Nitrogen in Grassland Ecosystems." *BioScience* 28:448–53.

World Commission on Environment and Development. 1987. *Our Common Future.* New York: Oxford University Press, 400 pp.

World Wildlife Fund. 1989. *Prairie Conservation Action Plan: 1989–1994.* Toronto: World Wildlife Fund.

World Wildlife Fund Canada. 1988. *Prairie Conservation Action Plan, 1989–1994.* Ottawa: World Wildlife Fund Canada, 38 pp.

Worster, D. 1979. *Dust Bowl: The Southern Plains in the 1930s.* New York: Oxford University Press.

————. 1992. "Cowboy Ecology." In *Under Western Skies: Nature and History in the American West.* New York: Oxford University Press.

————. 1994. "The Warming of the West." In *An Unsettled Country: Changing Landscapes of the American West, 91–120.* Albuquerque: University of New Mexico Press.

Wright, H. A. 1974. "Effect of Fire on Southern Mixed Prairie Grasses." *Journal of Range Management* 27:417–19.

Wright, H. A., and A. W. Bailey. 1982. *Fire Ecology: United States and Southern Canada.* New York: John Wiley and Sons.

Wright, H. A., et al. 1976. "Effect of Prescribed Burning on Sediment, Water Yield, and Water Quality from Dozed Juniper Lands in Central Texas." *Journal of Range Management* 29:294–98. Quoting Michler 1850.

Yahner, R. H. 1983. "Seasonal Dynamics, Habitat Relationships, and Management of Avifauna in Farmstead Shelterbelts." *Journal of Wildlife Management* 47:85–104.

Yeates, G. W. 1979. "Soil Nematodes in Terrestrial Ecosystems." *Journal of Nematology* 11:213–29.

Yoffe, E. 1992. "Silence of the Frogs." *New York Times Magazine,* 13 December.

Zerbe, R. O., Jr. and D. D. Dively. 1994. *Benefit-Cost Analysis: In Theory and Practice.* New York: HarperCollins College Publications.

Zimmerman, J. L. 1990. *Cheyenne Bottoms: Wetland in Jeopardy.* Lawrence: University Press of Kansas.

———. 1992. "Density-independent Factors Affecting the Avian Diversity of the Tallgrass Prairie Community." *Wilson Bulletin* 104:85–94.

Contributors

Michael G. Anderson is Assistant Director, Institute for Wetland and Waterfowl Research, Ducks Unlimited Canada, Manitoba, Canada.

Cody L. Arenz is a graduate student at the School of Biological Sciences, University of Nebraska, Lincoln, Nebraska.

Bruce D. J. Batt is Assistant Director, Institute for Wetland and Waterfowl Research, Ducks Unlimited, Memphis, Tennessee.

Richard K. Baydack is Associate Professor, Natural Resources Institute, University of Manitoba, Winnipeg, Manitoba, Canada.

Russel A. Benedict is a graduate student at the School of Biological Sciences, University of Nebraska, Lincoln, Nebraska.

Thomas B. Bragg is Professor, Department of Biology, University of Nebraska, Omaha, Nebraska.

Peter J. Buesseler is Prairie Biologist, Minnesota Department of Natural Resources, Fergus Falls, Minnesota.

Stephen J. Chaplin is Midwest Science Coordinator, The Nature Conservancy, Midwest Region, Minneapolis, Minnesota.

Jo S. Clark is Director of Programs, Western Governors' Association, Denver, Colorado.

Scott L. Collins is Associate Professor, Department of Botany and Microbiology, University of Oklahoma, Norman, and Director of the Ecological Studies Program, National Science Foundation, Arlington, Virginia.

Steve Corn is a zoologist at the Terrestrial Ecology Unit, Midcontinental Research Center, U.S. Department of the Interior, National Biological Service, Fort Collins, Colorado.

Penelope L. Diebel is Associate Professor, Department of Agricultural and Resource Economics, Oregon State University Agricultural Program, Oregon State University, LaGrande, Oregon.

Ian W. Dyson is Regional Environmental Coordinator, Prairie Region Alberta Environmental Protection, Lethbridge, Alberta, Canada.

Dan L. Flores is Professor, Department of History, University of Montana, Missoula, Montana.

Rod B. Fowler is Manager of Habitat Programs, Ducks Unlimited Canada, Manitoba, Canada.

Patricia W. Freeman is Curator of Zoology, Nebraska State Museum, and Associate Professor of Biology, School of Biological Sciences, University of Nebraska, Lincoln, Nebraska.

Hugh H. Genoways is Professor of Biology, School of Biological Sciences, University of Nebraska, Lincoln, Nebraska.

Daniel L. Gustafson is Assistant Research Professor, Department of Biology, Montana State University, Bozeman, Montana.

Anthony Joern is Professor of Biology, School of Biological Sciences, University of Nebraska, Lincoln, Nebraska.

James S. Kenney is Geographic Information Systems Specialist, The Nature Conservancy, Midwest Region, Minneapolis, Minnesota.

Fritz L. Knopf is Leader, Vertebrate Ecology, Midcontinent Ecological Science Center, U.S. Department of the Interior, National Biological Service, Fort Collins, Colorado.

Steven J. Kresl is Project Manager, Prairie Project and Wetland Management District, U.S. Department of the Interior, Fish and Wildlife Service, Mountain and Prairie Region, Woodworth, North Dakota.

Brian D. Ladd is Program Assistant, Migratory Bird Program, National Audubon Society, Boulder, Colorado.

James T. Leach is Coordinator, Regional Private Lands, U.S. Department of the Interior, Fish and Wildlife Service, Great Lakes and Big Rivers Region, Fort Snelling, Minnesota.

Carol A. Lively is Coordinator, North American Waterfowl Management Plan, U.S. Department of the Interior, Fish and Wildlife Service, Mountain and Prairie Region, Denver, Colorado.

Gene D. Mack is a wildlife biologist with the U.S. Department of the Interior, Fish and Wildlife Service, Kearney, Nebraska.

Jeffrey W. Nelson is Chief Biologist, Ducks Unlimited, Memphis, Tennessee.

Wayne R. Ostlie is a Great Plains Science Coordinator, The Nature Conservancy, Midwest Region, Minneapolis, Minnesota.

James H. Patterson is Director of International and Governmental Relations, Ducks Unlimited Canada, Ottawa, Ontario.

Elizabeth M. Payson is a graduate student at the Department of Biology, Montana State University, Bozeman, Montana.

Charles R. Peterson is Associate Professor, Department of Biology, and Curator of Herpetology, Idaho Natural History Museum, Idaho State University, Pocatello, Idaho.

Charles F. Rabeni is Professor of Fisheries, School of Forestry, Fisheries and Wildlife, University of Missouri, and Leader, Missouri Cooperative Wildlife and Fisheries Research Unit, U.S. Department of the Interior, National Biological Survey, Columbia, Missouri.

Ronald E. Reynolds is Project Leader, Habitat and Population Evaluation Team, U.S. Department of the Interior, Fish and Wildlife Service, Mountain and Prairie Region, Bismarck, North Dakota.

Paul G. Risser is President, Miami University, Oxford, Ohio.

Clayton D. Rubec is National Coordinator for Wetlands Conservation, Canadian Wildlife Service, Ottawa, Ontario.

Fred B. Samson is Regional Wildlife Ecologist, U.S. Department of Agriculture, Forest Service Northern Region, Missoula, Montana.

Rick Schneider is a Great Plains Community Ecologist, The Nature Conservancy, Midwest Region, Minneapolis, Minnesota.

Stanley E. Senner was Director, Migratory Bird Program, National Audubon Society, Boulder, Colorado, at the time his chapter was prepared; he is now Science Coordinator, Exxon Valdez Trustee Council, Anchorage, Alaska.

Ernest M. Steinauer is Visiting Assistant Professor, Department of Botany and Microbiology, University of Oklahoma, Norman, Oklahoma.

Allen A. Steuter is Director, Science and Stewardship-Nebraska, The Nature Conservancy Niobrara Valley Preserve, Johnstown, Nebraska.

Allen J. Tyrchniewicz is Research Officer, International Institute for Sustainable Development, Winnipeg, Manitoba.

T. Weaver is Professor, Department of Biology, Montana State University, Bozeman, Montana.

Ted W. Weins is Program Development Officer, Agriculture Canada, Regina, Saskatchewan.

Jeffery R. Williams is Professor, Department of Agricultural Economics, Kansas State University, Manhattan, Kansas.

Index